Joseph Victor Collins

Text-book of algebra

Through quadratic equations

Joseph Victor Collins

Text-book of algebra
Through quadratic equations

ISBN/EAN: 9783744745734

Printed in Europe, USA, Canada, Australia, Japan

Cover: Foto ©berggeist007 / pixelio.de

More available books at **www.hansebooks.com**

THROUGH

UADRATIC EQUATIONS.

BY

JOS. V. COLLINS, PH.D.,
PROFESSOR OF MATHEMATICS IN MIAMI UNIVERSITY.

CHICAGO:
ALBERT, SCOTT, & COMPANY.
1893.

C. J. Peters & Son,
Type-setters and Electrotypers,
Boston, U.S.A.

PREFACE.

THE usual treatment of algebra blends from the beginning the two distinct conceptions of the use of opposite numbers, and of letters to stand for numbers. In the following pages opposite numbers are presented first in the Arabic notation, all the rules for signs being given before letters are introduced. It is shown further, it is hoped in a simple and satisfactory way, how the operation symbols, + and —, can be used to distinguish positive and negative numbers. Careful attention is paid throughout to the interpretation of negative and imaginary results. Perhaps the most important feature is the persistent use of axioms in the solution of equations, so that the student can never forget that in handling an equation he is going at every step through a process of reasoning. This course secures to the student greater freedom in the choice of methods for the solution of all kinds of equations, and prepares for the examination into the validity of processes. Thus the treatment of algebra is made more like that of geometry.

Throughout the whole work every subject taken up is most carefully systematized. The demonstrations in nearly all cases are rigorous, and not mere illustrations of what is to be proved. Since algebra is one branch of analysis, one would naturally expect that its demonstrations and explanations would be given in analytic form. As a matter of fact, the synthetic method is often largely used to the student's great disadvantage. If the subject matter be first carefully arranged, and the demonstrations then put in the analytic form, the author believes a good presentation is attained. The different parts of the book are connected as closely as possible by cross references, so that each is made to throw light on the others. Aside from the exceptions noted, the book does not differ much in arrangement and method from other works of a similar scope in current use.

As the algebraic notation is made use of in all or nearly all of subsequent mathematical study, it is of great importance that the student should be thoroughly acquainted with its details and experienced in its use. The teacher should endeavor to make the student as familiar with it as possible. The interest awakened in the study of equations usually insures success in that part of the subject. Enough material is included here to constitute a good high school course, as also to fit for the best colleges.

Excepting a considerable number which are original, the exercises have been drawn from a great variety of sources, and are written in all legitimate notations. It would be very difficult to give proper credit for them. Special acknowledgment, perhaps, ought to be made for those taken from Heis's collection. The author is greatly indebted to his former instructor and esteemed friend, Professor S. J. Kirkwood, LL. D., of the University of Wooster, for valuable counsel and interest shown during the whole time of preparation. References to aid received from other sources are made in the text.

April, 1893.

SUGGESTIONS TO TEACHERS.

In addition to exercises in the text others may be assigned in drill books such as Perrin's or Ray's, or on algebra tablets. The latter are convenient for use, and are not very expensive. The rigorous classification of the subject matter has thrown some demonstrations and exercises in places where they should be used only on the second reading. It is thought that more is gained to the student in clearness of conception of the subject than is lost by the extra difficulty he finds in mastering such parts. Material which may be omitted is marked with a star. That which ought to be omitted by young students with a double star. The teacher is the best judge of what any particular class can do, and if any of the exercises are clearly too hard they should be passed over. With a very few exceptions the exercises will be found to be no more difficult than those of the best books in common use.

Not enough of the history of the different branches of mathematics has been taught in American schools. Much interest can be added to recitations now and then by brief accounts of the historical development of subjects under consideration. The author had thought of giving a short *résumé* of the history of algebra, but afterwards became convinced that it would be preferable to ask teachers themselves to go over this history pretty thoroughly, and then introduce the results of their reading in the class-room as occasion offered. For such preparation Ball's "Short

History of Mathematics" (Macmillan & Co., New York City) is the best book attainable, and it will be found to contain a list of other works on the subject should any one desire to prosecute the study further. Chrystal's Text-book contains many historical notes, and is besides one of the fullest and best treatises on algebra in the English language. In addition to such historical inquiries a study of methods of teaching algebra is also advised. For this purpose, Dr. Fr. Reidt's "Anleitung zum mathematischen Unterricht an hoeheren Schulen" (Berlin: Grote) is recommended. Circular of Information, No. 3, 1890, of the Bureau of Education, prepared by Florian Cajori, is on the teaching and history of mathematics in the United States.

TABLE OF CONTENTS.

(SYNOPSIS FOR REVIEWS.)

INTRODUCTION.

CHAPTER I. — ALGEBRA AS A BRANCH OF MATHEMATICS.

FIRST GENERAL SUBJECT. — THE ALGEBRAIC NOTATION.

CHAPTER II. — ON OPPOSITE NUMBERS.

SECTION I. — *Nature of Opposite Numbers. — Opposite Numbers and the Signs of Addition and Subtraction.*

CHAPTER III. — LETTERS USED TO SIGNIFY NUMBERS. — DEFINITIONS AND EXPLANATIONS.

ARTICLE. SECTION IV. — *On the Term.* PAGE.

CHAPTER IV. — EXERCISES IN NOTATION.

SECTION I. — *Exercise in Reading Algebraical Expressions.*

SECTION II. — *Exercise in Writing Algebraical Expressions.*

SECTION III. — *Numerical Values.*

CHAPTER V. — ADDITION.

CHAPTER VI. — SUBTRACTION.

CHAPTER VII. — SYMBOLS OF AGGREGATION. ALSO EXERCISES IN ADDITION AND SUBTRACTION.

SECTION I. — *On Compound Terms.*

CHAPTER XI. — COMMON FACTORS.

SECTION I. — *First Method. — The h. c. f. by factoring.*

SECTION II. — *Second Method. — By Continued Division.*

CHAPTER XII. — COMMON MULTIPLES.

SECOND GENERAL SUBJECT. — SIMPLE EQUATIONS.

CHAPTER XIV. — Simple Equations with One Unknown.

Section I. — *General Definitions.*

Section II. — *Algebraical Method of Treatment.*

I. — LOGICAL TERMS.

II. — AXIOMS.

III. — SOLUTION OF EQUATIONS.

ARTICLE. CHAPTER XX. — OF RADICALS. PAGE.

SECTION I. — *Reduction of Radicals.*

I. — SIMPLIFICATIONS.

II. — REDUCTION TO SURD FORM. CONVERSE OPERATION.

III. — REDUCTION OF RADICALS TO THE SAME INDEX.

SECTION II. — *Fundamental Operations with Radicals.*

SECTION III. — *Equations Involving Radicals.*

FOURTH GENERAL SUBJECT.—QUADRATIC EQUATIONS.

CHAPTER XXI.—QUADRATIC EQUATIONS CONTAINING ONE UNKNOWN QUANTITY.

CHAPTER XXIII. — QUADRATIC EQUATIONS AND EQUATIONS IN GENERAL.

FIFTH GENERAL SUBJECT. — TOPICS RELATED TO EQUATIONS. — INEQUALITY, PROPORTION, EXPONENTIAL EQUATIONS, LOGARITHMS, PROGRESSIONS, AND INTEREST.

ARTICLE. CHAPTER XXIV. — INEQUALITIES. PAGE.

CHAPTER XXV. — RATIO AND PROPORTION.

CHAPTER XXVI. — EXPONENTIAL EQUATIONS AND LOGARITHMS.

ARTICLE. CHAPTER XXVII. — THE PROGRESSIONS. PAGE.

CHAPTER XXVIII. — INTEREST, ANNUITIES, AND BONDS.

SECTION I. — *Interest.*

SECTION II. — *Annuities.*

SECTION III. — *Bonds.*

TEXT-BOOK OF ALGEBRA.

INTRODUCTION.

CHAPTER I.

ALGEBRA AS A BRANCH OF MATHEMATICS.[1]

By way of preparation for the study of algebra, it will be helpful to show its nature as one of the mathematical sciences, and to point out its relation to two other branches of mathematics, arithmetic and geometry.

1. **Mathematics** is the science of the exact relations of quantity as to magnitude and form.

a. The word "quantity" comes from a Latin adjective which means "how much," or "how many." Anything that has size or can be measured is a quantity. Any area, as 100 acres, any content, as 25 bushels, any length of time, as 10 hours, any number, as 15, is a quantity.

Quantity appears under one or other of two forms, number or extent. Thus we may say that a basket containing apples has 32 in it; or we may speak of the number of inhabitants in a town as, e.g., 2000. Or, on the other hand, we may speak of a square mile, or a cord of wood, and in this way denote the size of the object named. Now, many quantities have not only size but also shape; e.g., a house or a field; and so mathematics treats both of the size of objects and of their shapes.

[1] This chapter may be entirely omitted at the discretion of the teacher. It is too difficult for young pupils.

The following are intended as suggestive rather than exhaustive definitions of arithmetic and geometry.

2. Geometry treats of quantities in respect to their position, size, and shape.

3. Arithmetic treats of numbers with reference to the art of computation.

a. Arithmetic shows how to write numbers in the shortest and most convenient way; how to multiply and divide them; how to extract roots, and the like. Algebra, on the other hand, treats of numbers in the way of finding general truths in regard to them.

4. Algebra is that branch of mathematics which employs general characters as well as figures in the study of numbers, marking its numbers off into two opposite kinds.

5. The General Characters used.—In arithmetic we study numbers by using the arabic system of writing them. In algebra not only figures stand for numbers, but letters regarded as general characters are used for the same purpose, each letter standing for some number. However, when a letter is used to stand for a number, it is not like a figure, as 5 (which stands for 5 only, and can not mean anything else), but may stand for any number. This peculiarity the student will find one of the principal advantages algebra has over arithmetic.

Arithmetic teaches how to add and subtract, multiply and divide, extract roots, and the like, when figures stand for the numbers; algebra teaches how to perform the same operations when letters stand for the numbers.

6. The Two Kinds of Numbers.—The definition gives another difference between arithmetic and algebra. Arithmetic employs only one kind of numbers, using that kind to stand in one place for a debt, in another for a credit; sometimes for a gain, sometimes for a loss, and so on. Now, when both gain and loss, for example, appear in the same problem, contradictions may arise in the arithmetical language. Thus, if a man in business gain $500 in the first part of a year, and lose $200 in the latter part, the *amount* of his year's gain is found by *subtracting* $200 from $500. Contrariwise, if two men start in business with the same sum, and one loses $2000, while the other gains $3000, the *difference* in their fortunes will be obtained by *adding* $2000 to $3000. The same peculiarity presents itself in examples in longitude and time. We ask what is the difference in longitude between New York and Berlin, and expect the student to add to get the result. Algebra makes such questions clear by pointing out the opposite nature of the numbers and marking them in such a way as to show this. It also enables one to solve more difficult problems containing such numbers.

7. Opposite numbers will be investigated in the next chapter, before taking up the literal notation. All the laws governing the simultaneous use of such numbers will be developed while still using the familiar arabic system.

FIRST GENERAL SUBJECT.—THE ALGEBRAIC NOTATION.

CHAPTER II.

OPPOSITE NUMBERS.

8. In Algebra numbers are separated into two opposite kinds.

We will study the nature of opposite numbers first, and afterwards consider the changes necessary in the fundamental operations of addition, subtraction, multiplication, and division.

SECTION I.

Nature of Opposite Numbers.

9. Arithmetical and Opposite Numbers.—The science of arithmetic recognizes only real objects and will admit no element of unreality. It would say, for example, that a man can not lose more than he already has; that there is no such thing as a number less than zero. This may be described as a failure in arithmetical analysis. For it frequently happens that a man's debts exceed his credits, and he is worse off than if he had nothing at all. So stocks vary from premiums to discounts, and latitudes from north of the equator to south of the equator, and so on. As was pointed out and illustrated in Art. **6**, the operations of addition and subtraction often become confused in the solution of problems, because no distinctions were made at the outset between the two kinds of numbers.

10. The Arithmetical Series. — Arithmetic uses only one set of numbers, commencing with 1 and increasing without limit. We may write down an arithmetical series as follows:

1 2 3 4 5 6 7 8 9 10 11 12 13 14 . . . ∞.

(See Art. **66** for dots of continuation, and **62,** *b* for the meaning of ∞, which is used to denote an indefinitely large number.)

Fractional and irrational numbers are included between the whole numbers.

11. An Opposite Series. — Let us now commence at zero and write a similar series extending in the opposite direction, understanding that every number on the left has a signification opposite to that of the same number on the right. If numbers to the right hand mean credits, then numbers to the left hand mean debts; if numbers to the right mean north, then numbers to the left mean south; and so on. Combining the two, we have

∞ . . . . 7, 6, 5, 4, 3, 2, 1, 0, 1, 2, 3, 4, 5, 6, 7 . . . . ∞

12. The Series to be used. — For convenience merely we will imagine our double series of numbers to extend vertically. In order to distinguish between "above" and "below" numbers when removed from their places, some marks will be necessary.

13. Marking the Two Kinds of Numbers.[1] — The letter *a* (for above) might be written over all the upper numbers, and the letter *b* (for below) over all the lower ones. However, the notation adopted

+ ∞ . . . +13 +12 +11 +10 +9 +8 +7 +6 +5 +4 +3 +2 +1 0 −1 −2 −3 −4 −5 −6 −7 −8 −9 −10 −11 −12 −13 . . . − ∞

[1] The student is asked to prepare such a scale as is found in the margin on stiff pasteboard, and to make constant reference to it, verifying by the scale all the additions and subtractions given, until quite familiar with it.

is immaterial so long as two convenient marks are chosen and used consistently. We shall employ small *plus* and *minus* signs, not to denote addition and subtraction, but simply as marks for distinguishing the two kinds of numbers. A dagger and a dash might just as well be chosen.

14. Instances of Opposite Numbers. — This double series of numbers finds an application to problems involving (1) A merchant's gains or losses. (2) Income and outlay. (3) Latitude and longitude. (4) Scales, as the thermometer, etc. (5) Time, A.D. and B.C. (6) Attractive and repulsive forces, as positive and negative electricity, etc.

15. Definitions of Addition and Subtraction in Arithmetic. — Let us now examine addition and subtraction for a single series of numbers. Addition means literally *putting together*, and is performed by counting the total number of units in the different numbers added; and subtraction means finding a third number which added to the smaller of two numbers, will give the larger. These definitions are substantially the same as those given in the arithmetic. Now, how shall they be changed when both series are in use?

16. Algebraic Addition. — Remembering the opposite nature of the two kinds of numbers (**11**), let us understand that adding an "above" number to another means counting *upward* that many units, while adding a "below" number means the opposite, counting *downward* that many units.

To illustrate, let us take four simple examples:

$^{+}6 + {}^{+}5 = {}^{+}11$ Explanation. — When $^{+}5$ is added to $^{+}6$, we count 5 units *up* from $^{+}6$ and have $^{+}11$.

$^{-}7 + {}^{-}9 = {}^{-}16$ Explanation. — When $^{-}9$ is added to $^{-}7$, we count 9 units *down* from $^{-}7$ and have $^{-}16$.

Both these cases are just as in arithmetic. But, now, if adding $^{+}5$ carries us 5 units upward in the series, there is no reason why the starting-point (i.e., the number to which the other is added) should be one point in the double series rather than another. Let us suppose the starting-point is $^-3$, to which we are to add $^+5$.

$^-3 + {}^+5 = {}^+2$. EXPLANATION.—Starting at $^-3$ and counting 5 units *upward*, we have $^+2$.

Likewise, adding a "below" number to an "above," we write,

$^+6 + {}^-9 = {}^-3$. EXPLANATION.—Starting at $^+6$ and counting 9 units *downward*, we come to $^-3$.

17. Algebraic Subtraction—By the definition, Art. **15**, the subtrahend and remainder added must equal the minuend. We shall extend the application of this principle to algebraic numbers.

$^+7 + {}^+8 = {}^+15$. By the rule for addition.

$^+15 - {}^+8 = {}^+7$. Consequently, when $^+8$ is *subtracted* from $^+15$, in order to get $^+7$, one must count *downward* 8 units.

$^+13 + {}^-11 = {}^+2$. By the rule for addition.

$^+2 - {}^-11 = {}^+13$. Hence to *subtract* $^-11$ from $^+2$ we must count 11 units *upward* to get $^+13$.

Therefore to subtract an "above" number count from the minuend downward as many units as there are in the subtrahend, while to subtract a "below" number count from the minuend upward as many units as there are in the subtrahend. Evidently this is just the opposite of the rule for addition.

a. To subtract one number from another, or to find their difference, is the same as to *find the distance between them on the scale.* If the minuend is above the subtrahend, the difference is marked "above," +; if it is below, the difference is marked "below," −.

18. Exercise in Addition and Subtraction. — The student is expected to write down the answers and give the reasons as in example 1.

1. $^{+}8 + {}^{+}8 = {}^{+}16$. To add $^{+}8$ to $^{+}8$ count 8 units up from $^{+}8$.

2. $^{-}11 + {}^{+}16 = ?$
3. $^{-}9 + {}^{+}2 = ?$
4. $^{-}19 + {}^{-}1 = ?$
5. $^{+}32 + {}^{-}17 = ?$
6. $^{+}42 + {}^{-}55 = ?$
7. $^{+}8 - {}^{+}4 = ?$
8. $^{+}6 - {}^{+}9 = ?$
9. $^{-}3 - {}^{-}5 = ?$
10. $^{-}17 - {}^{-}12 = ?$
11. $^{+}26 - {}^{-}15 = ?$
12. $^{-}17 - {}^{+}29 = ?$

19. Meaning attached to Zero (0) in Algebra. — Zero has a real place in the double series, **(12)** — as real, for example, as the zero point in the thermometric or any other similar scale. It has a real place in the latitude scale; viz., the equator: and so in other instances. Consequently it can appear just like other numbers in algebra. Thus adding $^{+}5$ to 0 means going up from 0 to $^{+}5$. Adding $^{-}13$ to 0 means passing down from 0 to $^{-}13$. We have

1. $0 + {}^{+}6 = {}^{+}6$. Starting at 0 and moving up 6 units gives $^{+}6$.

2. $0 - {}^{+}5 = {}^{-}5$. Starting at 0 and moving down 5 units gives $^{-}5$.

3. $0 + {}^{-}3 = {}^{-}3$. Starting at 0 and moving down 3 units gives $^{-}3$.

4. $0 - {}^{-}4 = {}^{+}4$. Starting at 0 and moving up 4 units gives $^{+}4$.

20. The Use of Other Signs to Mark the Series. — From the first and second equations of the last article we see that instead of the sign "$^{+}$", "$0 + {}^{+}$" may be used; and instead of the sign "$^{-}$", "$0 - {}^{+}$" may be used.

But the latter notation has $^{+}$ before both kinds of numbers. Consequently this mark need not be retained. Hence instead of the sign "$^{+}$", "$0 +$" may be used, and instead of the sign "$^{-}$", "$0 -$" may be used.

NOTE.—Had the third and fourth equations been selected, the results would simply *duplicate* those found, the $^{+}$ and $^{-}$ merely being interchanged.

21. The 0 in the Series Sign just obtained is dropped.—Since 0 adds no value and can be readily supplied, it is plain that it need not be written, and that its absence will make no difference in either our operations or results. As in the first equation of Art. **19**, no sign, before a number and a $+$ sign signify the same thing.

To illustrate,

$$0 + 21 = +21 = 21;\ 0 - 21 = -21.$$

22. Conclusion.—By the last two articles it appears that we may distinguish between the two series by $+$ and $-$ signs. A $+$ sign prefixed to a number shows that it belongs to one, and a $-$ sign that it belongs to the other series.

23. From what we know of the nature of addition and subtraction, the sign of addition ($+$) refers to what may be called the **direct** series, and the sign of subtraction ($-$), to the other, which may be called the **reverse** series. Naturally no sign at all is direct, so that both $+$ and no sign refer to this series.

24. Two Views of the Signs.—The two signs $+$ and $-$ seem to have a double meaning: $+$ to denote the operation of addition, and to mark direct numbers; and $-$ to denote subtraction, and to mark reverse numbers.

In the thermometer scale $+$ means above zero. In accounts $+$ would appropriately refer to income, $-$ to outlay. But it would be better to regard them as always indicating the operations of addition and subtraction, 0 being understood when no other quantity precedes either of them.

25. Positive and Negative Numbers. — Numbers from the direct series are called positive or plus numbers, while those from the reverse are called negative or minus numbers.

26. Like and Unlike Signs. — When numbers or terms have the same sign preceding them they are said to have like signs; otherwise they are said to have unlike signs.

Thus 5, +7, +13 have like signs. Also −4, −3, −41, −50 have like signs. Again − 28 and + 4 have unlike signs; and 3, −6, +11, −23, −24 have unlike signs.

SECTION II.

The Fundamental Rules for Opposite Numbers.

27. Rules for all of the fundamental operations will now have to be investigated.

I. — ADDITION.

28. Addition of Positive and Negative Numbers. — There are two cases.

1. When the numbers added have like signs. The operations now to be considered are the same as those in **18**, but large + and − signs take the place of those there used to mark the series.

(1.		(2.	
	+ 4		− 3
	+ 6		− 7
	+ 7		− 5
	+ 9		− 12
	+ 26		− 27

Explanation of (2. — By **16** when − 7 is added to − 3, we count 7 units down to − 10; − 5 added to this result gives − 15; and − 12 added to − 15 gives the answer, — 27. Or, more simply, the numbers are added as in Example (1, and the − sign prefixed to the result.

(3. Set down as the above and add +9, +7, +6, +15, and + 81.

(4. Add −61, −23, −97, −5, −1, and −154.

(5. Add −83, −83, −83, −83, −83, −83, −83, −83.

2. When the numbers added have unlike signs.

(1. Add 5, − 4, + 15, − 9, and + 1.

5	EXPLANATION. — Following the method given in 16, we have,
− 4	− 4 added to + 5 gives + 1,
+ 15	+ 15 added to + 1 gives + 16,
− 9	− 9 added to + 16 gives + 7,
1	+ 1 added to + 7 gives + 8. *Ans.*
8	

(2. Add − 11, − 3, + 15, − 6, and − 14.

− 11	EXPLANATION.
− 3	− 3 added to − 11 gives − 14.
+ 15	+ 15 added to − 14 gives + 1.
− 6	− 6 added to + 1 gives − 5.
− 14	− 14 added to − 5 gives − 19. *Ans.*
− 19	

(3. Add 21, − 6, 3, − 49, − 25, 14, and 2.

(4. Add 602, 27, − 1003, 1292, − 359, − 1.

It is a well-known principle in arithmetic that the sum is the same in whatever order the numbers are added. If this be true in algebra there will be an obvious advantage in changing the order of the numbers added, so that all the positive numbers come first, followed by all the negative numbers. In the examples just given it is plain that starting with zero we count in the positive direction in all a certain distance, denoted by the sum of the positive numbers, and in the negative direction in all a certain distance denoted by the sum of the negative numbers: so that the sum obtained by adding the numbers *seriatim*, or one after another, is the same as that obtained by adding the sum of the positive numbers, and the sum of the negative. In other words, *the order in which numbers are added in algebra is indifferent.* This is called *the commutative law in addition.*

Thus, in examples (1 and (2 above

(1. $5 + 15 + 1 = 21$; $-4 - 9 = -13$; $21 - 13 = 8$,

answer as before.

(2. 15; $-11 - 3 - 6 - 14 = -34$; $15 - 34 = -19$,

answer as before.

Remark.—The student has no doubt observed that when two opposite numbers are added *the sum* is their *numerical difference.* It is positive if the positive number is the greater, and negative if the negative number is the greater.

The student may verify the truth of the commutative law for exercises (3 and (4.

(5. Add -6, -7, 11, 15, -3, -21, $+4$.

(6. Add 19, -3, -6, 0, -4, $+6$, -31, and 50.

29. Rule for Adding Positive and Negative Numbers.

1. If the numbers to be added have like signs, add them as in arithmetic, and prefix the common sign.

2. If the numbers to be added have unlike signs, add the positive numbers and the negative numbers separately, take the difference, and prefix the sign of the greater.

30. Exercise in the Addition of Positive and Negative Numbers. Problems.

1. Add 51, 96, -37, -72, 101, -49, $+237$.
2. Add -79, -106, -304, $+40$, -9, $+1$, and 362.
3. $142 - 16 - 50 - 31 - 199 + 777 = ?$

4. $18 \times 47 - 39 \times 47 - 21 \times 47 + 47 = ?$

5. Add $5\frac{1}{8}$, $-3\frac{13}{16}$, $8\frac{1}{2}$, $-19\frac{3}{4}$, $-20\frac{7}{8}$.

6. $29.4 - 3.3 + .079 - 1.001 - 20 + 43.05 + 11 = ?$

7. A boy has 15 cts., his father owes him 12 cts., and one of his playmates 7 cts. If credits be marked +, how much has the boy, and how shall the amount be marked?

8. William has no money, and he owes a grocer 8 cts., a pieman 4 cts., his sister 3 cts., and a playmate 10 cts. What is he worth, and how shall we mark the amount?

9. A ship starting from 17° north latitude goes one day 2° north, the next 3° north, and a third 4° south; what latitude is she now in, north latitudes being marked + and south latitudes −?

10. A vine that was 30 inches long grew in one month 9 in., in the next 12 in., and in the next 15 in. It was pruned back at one time 11 in., and at another time 5 in. What is its length?

11. A boy has 25 cts. and owes 10; what is he worth? How shall 25 be marked, how 10, and how the result?

12. George has 16 cts. in his money box, 10 in his pocket, and William owes him 10 cts. He owes his brother John 7 cts., and a confectioner 14. What is he worth?

13. A thermometer which stood at 4 P.M. at 70°, and which by 8 o'clock had fallen 14°, and by midnight 8° more, and by 4 A.M. 3° more, rose from 4 o'clock to 8 o'clock 12°, and from 8 o'clock to 12 o'clock noon 25°. Where did the mercury stand at the last-named hour?

14. A man who has $300 in the bank and $125 in his pocket owes A $550 and B $1000. He has due him from one party $400, and from another $640. But he owes as surety on a note $190. What is he worth?

15. A rope which was 40 ft. long had at one time 8 ft. cut off, at another 6 ft.; it was spliced with a piece 15 ft. long, after which 9 ft. was first broken off, then 11 ft., then 3 ft., when it was again spliced with a 20 ft. piece. How long was the rope then?

QUERY.—If a question like this were propounded, in which the parts broken off taken together exceeded the original length of the rope added to the sum of the splices, what would the student judge concerning the nature of the problem? What then will a negative result sometimes indicate?

II.—SUBTRACTION.

31. Subtraction of Positive and Negative Numbers.—See **17**.

1. $\begin{array}{r} 18 \\ 7 \\ \hline 11 \end{array}$ 2. $\begin{array}{r} 15 \\ -\ 8 \\ \hline 23 \end{array}$

EXPLANATION.—To *subtract* -8, since it is a negative number, count from 15, 8 units *up* to $+23$.

3. $\begin{array}{r} -14 \\ 10 \\ \hline -24 \end{array}$ 4. $\begin{array}{r} -33 \\ -17 \\ \hline -16 \end{array}$

EXPLANATION.—To subtract $+10$, count 10 units down from the minuend. To subtract -17, count 17 units *up* from the minuend.

5. $\begin{array}{r} 6 \\ 11 \\ \hline -5 \end{array}$ 6. $\begin{array}{r} 61 \\ -89 \\ \hline 150 \end{array}$ 7. $\begin{array}{r} -3 \\ +7 \\ \hline -10 \end{array}$ 8. $\begin{array}{r} -1 \\ -12 \\ \hline +11 \end{array}$

By Arts. **16** and **17** we perceive that subtracting a number of one series is always the same as adding the corresponding number of the other series. Thus:

$$^{+}18 - {}^{+}10 = {}^{+}8, \quad \text{as also,} \quad {}^{+}18 + {}^{-}10 = {}^{+}8;$$
$$^{+}21 - {}^{-}12 = {}^{+}33, \quad \text{as also,} \quad {}^{+}21 + {}^{+}12 = {}^{+}33;$$

and so in every case.

Now, since this is true, if we choose to do so, we can subtract by first changing the series to which the subtrahend belongs, and then *adding* it to the minuend. But to change the series of a number, we merely change its sign. Consequently we may use this rule:

To Subtract, change the sign of the subtrahend and add.

a. Instead of actually changing the sign of the subtrahend, it is more convenient to *imagine* it to be done.

Thus, in Ex. 3, imagine the sign of 10 to become −; and then, if − 10 is added to − 14, the sum is − 24. In Ex. 6, conceive − 89 to become + 89, and adding + 89 to 61, we have + 150; and so in the other examples.

9. From 16 take 23.
10. From − 15 take − 38.
11. 44 − 100 = ?
12. From − 399 take 183.
13. From − 1000 take 1001.
14. From 436 take 591.

32. Rule for Subtracting Positive and Negative Numbers. — Conceive the sign of the subtrahend number to be changed, and add it to the minuend.

33. Exercise in Subtracting Positive and Negative Numbers. —

1. From 1428 take − 794.
2. From − 37 take − 56.

3. From the sum of 16, 32, − 79, − 37, − 109, 2, and 9, take the sum of − 19, − 108, − 42, 83, 29, − 60, and 41.

4. From 19 × 99 take 73 × 99.

5. From 16 − 31 + 172 − 200 less 169 − 508 − 263 take 396 − 243 + 27 less 100 − 6 + 23.

6. On a certain day the mercury in a thermometer stood at 60°, and on the next it stood at 77°. What was the difference of temperature of the two days if + denote upward movement and − downward?

7. A man who has property valued at $3000 owes, in various amounts, to the extent of $3355. What is he worth, and how shall we mark the number?

8. What is the difference in longitude between New York and Berlin if New York is 74° 0′ 24″ west and Berlin is 13° 23′ east? How shall we mark the result if east be + and west be −, the answer thus showing how far and in which direction New York is from Berlin?

9. If north latitude be marked + and south −, what is the difference in latitude between two ships, one at 19° south and the other at 61° north?

10. What is the difference between the average temperatures of January and December if that of December be 15° above zero and that of January 4° below?

III. — MULTIPLICATION.

34. Multiplication in Algebra is the process of taking one number as many times as there are units in another, giving the product the same sign as the multiplicand's, or the opposite, according as the multiplier is a direct or reverse number.

To illustrate (we suppose the first factor to be the multiplier):

$$+5 \times +6 = 30, \qquad -3 \times +7 = -21;$$
$$+4 \times -9 = -36; \qquad -2 \times -8 = +16.$$

a. The definition of the multiplication of algebraic numbers given above is based on the nature of such numbers. One series (10) attaches a direct meaning to all its numbers, and is the first and natural series. The other is joined to the first, and obtains its meaning from the first by always using its numbers in a sense contrary to that which is applied to those of the first. Looking at the two series as written in 11, the two parts are the same, or, rather, they are symmetrical, like the two arms of a balance. Nevertheless, the meanings attached to them place them on a different footing. So we find, while the product of two positive numbers is a positive number, that, owing to the difference in their significations, the product of two negative numbers is not a negative, but a positive number. This last is not unlike the grammatical rule which says two negatives make an affirmative. Thus, I said "not unlike," when I meant "alike," at least, in some respects. The problems in 41 illustrate the definition.

35. Investigation of the Rule of Signs in Multiplication.— It can be shown from the definitions of addition and subtraction that if a positive multiplier gives to the product the same sign as the multiplicand's, a negative multiplier will give opposite results.

1. When the sign of the multiplier is +:

(1. To multiply + 15 by + 7.

Once + 15 is + 15; twice + 15 is + 30; three times + 15 is + 45, and so on. 7 × + 15 is + 105. And so in general.

(2. To multiply − 9 by + 8.

Once − 9 is − 9; twice − 9 is − 18; three times − 9 is − 27, and so on. 8 × − 9 is − 72. And so in general.

2. When the sign of the multiplier is −:

In the expression 6 × 12 − 13 × 12 it is evidently intended that 13 times 12 shall be subtracted from 6 times 12, and the difference is therefore − 7 times 12.

Now,

$$6 \times 12 - 13 \times 12 = 72 - 156 = -84. \quad \text{(Art. 32.)}$$

Hence, $\quad -7 \times 12 = -84.$

Again, let us evaluate the expression 3 × − 14 − 8 × − 14, which plainly means that 8 times − 14 is to be taken from 3 times − 14, and the remainder is, therefore, − 5 times − 14.

But,

$$3 \times -14 - 8 \times -14 = -42 \text{ minus } -112 = +70. \quad \text{(Art. 32.)}$$

Hence, $\quad -5 \times -14 = +70.$

This reasoning is plainly applicable to any other numbers. If, now, $+$ refer to the direct series and $-$ to the reverse, these four products agree with the definition (**34**). But the last two have been obtained without any further reference to that article. Consequently, we are assured that the definitions of multiplication and addition and subtraction are consistent.

36. Rule for Signs in Multiplication. — Like signs give $+$, and unlike, $-$. Or, more specifically, $+$ by $+$ and $-$ by $-$ give $+$; while $+$ by $-$ and $-$ by $+$ give $-$.

37. Exercise in Multiplication of Positive and Negative Numbers.

1. $8 \times -2 = ?$
2. $6 \times +36 = ?$
3. $-19 \times 3 = ?$
4. $42 \times -1 = ?$
5. $-72 \times -4 = ?$
6. $2\frac{1}{2} \times -3\frac{1}{3} = ?$
7. $14\frac{2}{7} \times -\frac{2}{25} = ?$
8. $4.6 \times -.23 = ?$

38. The Commutative and Associative Laws in Multiplication. — We knew in Arithmetic that it made no difference in what order numbers were multiplied; for example, $7 \times 9 = 9 \times 7$, and $3 \times 4 \times 5 = 3 \times 5 \times 4 = 4 \times 3 \times 5$, and so on. We will show that this is true in ordinary Algebra as well.

1. *The Commutative Law.* — That the numerical value of the product of two factors is the same, whichever be the multiplier, is known from Arithmetic. We are to show that interchanging the factors will not change the ***sign*** of the product.

(1. *If the Factors are both Positive or both Negative.* — In either case the product would be $+$, both before and after changing the factors (**36**).

(2. *If one Factor is Positive and the other Negative.* — Here the product would be negative, both before and after changing (**36**).

Hence, the multiplier and multiplicand may change places in Algebra, and the product remains the same.

2. *The Associative Law.* — To see whether the same law will hold for three or more factors, let us first take three factors. Now, by uniting any two of them into *one factor*, we have this product multiplied by the third, in which multiplication, as well as in the first partial one, the order may be disregarded by the commutative law. And so also for a larger number of factors.

Hence, it may be inferred that the product of any number of factors is the same in whatever order they are associated.

39. The Sign of the Product for Three or more Factors.

The rule for signs, where two factors are multiplied, has just been given. The rule for a greater number of factors is derived from that for two in the following manner:

1. The product of any number of positive factors is positive.

For, any two multiplied together give a positive product (**36**), and this product by a third factor is positive; and so on for any number of factors, $+$ by $+$ always giving a positive product.

Thus, $+6 \times +9 \times +2 \times +1 = +6 \times 9 \times 2 \times 1 = +108.$

For, $+6 \times +9 = +54$; $+54 \times +2 = +108$; $+108 \times +1 = +108.$

2. The product of any *even* number of negative factors is always positive.

If there be positive factors, change the order so as to bring them all together (**38**). Then, by taking the negative factors in pairs, since two negative factors give a positive product, these positive products with the positive factors, if any, give a positive product by 1 above.

Thus, $-6 \times -3 \times -2 \times -5 \times +4 \times +1 = +18 \times +10 \times +4 = +720.$

3. The product of an *odd* number of negative factors is negative.

For, if one negative factor be withdrawn (since one less than an odd number is an even number), the product then obtained from the others will be positive by the preceding case. This positive product multiplied by the negative factor withdrawn gives a negative result (**36**).

Thus, $+8 \times +2 \times -7 = 16 \times -7 = -102.$

Also, $+5 \times -3 \times -9 \times -4 \times -2 \times -4 = +5 \times +27 \times +8 \times -4 = +1080 \times -4 = -4320.$

40. Rule. — If the number of negative factors in a product be *odd*, the product is negative. Otherwise, it is positive.

41. Exercise in the Multiplication of Opposite Numbers.

1. $6 \times 11 \times -25.$
2. $19 \times -1 \times -1 \times -2.$
3. $72 \times 2 \times -1 \times -40.$
4. $1\frac{1}{2} \times -4 \times -\frac{2}{3} \times -\frac{4}{3}.$
5. $-.6 \times -.6 \times -.3 \times -.02.$
6. $\frac{2}{3} \times -\frac{4}{9} \times -\frac{3}{8} \times -\frac{6}{5}.$
7. $11 \times -3 \times -4 \times -7 \times -21 \times 2.$
8. $-1 \times -1 \times -1 \times -1 \times -2 \times 2 \times 2 \times -3 \times -3 \times 3.$
9. $-6 = -5 \times -4 \times -3 \times -2 \times -1 \times 1 \times 2 \times 3.$
10. $6 \times 7 \times 2 \times 3 \times 4 \times 17 \times 2 \times -1.$

11. A boy engaged in catching fish secures each day 50 fish for 10 successive days. How many did he catch in all?

12. A hunter used 24 charges of ammunition without securing any game, each charge costing him $2\frac{1}{2}$ cents. If waste be marked $-$, how shall we write the factors and product indicating the expense connected with his sport?

13. A manufacturer of telescopic object glasses, the casting of which costs $150 each, out of a certain number cast breaks 11 in the grinding. If the number of those broken be marked $-$, how shall we indicate his loss in the factors and product?

14. * * The ten hindmost cars of a train which is going directly across a valley are still directed down hill, and besides overcoming friction are exerting a forward tendency equal to two horse-power each. How shall we represent their combined effect on the train by using numbers from the double series?

Solution.—On the level the force exerted by the locomotive in a forward direction would appropriately be marked $+$, while the resistance of each car would be marked $-$. On this supposition we mark every car's effect $-$, and in the case of the ten hindmost each -2. But the ten hindmost cars act contrary to the original supposition. Therefore we write

$$-10 \times -2 \text{ horse-power} = +20 \text{ horse-power};$$

i.e., 20 horse-power in the positive direction.

15. * * A book-agent selling a book at $6.50 which costs him $4 finds himself at the end of a certain year in debt to his company for 24 copies, the pay for which he never expects to be able to collect. The next year he sells 150 copies of a book at $5 a copy, the cost of which is $3. But at the end of this year he finds he has been able to

collect for only 148 copies. During the year, however, 9 copies of the first book have been paid for leaving only 15 still unpaid. It is required to represent his gains and losses by factors and products from the double series.

Solution. — In his accounts the agent would naturally keep the old and the new items separate. The former as supposed loss would appropriately be marked $-$, and the latter as supposed gain $+$. Again the number of books paid for and the number of books not paid for *in accordance with his expectations,* would be direct series numbers, while the number of books not paid for and the number of books paid for *contrary* to his expectations would be reverse series numbers

$+15 \times -\$4 = -\$60$ loss; $+148 \times \$2 = +\$296$ gain; $-12 \times \$3 = -\$36$ loss; $-9 \times -\$2.50 = +\$22.50$ gain.

Adding, $-\$60 + \$296 - \$36 + \$22.50 = \$222.50$ net profit.

42. Powers of Algebraic Numbers. — The term power of a number is used in the same sense in algebra as in arithmetic.

Thus, $6^2 = 6 \times 6 = 36$; $(-2)^4 = -2 \times -2 \times -2 \times -2 = +16$.

a. In algebra it is often convenient to place the exponent (i.e., the small figure written to the right and above the number) outside a parenthesis, as in the second example just given. This means that the number inside with its proper sign is to be taken as a factor as many times as the exponent has units.

1. The Signs of Powers.

(1. If we use a positive number continuously as a factor, we always get a positive product (**39**). Hence any power of a positive number is also positive.

(2. Even powers of negative numbers are positive (**39**, 2).

(3. Odd powers of negative numbers are negative (**39**, 3).

2. Exercise in raising numbers to powers.

(1. Square 16.

(2. Raise $+4$ to the fourth power.

(3. Cube 16.

(4. Raise -3 to the fifth power.

(5. Cube -14.

(6. Square -11.1.

(7. $(-1.3)^3 = ?$

(8. $(-2)^6$.

(9. $(3\frac{1}{2})^3$.

(10. $(-5 \times -3)^4$.

(11. $(3 \times -2\frac{2}{3})^5$.

IV.—DIVISION.

43. Division in Algebra is the process of finding one factor when the product and the other factor are given.

The product is the dividend, the given factor is the divisor, and the required factor is the quotient.

Since division is thus seen to be the converse of multiplication, the law of signs in division may be inferred from that in multiplication.

By the definition of division, if

$+6 \times +5 = +30$, then $+30 \div +6 = +5$; i.e.,
+ divided by + gives +.

$+6 \times -5 = -30$, then $-30 \div +6 = -5$; i.e.,
− divided by + gives −.

$-6 \times +5 = -30$, then $-30 \div -6 = +5$; i.e.,
− divided by − gives +.

$-6 \times -5 = +30$, then $+30 \div -6 = -5$; i.e.,
+ divided by − gives −.

Thus the rule of signs in multiplication holds in division also; viz., like signs give plus, and unlike, minus.

44. Exercise in the Division of Algebraic Numbers.

1. Divide $+10$ by -2.
2. Divide -15 by 3.
3. Divide -50 by 5.
4. $\frac{30}{5} \div -6 = ?$
5. $-3.1 \div 4$?
6. $1.06 \div -9$?
7. $-256 \div -8000$?
8. $\frac{-39.6\frac{1}{3}}{-2\frac{1}{2}}$
9. $\frac{2}{-3} \times \frac{6.5}{-13} \times -\frac{29}{2} \times -.007 \div -\frac{2}{.3} = ?$

10. If in 11 days the mercury falls from 15° above zero to 18° below, how much is that a day, and how shall we mark the terms of the division?

11. A man losing 5 cents on every bushel of wheat he sold found he had fallen short \$400. How many bushels did he sell? Mark the terms.

45. Roots of Opposite Numbers. — The term root of a number is used in the same sense in algebra as in arithmetic.

Thus, $\sqrt{81} = 9$; $\sqrt[3]{-64} = -4$, etc.

a. The radical sign ($\sqrt{}$) is used to show that a root is to be taken. What root is desired is indicated by the small figure above and to the left. When no figure is written the square root is understood.

It will be convenient to first take up square roots, then cube roots, and lastly other roots. As in Powers, the *signs* of the roots is the main point considered.

1. Square Roots.

(1. The square root of a positive number *may be either plus or minus.*

Thus, $\sqrt{4} = +2$ or -2. For, $+2 \times +2 = +4$; also $-2 \times -2 = +4$.

$\sqrt{28} = +5.29+$, or $-5.29+$. For, $+5.29 \times +5.29 = +28$; also $-5.29 \times -5.29 = +28$;

and so for any other number.

b. To mark the two roots without rewriting the number the sign $\pm$ (read "plus or minus") is used. It indicates two different numbers.

(2. The square root of a negative number is *impossible*. For, the square root of any negative number, say of -36, to be an algebraic number must be either $+6$ or -6. Now it is *neither*, since $(+6)^2 = +36$, and $(-6)^2 = +36$ also. Hence -36 has no algebraic square root.

2. Cube Roots.

(1. The cube root of a positive number is positive.

Thus, $\sqrt[3]{27} = +3$, since $(+3)^3 = +27$.

(2. The cube root of a negative number is negative.

Thus, $\sqrt[3]{-64} = -4$, since $(-4)^3 = -64$.

3. Other Roots.

(1. Roots whose index is a prime number. — Of these, positive numbers have a positive root, and negative numbers a negative root, as in the cube root.

(2. Roots whose index is a composite number. — Such roots may be found by extracting the roots denoted by the factors of the index, the one after the other. Thus, the fourth root can be derived by extracting the square root twice, the sixth root by extracting the square root first and then the cube root of the result. The fifteenth, by extracting the cube root and the fifth root, and so on.

4. Rules for signs of roots. — We saw above that the square root of negative numbers is impossible. Hence we have

(1. Even roots of negative numbers are impossible.

(2. Even roots of positive numbers are to be written $\pm$.

(3. Odd roots of numbers have the same sign as the powers themselves.

NOTE. — These rules have reference only to what are called *real roots*. This subject will be treated more fully in Chapter XVIII.

46. Exercise in extracting roots of algebraic numbers.

1. Extract the square root of 36, of 169, of 49, of 225,[1] and of 121.

2. Extract the cube root of 8, of − 27, of − 64, of 125, and of − 343.

3. Extract the fourth root (i.e., the square root twice) of 16, 81, and of 256.

4. Extract the sixth root of 64, of 729.

5. What algebraic numbers are those whose squares are each + 144?

6. What algebraic number is it whose cube is − 512?

7. What is the length of one side of a cubical box whose content is 216 cubic inches?

[1] If the student is not familiar with the rules for finding square and cube roots, the numerical results will have to be found by trial multiplications.

CHAPTER III.

LETTERS USED TO REPRESENT NUMBERS, DEFINITIONS, AND EXPLANATIONS.

47. The Use of Letters to Stand for Numbers constitutes the second distinguishing characteristic of algebra. We proceed to give some illustrations of this use before taking up the next topic. The solutions of the following questions show how a letter (x) may represent different numbers in different problems.

1. What number is that which being added to twice itself the sum is 42?

Solution. — Let $x =$ the number,
then $2x =$ twice the number, ($2x$ means 2 times x)
and $x + 2x =$ the sum.
But the sum is 42.

Therefore $x + 2x = 42$,
or, $3x = 42$ (for $x + 2x = 3x$).
Now if $3x = 42$, $1x$ or $x = \frac{1}{3}$ of $42 = 14$.
Therefore $x = 14$, which is the number.

2. What number is that to which if we add its half and ten more the sum is 43?

Let $x =$ the number; then $\frac{1}{2}x =$ one-half the number, and $\frac{3}{2}x =$ the number $+ \frac{1}{2}$ of the number.

But the number $+ \frac{1}{2}$ the number $+ 10 = 43$.

Therefore $\frac{3}{2}x + 10 = 43$.

Now, if 10 added to $\frac{3}{2}x = 43$, $\frac{3}{2}x = 43$ *less* $10 = 33$;

$$\text{If } \frac{3}{2}x = 33, \text{ then } \frac{1}{2}x = \frac{1}{3} \text{ of } 33 = 11;$$

$$\text{and } \frac{2}{2}x, \text{ i.e., } x = 2 \times 11 = 22.$$

3. In a store-room containing 40 barrels the number of those that are filled exceeds the number that are empty by 16. How many are there of each?

Let x equal the number filled, then $x - 16$ equals the number empty.

Hence, $x + x - 16 = 40$,
or, $2x - 16 = 40$.

Now, if 16 has to be subtracted from $2x$ to give 40, then

$$2x = 40 + 16 = 56$$
$x = 28$ barrels filled,
$x - 16 = 12$ barrels empty.

4. Three pieces of lead together weigh 47 lbs.; the second is twice the weight of the first, and the third weighs 7 lbs. more than the second; what is the weight of each piece?

Let $x =$ the number of lbs. the first piece weighs;
then $2x =$ the number of lbs. the second piece weighs,
and $2x + 7 =$ the number of lbs. the third piece weighs.

But the sum of these weights is 47 lbs.;
therefore, $x + 2x + 2x + 7 = 47$,
or, $5x + 7 = 47$.

Now, if 7 added to $5x$ is 47, $5x = 40$. Hence,

$$x = 8$$
$$2x = 16$$
$$2x + 7 = 23$$

5. A boy bought a certain number of lemons and twice as many oranges for 40 cts., the lemons costing 2, and the oranges 3 cts. apiece; how many were there of each?

Let $x =$ the number of lemons,
$2x =$ the number of oranges.

Now, x lemons at 2 cts. apiece amount to $2x$ cts., and $2x$ oranges at 3 cts. apiece amount to $6x$ cts. Then, to find the cost of both,

$$2x + 6x = 40.$$
$$8x = 40.$$
$$x = 5$$
$$2x = 10$$

6. A father gave his boy three times as many cents as he had, his uncle then gave him 40 cts., when he found he had nine times as many as he had at first? How many had he at first?

Let $x =$ the number of cents he had at first, then $3x =$ the number of cents his father gave him, so that $4x =$ the number he now had.

His uncle gave him 40 cts. more, when the statement is made that $9x =$ his total amount.

Hence, $$4x + 40 = 9x.$$

It is plain now, that if 40 added to $4x$ makes $9x$, then $40 = 5x$.

If $5x = 40$,
$x = 8$ the number he had at first.

Exercises like the preceding will be found in Art. 85.

SECTION I.

Letters with the Signs.

48. An Algebraic Expression is anything written in the algebraic notation.

Thus, a, $-5b$, $a^2 + ab$, and $\dfrac{abc - 3mn}{2b}$ are all algebraic expressions.

49. As suggested by the expressions just written, and by others already given, the signs appearing in arithmetic are used in algebra also, and retain the same meaning. Of them the most important are

$+$, $-$, $\times$, $\div$, $(\ \ )$, $=$, and the signs of powers and roots.

50. The Sign $+$ (read "plus") denotes addition, and is used to show that the numbers[1] between which it is placed are to be added.

Thus, $a + b$ means the sum of the numbers denoted by a and b.

51. The Sign $-$ (read "minus") denotes subtraction, and is used to show that the number following the sign is to be subtracted from the other.

Thus, $c - d$ means the number denoted by c less the number denoted by d.

52. The Sign $\times$ (read "multiplied by") is used to show that the numbers between which it is placed are to be multiplied.

Thus, $a \times c$ means the product of the numbers denoted by a and c.

a. The sign "$\cdot$" is sometimes used instead of $\times$ to denote multiplication. Thus, $5 \cdot b \cdot c$ means that the numbers denoted by 5, b, and c are to be multiplied together.

b. In algebra no sign indicates multiplication.

Thus, $15abc$ denotes the product of the numbers 15, a, b, and c.

[1] **Letters and Figures Representing Numbers.** — A word is the sign of an idea. Now, a word (or figure or letter) that *stands for* a number is as different from the number itself as the written word *wagon* is different from a real wagon. A number as defined in arithmetic is a unit or collection of units, and 5, or five, for example, is nothing but a symbol which refers to this particular number. In ordinary language we drop this distinction. To illustrate: an older person pointing to the picture of a horse in a book asks "What is this?" And the child immediately replies "It is a horse." Neither of the two refers to its being only the *picture* of a horse. So to the class in arithmetic the teacher says, "Put down the number fifty," when a more correct statement would be "Put down the figures which denote fifty. Such ellipses are very frequent in all discourse. Since much is gained in clearness and brevity and nothing is lost if this explanation be remembered, it seems best to use the abridged expressions. Thus in the above a correct statement would be "the sign $+$ denotes addition, and is used to show that the numbers between whose *symbols* it is placed are to be added. The difference between numbers and symbols of numbers, it is hoped, is here plainly emphasized, but circuitous language in definitions to insist on this distinction is avoided.

However, it must be understood that this does not apply to the arabic symbols. Thus, 2524 means two thousand five hundred twenty-four in algebra as well as in arithmetic.

53. The Sign $\div$ (read "divided by") denotes that the number preceding the sign is to be divided by the number following it.

Thus, $6\,ab \div 4\,c$ means that $6\,ab$ is to be divided by $4\,c$.

a. The fractional form $\frac{a}{b}$ is also used to denote a divided by b.

54. The Parentheses, ( ), are used to show that all inside is looked on as one number. Thus, $\left(\frac{6+4}{7}-\frac{8}{9}\right)$ would be treated just as if it were a simple fraction $\frac{34}{63}$. In the same way $\left(a-3\,b+\frac{c}{2}\right)$ would also be regarded as one number.

55. The Sign $=$ (read "equals to") is used to indicate that the algebraic expression on its left has the same numerical value as the expression on its right.

a. To explain further: The sign $=$ denotes that the numbers or combinations of numbers on its two sides reduce, when simplified, to the same number. We proceed to illustrate this: —

$$\frac{6 \div 2 + 4}{7 \times 3 \times 4} = \frac{11 - 2 \times 5}{3 + 18 \div 2}.$$

For, $\frac{6 \div 2 + 4}{7 \times 3 \times 4}$ reduces to $\frac{3+4}{84}$, which reduces to $\frac{1}{12}$;

and $\frac{11 - 2 \times 5}{3 + 18 \div 2}$ reduces to $\frac{11-10}{12}$, which reduces to $\frac{1}{12}$;

that is, when reduced, both have the same value, and therefore they are equal.

Likewise, $ab = 8\,c + a$ when $a = 8$, $b = 6$, and $c = 5$.

For, ab, or 8×6 is 48; and $8\,c + a = 8 \times 5 + 8$ is 48 also.

Hence, $ab = 8\,c + a$, when $a = 8$, $b = 6$, $c = 5$.

56. **An Exponent**[1] is a small symbol of number written to the right and above another number. When a whole number, it shows how many times the other is used as a factor.

Thus, $5 \times 5 = 5^2$; $a \times a \times a \times a = a^4$; $2a \times 2a \times 2a = (2a)^3$. b^c means c b's multiplied together.
If $c = 6$, then $b^c = b \times b \times b \times b \times b \times b$.

a. When no exponent is written, 1 is understood:—

Thus, a, i. e., a used once equals a^1.

b. The meaning and treatment of fractional exponents it is preferable to give further on, in Chapter XIX.

57. The **Product** arising from taking a number as a factor a certain number of times is called a **Power of the Number**.

Thus, $a \times a = a^2$ is the second power, or square of a.
$5 \times 5 \times 5 = 5^3$ is the third power, or cube of 5.
$(a + b) \times (a + b) \times (a + b) \times (a + b) = (a + b)^4$ is the fourth power of $a + b$.

a. The exponent is called the *index* of the power.

b. The second power of a number is usually styled its *square*; and the third power of a number, its *cube*.

58. **The Sign** $\sqrt{\ }$ (called a radical sign) is used to indicate a factor which multiplied by itself some number of times will produce the number under the sign.

59. **An Index of a Radical Sign** is a small figure or letter placed to the left and above it, to show into how many equal factors the number is to be separated.

Thus, $\sqrt{25} = \pm 5$; $\sqrt[3]{64} = 4$; $\sqrt[4]{81} = \pm 3$; $\sqrt[5]{32} = 2$; $\sqrt[a]{1} = 1$, in which a can have any value; for any root of 1 is 1.

a. When no index is written, 2 is understood. Thus $\sqrt{16} = \pm 4$; $\sqrt{a^2 + b^2}$ means a number, which, multiplied by itself, equals $a^2 + b^2$.

[1] From Latin *ex*, "out of," and *pono*, "to place;" i. e., it is placed *out from* the number to which it belongs.

60. One of its equal factors is called a **Root of a Number.**

Thus, $36 = 6 \times 6$, therefore 6 is a root of 36; $243 = 3 \times 3 \times 3 \times 3 \times 3$, therefore 3 is the fifth root of 243.

a. One of two equal factors is called a *square root*, and one of three equal factors is called a *cube root*.

Thus, $4a^2 = 2a \times 2a$, and $2a$ is the square root of $4a^2$. -6 is one of the square roots of 36; $\sqrt{49x^2} = \pm 7x$, read, "the square root of $49x^2$ equals plus or minus $7x$." $\sqrt[3]{27} = 3$, read, "the cube root of 27 equals 3."

SECTION II.

Classification of Symbols.

61. A Symbol, as used in Algebra, is a letter or sign with a distinct meaning.

Thus, a letter, as a, stands for a number; $+$ stands for the operation of addition; and "( )" indicates that an expression inside is to be looked upon as one number.

a. The Symbols used in Algebra are (1) Symbols of Number, (2) Symbols of Operation, (3) Symbols of Relation, (4) Symbols of Aggregation, (5) Symbols of Omission, (6) Logical Symbols.

62. Symbols of Number. — Besides the use of letters and figures to stand for numbers, two other characters are used, 0 (see **19**), and ∞.

a. The sign 0 (read "nought" or "zero") either denotes zero, as in Arithmetic, or an infinitesimally small number. It marks the beginning point for the two series.

b. The sign ∞ (read "infinity") stands for an exceedingly large number, greater than any that can be named.

63. Symbols of Operation. — These are $+$, $-$, $\times$, $\cdot$, $\div$, the exponent, and the radical sign. Division is also indicated by writing the dividend over the divisor.

64. Symbols of Relation. — These are $=$, $> <$, $\propto$, called, respectively, the signs of equality, inequality, and variation.

a. A horizontal V, $>$, or $<$ is used to show that the numbers between which it is placed are unequal. The opening is always turned toward the larger number. $>$ is read "greater than;" $<$ is read "less than."

Thus, $8 < 7$; $8 + 4 > 3 \times 3$; $6 < 10$.

b. The symbol $\propto$ means "varies as." As the number on its right increases, the number on its left increases, and as the number on its right decreases, the number on its left decreases in the same ratio.

Thus, a man's monthly wages $\propto$ the number of days he works.

c. A line drawn across any of these signs, thus, $\neq$, or $\not<$ is used to deny that which the symbol unmarked expresses; e.g., $6 \neq 5$, $11 \not< 10$.

65. Symbols of Aggregation. — These are ( ), { }, [ ], $\overline{\quad}$, $|$. Named in order, they are parenthesis, brace, bracket, vinculum, and bar. They are all used for the same purpose; viz., to show that an included expression is to be looked upon as one number. See **54**.

66. Symbols of Omission. — Dots or dashes, usually called symbols of continuation, are used to show that certain expressions have not been written, and are to be filled out from those that are given.

Thus, $a, a^2, a^3 \ldots a^{15}$; or, $x + x^2 + x^3 + x^4 \ldots x^{20}$.

They may be read "and so on to."

67. Logical Symbols. — The logical symbols are those of reason and conclusion The sign $\because$ (read "since" or "because") is used to mark a reason. The sign $\therefore$ (read "therefore," "hence") is used to mark an inference or conclusion.

SECTION III.

Classification of Algebraic Expressions.

68. **A Term** is an algebraic expression the parts of which are not separated by plus and minus signs.

Thus, $3\,ab$, $9\,a^2bc$, $10\,x$ are terms. Also $4\,ab + 2\,ac - 3\,bc$ is an expression consisting of three terms: the first $4\,ab$, the second $2\,ac$, and the third $3\,bc$.

a. A *simple* term is a single expression, as $2\,ax$, $5\,m$, $3\,p^2\,q$.

b. A *compound* term is any combination of simple terms *looked upon* as one expression. This is usually indicated by a symbol of aggregation. Thus, $(5\,a + 2\,b)$, $\left(\frac{a}{b} + \frac{c}{d}\right)$.

69. An algebraic expression consisting of but one term is called a **monomial.**

Thus, $5\,abc$, $11\,m^2$, $16\,ax$ are each monomials.

a. It is plain that both words *monomial* and *term* can be applied to the same expression and with the same intent.

To illustrate: $3\,abc$ can be called either a term or a monomial. Term is the more general word of the two, and may cover compound expressions. Monomial is more explicit in referring to a single term.

70. **A Binomial** is an algebraic expression consisting of two terms.

Thus. $a + b$. $2\,a - c$. $a^2b - c^2$.

71. **A Trinomial** is an algebraic expression consisting of three terms.

As, $a + b + c$, $ab + bc + ac$, $2\,abc - ab^2 + bc^2$.

72. **A Polynomial** is an algebraic expression consisting of many terms, usually taken, however, to mean two or more.

Thus. $a + b + c + d$; $a^2 - b^2 - c^2 - d^2 - e^2$; $3\,mn - 4\,m^2 + 5\,n^2 + 10\,m^2n^2$.

SECTION IV.

ON THE TERM.

73. A simple term in algebra, such as $5a^2x$, usually consists of a *coefficient* and a *literal part*, each letter having an exponent expressed or understood.

74. The Coefficient [1] shows how many times the number denoted by the product of the other factors is taken.

In $5a^2x$, (i. e., $a^2x + a^2x + a^2x + a^2x + a^2x$), 5 is the coefficient, and shows how many times the number denoted by a^2x is taken. In $6ax$, $6a$ may be regarded as the coefficient, showing how many times x is taken. Of course, the a is a literal factor; but we may, if we choose, regard some of the literal factors as part of the coefficient. In $3(a+b)$, [i. e., $(a+b)+(a+b)+(a+b)$], 3 is the coefficient of $(a+b)$.

a. The coefficient standing in front of its literal part shows how many times that part is to be added; while the exponent of an expression, written to the right and above it, shows how many times it is to be used as a factor in a multiplication.

To illustrate: $5ab$ means $ab + ab + ab + ab + ab$; while $(ab)^5 = ab \times ab \times ab \times ab \times ab$.

b. The coefficient 1 is not written, although, of course, it may be if desired. Thus, bc and $1bc$ are the same thing.

75. The Literal Part of a term is the part containing the letters.

In $10a^2b$, a^2b is the literal part.

a. One letter, as a in a^2b, is sometimes called a *dimension* of the term.

[1] From Latin *co* or *con*, "with," and *efficio*, "to effect, or make;" i. e., the coefficient is that which with the literal part makes up the term.

76. Similar Terms or Like Terms are those which have the same letters affected by the same exponents.

$7\,ab^2$ and $9\,ab^2$ are similar terms. So also are $11\,x^3y^2$ and $14\,x^3y^2$. Again, ax^2z and mx^2z are similar with respect to the literal part x^2z.

a. In arithmetic we could add and subtract like numbers, as $7 − $5 = $2 ; 10 qts. + 16 qts. = 35 qts. Likewise we can add 8 times a certain number to 5 times the same number and get 13 times the number. In the same manner *we can add and subtract what have been defined as similar terms.*

Thus, $7\,ab^2 + 9\,ab^2 = 16\,ab^2$.
For, if $a = 4$ and $b = 3$, $ab^2 = 4 \times 3^2 = 36$.
And $7 \times 36 + 9 \times 36 = 16 \times 36$.
Likewise, $11\,x^3\,y^2 - 8\,x^3\,y^2 = 3\,x^3\,y^2$.

b. For like and unlike *signs* of terms, see 26.

77. The Degree of a Term is the sum of the exponents of its literal factors. (Remember **56**, *a.*)

The expression a^2bc is of the fourth degree.
$6\,m^2n$ is of the third degree.
$9\,x^2y^3z^4$ is of the ninth degree.

78. Homogeneous Terms are those which are of the same degree. The following sets are homogeneous:

$6\,a^2bc^2$, $7\,ab^2c^2$, $9\,a^2b^2c$ all of the fifth degree.
$9\,x^2y^4$, $11\,x^6$, $4\,x^3z^3$, yz^5 all of the sixth degree.

a. A polynomial is said to be homogeneous when all its terms are of the same degree.

To Illustrate: $6\,m^3n^3 - 5\,m^2n\,p^3 - p^2q^4 - 12\,m^2p^2q^2$ is homogeneous.
So also is $a^2b - ab^2 - abc - a^2c + ac^2 + bc^2$

CHAPTER IV.

EXERCISES IN THE NOTATION.

SECTION I.

Exercise in Reading Algebraic Expressions.

79. **Read** the following expressions:

1. $a + b - c.$ 2. $x + x\,y - y.$

Remark.—If the first were to be read as suggested in **50** and **51** we should say "the sum of the numbers denoted by a and b, diminished by the number denoted by c." The second should read "the sum of the numbers denoted by x, and the product of the numbers denoted by x and y, less the number denoted by y." It is assumed, however, that by this time the student is familiar with the idea of a letter standing for a number, and it will be sufficient to read the first "a plus b minus c," and the second "x plus x times y minus y."

3. $12\,a - bc + d$ (**52**, b). 5. $a^2 + 3\,a$ (**57**).

4. $2\,a + 3\,b - c.$ 6. $2\,a^2 + 4\,b^3 - 3\,c^2.$

7. $a + b + c + d.$

Remark.—This may be read "a plus b, plus c, plus d;" but it is regarded as more elegant to say "the sum of a, b, c, and d." Also to use this mode of expression in reading binomials particularly. Thus, ex. 5 would be read, "the sum of a squared, and three times a." Likewise, a residual binomial, such as $a^2 - 4\,b^2$ would be read "the difference of a squared and four times b squared.

8. $6\,ab + 9\,ac + 14\,bc.$ 10. $4\,x^2 - 9\,y^2.$

9. $44\,abc - 11\,bcd.$ 11. $24\,a - 5\,b + 14\,c - 7\,d.$

12. $8\,abc - bcd + 9\,cde - def$.

13. $\sqrt{a+b}$ (59, a).

14. $c^2 \div d^2$, or $\dfrac{c^2}{d^2}$ (53, a).

15. $7\,a^3 - 3\,a^2b + c^3$.

SECTION II.

Exercise in Writing Algebraic Expressions.

80. Exercise in **writing** expressions: —

1. Express the sum of a, b, and c.

2. Express the double of b.

3. Express a, plus b divided by c.

4. By how much is a greater than 5?

5. Write the sum of a cubed, three times c squared, and the product of b, c, and d.

6. Write cd over b, plus four times b divided by three times a, minus cd divided by 24.

7. Write the sum of b squared and c squared, divided by the difference of two times c and three times a.

8. Express c to the fourth power, less four times c cubed, plus three times c less 6.

9. Write the sum of the sixth powers of a and b.

10. Write the difference of the fifth powers of x and y.

SECTION III.

Numerical Values.

81. The Numerical Value of an algebraic expression is the number obtained by giving a particular value to each letter and then performing the operations indicated.

82. Evaluate the following expressions:

1. $c + d - b$ when $c = 5$, $d = 10$, $b = 3$.
2. $6a^2x - 9y^2$ when $a = 4$, $x = 2$, $y = 3$.
3. $3a^2 + 2cx - b^3$ when $a = 5$, $b = 7$, $c = 15$, $x = 3$.
4. $\dfrac{b + c + d - x}{a}$ when $a = 4$, $b = 3$, $c = 5$, $d = 10$, $x = 2$.
5. $\sqrt{a^2 + b^2 + c}$, when $a = 9$, $b = 7$, $c = 14$.
6. $\sqrt{a} + \sqrt{b} + \sqrt{c}$, when $a = 9$, $b = 121$, $c = 64$.
7. Obtain the numerical values of the expressions in **80** when $a = 1$, $b = 2$, $c = 3$, $d = 4$, $e = 5$, $f = 6$, $x = 10$, $y = 5$.

83. It was pointed out in **76**, *a*, that **similar terms can be united into a single term by combining their coefficients.**

Thus, $7abc + 3abc - 2abc = 8abc$. And if $a = 2$, $b = 3$, $c = 1$, then $8abc = 48$, which is the same result as would be obtained by substituting in each term, and then uniting the separate results. $7abc = 42$; $3abc = 18$; $2abc = 12$, and $42 + 18 - 12 = 48$.

But if the literal parts were not the same, they would not have the same numerical value, and could not be added.

Thus, x^2z and xz^2 are not similar, and so we find when $x = 2$ and $z = 3$, that the first, $x^2z = 12$, and the second, $xz^2 = 18$.

Again, if we regard part of the literal factors as belonging to the coefficients, we can evaluate certain kinds of exercises more readily.

Thus, $ax^2y + bx^2y - 3cx^2y + 4dx^2y$, when $a = 5$, $b = 3$, $c = 2$, $d = 4$, and $x = 3$, $y = 7$ becomes $(a + b - 3c + 4d)x^2y = (5 + 3 - 6 + 16) \times 3^2 \times 7 = 18 \times 63 = 1134$. *Ans.*

1. $8x^2 + 6x^2 - 5x^2 = ?$ when $x = 4$.

2. $9x^2yz - 6x^2yz + 11x^2yz = ?$ when $x = 2$, $y = 3$, and $z = 5$.

3. $a^2x^2z + 3bx^2z - 2abx^2z = ?$ when $a = 5$, $b = 2$, and $x = 3$, $z = 2$.

4. $2m^2y^2z^2 - 2n^2y^2z^2 + 3mny^2z^2 = ?$ when $m = 4$, $n = 3$, and $y = 1$, $z = 2$.

84. Numerical Values introduced to verify Equations.

Show that the condition of an equality is fulfilled in the following:

1. $\dfrac{a^2 - x^2}{a + x} = a - x$, when $a = 5$, $x = 3$; when $a = 6$, $x = 1$.

2. $\dfrac{x^2 - x - 30}{x + 5} = x - 6$, when $x = 8$; also when $x = 9$.

3. $\dfrac{x^3 - y^3}{x - y} = x^2 + xy + y^2$, taking $x =$ any number, $y =$ any number.

85. To find the Numerical Value of a letter in an equation.

Take the solutions in Article **47** as models.

1. $x + 6 = 20$, $x = ?$ If 6 added to x equals 20, x must equal 14.

2. $2x + 6 = 20$.

3. $2x - 4 = 20$.

4. $3x + 6 = 24$.

5. $3x - 6 = 24$.

6. $3x + 10 = 5x$ ($\therefore 2x = 10$).

7. $9x - 16 = 5x$.

8. $8x - 35 = x$.

9. John is three times as old as James and the sum of their ages is 36; how old is each?

Let $x =$ James' age,
then $3x =$ John's age,
and $x + 3x =$ the sum of their ages.
But the sum of their ages is 36.

$$\therefore x + 3x = 36$$
$$4x = 36$$
$$\left.\begin{aligned} x &= 9 \\ 3x &= 27 \end{aligned}\right\} \textit{Ans.}$$

10. There are four times as many girls as boys in a party of 60. How many were boys and how many were girls?

Let $x =$ the number of boys, then $x + 4x = 60$ is the equation.

11. Divide a line 21 inches long into two parts, such that one part may be $\frac{3}{4}$ of the other.

SUGGESTION. $x + \frac{3}{4}x = 21$; $\frac{7}{4}x = 21$, $\frac{1}{4}x = \frac{1}{7}$ of $21 = 3$; $\frac{4}{4}x$, or $x = 4 \times 3 = 12$: $21 - 12 = 9$.

12. A stick of timber 40 feet long is sawed in two, so that one part is $\frac{2}{3}$ as long as the other. Required the length of each.

13. A farmer sold a sheep, a horse, and a cow for \$105. For the horse he received 10 times, and for the cow 4 times, as much as for the sheep. How much did he get for each?

14. A man being asked how much he gave for his watch replied: If you multiply the price by 4, to the product add 70, and from the sum subtract 50, the remainder will be 220. What did his watch cost?

15. Three boys together spend 50 cts.: the second spends 5 cts. more than the first, and the third three times as much as the first. How many cents did each spend?

16. Says A to B, "Good-morning, master, with your hundred geese." Says B, "I have not one hundred, but if I had twice as many as I now have and ten more I should have one hundred." How many geese had he?

17. An apple, a pear, and a peach cost 19 cts.: a pear costs 2 cents more than an apple and a peach 3 cents more than a pear. How much did each cost?

18. Distribute $3.50 among Thomas, Richard, and Henry, so that Richard shall have twice as much as Thomas, and Henry twice as much as Richard.

19. Divide the number 60 into three parts, so that the second may be three times the first, and the third double the second.

20. What number is that from which we obtain the same result whether we multiply it by four or subtract it from 100?

21. A boy bought the same number of tops, of marbles, and of balls for 65 cts.: for the marbles he gave 1 ct. each, for the tops 2 cts. each, and for the balls 10 cts. each. How many of each did he buy?

22. Out of 77 books a certain number were sold, when there remained three more than were sold? How many were sold?

23. John bought a certain number of tops and fifteen times as many marbles. After losing 14 of the marbles and giving away 30, he had only 16 left. How many tops did he buy?

24. A man bought 3 horses and 4 cows for $600. Each horse cost twice as much as a cow. How much did he give for each?

SUGGESTION. — Let x represent the price of a cow, and $2x$ that of a horse. How much did all the cows cost, and all the horses?

25. Seven men and three boys were hired for a week (6 days), a man receiving three times as much as a boy. Altogether their wages amounted to $72. What did each receive?

Additional problems may be found in Tower's Intellectual Algebra.

CHAPTER V.

ADDITION.

86. Addition[1] is the process of uniting two or more expressions into one called their sum. The sum is usually obtained in its simplest form.

a. **Meaning of Algebraic Addition.** — An algebraic expression was defined (48) as anything written in algebraic language. In Chapter II. the rules for adding algebraic numbers were given. We now proceed to apply the same rules for signs to literal expressions.

87. Examples in Addition. — It is convenient to classify examples in addition into four cases:

1. When the terms added are similar and have like signs.

(1. Monomials. (See **28.**)

(1)		(2)	(3)	(4)
$2a$		$-\ b$	$8x^2y^2$	$-18a^2y$
$8a$		$-10b$	$6x^2y^2$	$-2a^2y$
$9a$		$-4b$	$9x^2y^2$	$-20a^2y$
$12a$		$-3b$	$15x^2y^2$	$-a^2y$
$31a$	(See **76**, *a*.)	$-18b$	$38x^2y^2$	$-41a^2y$

(2. Polynomials. (**72.**)

(1) Add the three trinomials given below.

$$\begin{array}{rrr} 4a^2x & -\ 6a^2y & +\ 9y^2 \\ 5a^2x & -14a^2y & +17y^2 \\ 2a^2x & -\ \ a^2y & +10y^2 \\ \hline 11a^2x & -21a^2y & +36y^2 \end{array}$$

[1] The essential definition was given in Art. 16.

This solution evidently proceeds on the assumption that the three trinomials can be added *part at a time*. Thus, we first add the quantities in the left column, next those in the middle column, and lastly those in the right column. Is this justifiable by the nature of addition? See Commutative Law, **28**.

2. When the terms added are similar, but have unlike signs.

(1. Monomials. To add similar terms, add their coefficients (**76**, *a*). Consequently, in the examples now to be given the coefficients are added by the rules of **29**, and the resulting coefficient is prefixed to the common literal part.

(1) $\begin{array}{r} -6a^2 \\ +2a^2 \\ +5a^2 \\ +4a^2 \\ -3a^2 \\ \hline 2a^2 \end{array}$

EXPLANATION. — The sum of the positive coefficients is 11; the sum of the negative coefficients is 9; the difference is 2, and the sign of the greater is +. Hence, $+2$ is the coefficient of a^2 in the sum.

(2) $\begin{array}{r} 5xy \\ -11xy \\ -14xy \\ +7xy \\ -xy \\ \hline -14xy \end{array}$

(4) $\begin{array}{r} 8mn^2 \\ -2mn^2 \\ +21mn^2 \\ -72mn^2 \\ -11mn^2 \\ \hline \end{array}$

(3) $\begin{array}{r} 9xz^2 \\ -xz^2 \\ -7xz^2 \\ +3xz^2 \\ +10xz^2 \\ \hline +14xz^2 \end{array}$

(5) $\begin{array}{r} 100abc \\ 4abc \\ -92abc \\ +42abc \\ -9abc \\ \hline \end{array}$

(2. Polynomials.

The columns are added separately by the rule for monomials.

(1) $\begin{array}{rrrr} a & -2b & +3c & -4d \\ -2a & +3b & -4c & +5d \\ -4a & +5b & -4c & +7d \\ 3a & -4b & +5c & -6d \\ \hline -2a & +2b & & +2d \end{array}$

REMARK. — The third column cancels out of the sum.

3. When the terms to be added are not similar, or some similar and others dissimilar.

(1. Add $8a^3x^2 - 3ax$, $7ax - 5xy^4$, $9xy^4 - 5ax + 2x^2$, and $2a^3x^2 + xy^4 + 8y^2$.

OPERATION.
$$\begin{array}{rrrrr} 8a^3x^2 & -3ax & & & \\ & 7ax & -5xy^4 & & \\ & -5ax & +9xy^4 & +2x^2 & \\ 2a^3x^2 & & +\ xy^4 & & +8y^2 \\ \hline 10a^3x^2 & -\ ax & +5xy^4 & +2x^2 & +8y^2 \end{array}$$

EXPLANATION. — Similar terms are placed in the same column, changing the order in the expressions if necessary. Thus, the $-5ax$ of the third expression is written down to the left of the $9xy^4$ which came before it, as given in the problem. Moreover, as fast as new terms are obtained, they are written in new columns at the right.

REMARK. — In connection with this case it is proper to say that since addition is merely *uniting* algebraic expressions to obtain the sum, it is sufficient in all cases to write the expressions one after another with their proper signs. But the definition says the sum is usually obtained *in its simplest form.* The processes explained above give the results in their simplest forms provided that polynomials which have similar terms are first simplified as explained in the next case.

4. When the terms to be added constitute a single polynomial having some of its terms similar.

Simplification of a polynomial.

(1. Simplify $3a^2b + 9a^2c - 7a^2b + 4a^2c^2 - 2a^2c + 4a^2b$.

$3a^2b - 7a^2b + 4a^2b = 0$; $9a^2c - 2a^2c = 7a^2c$.

Therefore the polynomial reduces to $7a^2c + 4a^2c^2$.

From the examples and explanations here given we derive the rule in the next article.

88. Rule for Algebraic Addition.

1. Write the expressions to be added so that similar terms (**76**), shall stand in the same column. If none of the terms are similar, unite them just as they stand, case 3.

2. Add the coefficients of each column and annex the common literal part to the sum. In adding observe the rule in **29**.

3. Any terms standing alone are to be brought down with their signs unchanged.

4. To simplify a polynomial regard each term as one quantity, and proceed to add such as are similar as in the first part of this rule.

89. Exercise in Addition.

1. Add the following sets of monomials:

(1. $3a$, $5a$, $-2a$, $7a$, and $-4a$.
(2. $10am$, $-6am$, $4am$, $7am$, $-9am$, and am.
(3. $-3dy^3$, $4dy^3$ $-8dy^3$ $-13dy^3$, $2dy^3$ and $18dy^3$.
(4. $-6a^2$, $+2a^2$, $-5a^2$, $4a^2$, $-3a^2$ and a^2.
(5. $13m^2n$, $-10m^2n$, $-6m^2n$, $5m^2n$, and $-4m^2n$.
(6. axy, $-7axy$, $+8axy$, $-axy$, $-8axy$, and $+9axy$.

2. Add $3ab+5a^2b$, $6ab-8a^2b$, $8ab-a^2b$, and $3ab+2a^2b$.

3. Add $5x-3a+b+7$, and $-4x-3a+2b-9$.

4. Add $3a+7b-8c+d$, $3a-2b+c-e$, and $-a$, $-b$, $-c$, $-d$.

5. Add $7x^2-2x-5$, $2x^2-3x+8$, and $-9x^2+5x+3$.

6. Add $3a^2b^3-7ab^4+5axy$, $-7a^2b^3-2ab^4-axy$, $ab^4-7axy+8a^2b^3$, $-10ab^4+a^2b^3+3axy$, and $-5a^2b^3+18ab^4$.

NOTE. — Here it will be necessary to change the order of the terms as in 87, 3. In the review, the student may solve these problems mentally; i. e., by simply running the eye along, picking out similar terms, and writing down the sums thus obtained with their proper signs. By this plan it will not be necessary to write the problem or to arrange the terms.

7. Add $2ab - 3ax^2 + 2a^2x$, $12ab - 6a^2x + 10ax^2$, and $ax^3 - 8ab - 5a^2b$.

8. Add $x^4 - 2x^3 + 3x^2$, $x^3 + x^2 + x$, $4x^4 + 5x^3$, $2x^2 - 3x - 4$, and $-3x^2 - 2x - 5$.

9. Simplify $3xy - 5a + 6c - 3m + 5 + xy + 12 - 2m$.

10. Simplify $a + b + c - d - 2e - 2a + 3b - 5c - 5d + a - 4b + 6d - 3c + 4c + 5c$.

11. Simplify $8x^3 - 4x^2 + 2x^3 - 5x - 6x^2 + x^3 - 2 - x + x^2$.

12. Simplify $9b - \frac{3}{4}c + 15d - \frac{1}{8}b + \frac{7}{8}c - 2d + 4c - e$.

13. Add $2x^a + 3y^b$, $4x^a - 6y^b$, and $7x^a - 5y^b$.

14. Add $10a^2b - 12a^3bc - 15b^2c^4 + 10$, $8a^3bc - 10b^2c^4 - 4a^2b - 4$, $20b^2c^4 - 3a^3bc - 3a^2b - 3$, and $2a^2b + 12a^3bc + 5b^2c^4 + 2$.

15. Add $a + b$, $c - d$, and $e + f - g$.

16. Add $5ca^2x^2 + 4ba^2x^3 + mx^2y^2$, and $10ca^2x^2 - 2ab^2x^3 + 6mx^2y^2$.

17. Of two farmers the first had $2x - 3y$ acres, and the second had $x - y$ acres more than the first. How many acres had the second? How many acres had both together?

18. One man is worth $a + c$ dollars, a second is worth $a + 2b + c$ dollars, a third is worth $2d - e$ dollars, while a fourth is worth $a + b + f$ dollars. What is the sum of their fortunes?

CHAPTER VI.

SUBTRACTION.

90. Subtraction is the process of finding from two given expressions a third which added to the second will give the first.

The two given expressions are called respectively minuend and subtrahend, and the third, difference or remainder.

91. Examples in Subtraction.

1. Monomials.

To subtract one term from another similar to it, take the difference of their coefficients. (See **31** and **32**.)

EXPLANATION OF (2. — By **31** and **32**, we conceive the sign of the subtrahend coefficient -6 to be changed to $+6$, and we then add it to 11. So in the subsequent examples.

$$\text{(1.}\quad \begin{array}{r} 17\,a \\ 9\,a \\ \hline 8\,a \end{array} \qquad\qquad \text{(2.}\quad \begin{array}{r} 11\,x^2 \\ -\,6\,x^2 \\ \hline 17\,x^2 \end{array}$$

$$\text{(3.}\quad \begin{array}{r} -\,13\,xy^3 \\ 10\,xy^3 \\ \hline -\,23\,xy^3 \end{array} \qquad \text{(5.}\quad \begin{array}{r} 8\,a \\ 14\,a \\ \hline -\,6\,a \end{array} \qquad \text{(7.}\quad \begin{array}{r} -\,5\,xy^4 \\ 9\,xy^4 \\ \hline \end{array}$$

$$\text{(4.}\quad \begin{array}{r} -\,15\,y^4 \\ -\,12\,y^4 \\ \hline -\,3\,y^4 \end{array} \qquad \text{(6.}\quad \begin{array}{r} 16\,x^2 \\ -\,20\,x^2 \\ \hline \end{array} \qquad \text{(8.}\quad \begin{array}{r} -\,4\,y^4 \\ -\,12\,y^4 \\ \hline \end{array}$$

2. Polynomials. (Will it be correct to subtract *part at a time* just as we added part at a time in addition?)

(1. From $6a^2b^3c - 7abc^3 + 9ac - 3bc^4$ take $4a^2b^3c + 3abc^3 - 8ac - 10bc^4$.

Operation.

$$\begin{array}{rrrrl} 6a^2b^3c & -\ 7abc^3 & +\ 9ac & -\ 3bc^4 & \text{minuend.} \\ 4a^2b^3c & +\ 3abc^3 & -\ 8ac & -\ 10bc^4 & \text{subtrahend.} \\ \hline 2a^2b^3c & -\ 10abc^3 & +\ 17ac & +\ 7bc^4 & \text{remainder.} \end{array}$$

Explanation. — Conceiving $4a^2b^3c$ to have its sign changed to $-$, and adding it to $6a^2b^3c$ we get $2a^2b^3c$; conceiving $3abc^3$ to become $-3abc^3$ and adding this to $-7abc^3$, we have $-10abc^3$; conceiving $-8ac$ to become $+8ac$ and adding to $+9ac$ we have $+17ac$; finally conceiving $-10bc^4$ to become $+10bc^4$ and adding it to $-3bc^4$, we have $+7bc^4$. The difference is $2a^2b^3c - 10abc^3 + 17ac + 7bc^4$.

(2. From $c^2x^2 + 3cx^2 - 5cx - 4x^2$ take $c^2x^2 - 2cx + 3c^2x - 6c^3$.

$$\begin{array}{llllll} c^2x^2 & +\ 3cx^2 & -\ 5cx & -\ 4x^2 & & \\ c^2x^2 & & -\ 2cx & & +\ 3c^2x & -\ 6c^3 \\ \hline & 3cx^2 & -\ 3cx & -\ 4x^2 & -\ 3c^2x & +\ 6c^3. \end{array}$$

Explanation. — Conceiving c^2x^2 to become $-c^2x^2$ and adding we have 0, or simply c^2x^2 from c^2x^2 leaves nothing. Next we bring down $3cx^2$ since there is nothing to be subtracted from it; and so in the fourth column. Conceiving the sign of $3c^2x$ to become $-$, and adding $-3c^2x$ to 0, we have $-3c^2x$. Lastly, changing $-6c^3$ to $+6c^3$ and adding to 0 (*which amounts to bringing it down with its sign changed*), we get $+6c^3$.

92. Rule for Algebraic Subtraction.

1. Write the subtrahend under the minuend so that similar terms shall stand in the same column. If none of the terms are similar, unite the expressions after changing the signs of all the terms of the subtrahend.

2. Conceive, in turn, the sign of each term of the subtrahend to be changed, and add the result to the term above in the minuend, writing the sum beneath. Dissimilar terms in the minuend are brought down with their signs unchanged.

93. Exercise in Algebraic Subtraction.

1. Find the difference between the following sets of monomials:

(1. $8a$ and $6a$.
(2. $3b$ and $-2b$.
(3. $-6c$ and $11c$
(4. $-14x^2$ and $-13x^2$.
(5. $16a^2b^2$ and $-15a^2b^2$.
(6. a^2b and ab^2.
(7. $-11ad^3$ and $4ad^3$.
(8. $43a^2b^4$ and $19a^2b^4$.
(9. $5m$ and $-m$.

2. From $2x^2 - 3a^2x^2 + 9$ take $x^2 + 5a^2x^2 - 3$.

3. From $3ax - 4by + 3cz$ take $ax + 3by - 5cz$.

4. From $19ab - 2c - 7d$ take $2x^3 - 3x^2 + x + 1$.

5. From $a^2 + 2ax + x^2$ take $a^2 - x^2$.

6. From $x^3 - 3x^2y + 3x^2y^2$ take $-x^2y + 5x^2y^2 - y^3$.

7. Simplify the following expressions, and then take the second from the first: $4x^2y^3 - 5cz + 8m - 4cz - 2m$, and $cz + 2x^2y^3 - 4cz$.

8. From $a^3 - 6a^2c + 9ac^2$ take the sum of $-2a^2c - 4a^3 + 2ac^2$, and $-2a^2c + 3a^3 - ac^2$.

9. From $a^2b^2 + 12abc - 9ax^2$ take $4ab^2 - 6acx + 3a^2x$.

10. From $x^4 + 4x^3 - 2x^2 + 7x - 1$ take $x^4 + 2x^3 - 2x^2 + 6x - 1$.

11. A man bought two storerooms: for one he paid $3ax^2 + 2ar^2 - 3as^2 - 4at^2$ dollars; and for the other $ax^2 - 9as^2 + 3at - a^2$ dollars. How much more did he pay for the first than for the second?

12. A man who had four sons gave to the youngest $2x$ dollars, to the next older $2a + 100$ more than to the youngest; to the third he gave $2b + 50$ dollars more than he gave the seccnd; and to the eldest as much as to the three younger sons together. How much more did the eldest get than the next to the youngest?

CHAPTER VII.

SYMBOLS OF AGGREGATION, WITH EXERCISES IN ADDITION AND SUBTRACTION.

94. The Symbols of Aggregation (**54, 65**) are used to show that an enclosed expression is to be regarded as one number.

Thus $(3a^2 + 2ab + b^2)$ is no longer looked upon as three terms to be added together, but as a single compound term.

We might illustrate this use still further by an example, letting $\$3a^2$ stand for the value of a man's real estate, $\$2ab$ for his notes and cash, and $\$b^2$ for his personal property. Then $\$(3a^2 + 2ab + b^2)$ indicates how much he is worth without any reference to the values of the different terms.

Since the order in which terms are added is indifferent (**28**), a polynomial may be separated into any sets of different terms by means of parentheses, provided each term is included. This is called the *associative law in addition.*

95. The Word Quantity as used in algebra means an algebraic expression. Consequently the word quantity may be used to describe simply a positive or negative number; or, on the other hand, the most complicated expression.

a. Since *quantity* and *algebraic expression* are synonymous terms, see again the definition of the latter in **48**. Algebraists, and mathematicians generally, use the word quantity constantly. It is therefore very important that the student should have a just notion of its signification in algebra as explained above, as well as to know its original meaning given in **1**, *a*. The word quantity is not so suggestive as either of the words, number, or algebraic expression, and for this reason its use has hitherto been withheld. It is hoped that by this time the student has sufficiently clear ideas concerning numbers and algebraic expressions that he will be able to use intelligently instead of them the word quantity.

SECTION I.

ON COMPOUND TERMS.

96. Addition and Subtraction of Similar Compound Terms with numerical coefficients.

1. Add $7\ (a+b-c)$, $4\ (a+b-c)$, $-2\ (a+b-c)$, and $-6\ (a+b-c)$.

EXPLANATION. — As in **52**, *b*, the multiplication sign between 7 and the quantity $(a+b-c)$ is dropped. Adding the coefficients as in simple terms, $7+4-2-6=3$; $\therefore$ the answer is $3\ (a+b-c)$.

2. Add $2\ (a+b)$, $3\ (a+b)$, and $(a+b)$.

3. Simplify $4\ (x+z)+5\ (x+z)-11\ (x+z)$.

4. Simplify $9\ (a+b)^2+10\ (a+b)^2-(a+b)^2-2\,(a+b)^2$. (**57**).

5. Add $2\ (x+y)+3\ (x-y)$ and $3\ (x+y)-2\ (x-y)$.

6. Add $4\ (a+2\,b)^2-4\ (a-m)$, $-20\ (a+2\,b)^2+5\ (a-m)$, $12\,(a+2\,b)^2-14\ (a-m)$, and $-5\ (a+2\,b)^2+5\ (a-m)$

7. Reduce $17\,abc\ (a^4+a^3b+a^2b^2+ab^3+b^4)-6\,abc\ (a^4+a^3b+a^2b^2+ab^3+b^4)$ to its simplest form.

97. The Formation of Compound Coefficients for a Common Literal Part.

1. Add ax, bx, and cx with respect to x.

EXPLANATION. a times x plus b times x plus c times x equals the sum of a, b, and c times x; or $(a+b+c)\ x$. Otherwise, $ax+bx+cx=(a+b+c)\ x$.

2. Add axy and $3\,xy$ with respect to xy.

3. Add $ax + by$, $cx + dy$, $ex + fy$, and $gx + hy$ with respect to x and y.

4. Add ax, $2\,cx$, and $4\,dx$.

5. Add $am + bn$, and $bm + an$.

6. Add $ay + cx$, $3\,ay + 2\,cx$, and $4\,y + 6\,x$.

7. Simplify $ax + ay + az$ with respect to a.

EXPLANATION. x times a plus y times a plus z times a equals the sum of x, y, and z times a. Or, $ax + ay + az = a\,(x + y + z)$.

8. Add $3\,mx$, $7\,my$, $-\,7\,mz$, and $4\,m$ with respect to m.

SUG. $3\,mx = 3\,x \,.\, m$.

98. The Formation of Compound Coefficients for Similar Compound Terms.

1. Add $a\,(x + y)$ and $b\,(x + y)$. Answer $(a + b)\,(x + y)$.

NOTE. — As in other places the multiplication sign between parentheses is omitted.

2. From $a\,(x + y + z)$ take $b\,(x + y + z)$.

3. Add $a\,(x + y) + b\,(x - y)$, and $m\,(x + y) - n\,(x - y)$.

SECTION II.

PRINCIPLES AND RULES CONNECTED WITH THE SYMBOLS OF AGGREGATION.

99. Explanation of the Use of the Symbols of Aggregation.

1. By using parentheses the addition or subtraction of polynomials can be indicated.

Thus $(+\,5\,a^2cx^2 + 4\,a^2bx^3 + mx^2y^2) + (40\,a^2cx^2 - 2\,a^2\,bx^3 + 6\,mx^2y^2)$ indicates the sum of the two enclosed polynomials.

So, also, $(a^2b^2 + 12\,abc - 9\,ax^2) - (4\,ab^2 - 6\,acx + 3\,a^2x)$ indicates the difference required in ex. 9, Art. **93**.

By the remark in **87** the two quantities given first above can be added by simply uniting their terms. We have $(5\,a^2cx^2 + 4\,a^2bx^3 + mx^2y^2) + (40\,a^2cx^2 - 2\,a^2bx^3 + 6\,mx^2y^2)$ $= 5\,a^2cx^2 + 4\,a^2bx^3 + mx^2y^2 + 40\,a^2cx^2 - 2\,a^2bx^3 + 6\,mx^2y^2$.

It is plain from this that the only difference between the two modes of writing the sum is that the first indicates that it is made up of two parts, while the second represents it as one quantity. Consequently if no sign or a plus sign precedes a parenthesis, or other symbol of aggregation, the marks of parenthesis may be removed without altering the value of the quantity.

Again, the rule for subtraction directs to proceed as in addition *after having changed the sign of the subtrahend.* Hence, after changing the signs of the subtrahend, it may be united to the minuend.

$$\text{Thus } (a^2b^2 + 12\,abc - 9\,ax) - (+4\,ab^2 - 6\,acx + 3\,a^2x)$$
$$= a^2b^2 + 12\,abc - 9\,ax^2 - 4\,ab^2 + 6\,acx - 3\,a^2x.$$

This time when the parenthesis preceded by a minus sign was taken away, all the signs within were changed. Of course the reasons given apply to any other polynomials as well as to the trinomials used in illustration.

Restating the foregoing, it is briefly this: to add a quantity (indicated by a plus sign preceding the parenthesis enclosing it) its terms are joined just as they stand; but to subtract a quantity the sign of each of its terms must be changed according to the rule for the subtrahend in subtraction. It follows from this that any terms in a sum can be enclosed in a parenthesis with a + sign before it just as they stand, but must have their signs changed if enclosed in a parenthesis preceded by a − sign.

2. The parentheses are also sometimes used to enclose the *series sign* of a number.

Thus, $a - (-b)$ means that b from the negative series is to be subtracted from a.

To remove such parentheses, we follow the same course as in the preceding case.

$$a + (+b) = a + b;\ a + (-b) = a - b;$$
$$a - (+b) = a - b;\ a - (-b) = a + b.$$

100. Rules for the Symbols of Aggregation.—The previous article shows:

1. Any terms can be placed within a symbol of aggregation having a + sign before it just as they stand.

2. Any terms can be placed within a symbol of aggregation having a − sign before it, if the sign of each term be changed from + to −, or from − to +.

3. A symbol of aggregation having a + sign before it can be removed without changing the quantity.

4. A symbol of aggregation with a − sign before it can be removed if the sign of each term within it be changed.

NOTE.—The terms inside a symbol of aggregation are generally arranged so that a positive term comes first, when, following the usual custom, *no sign is written. The sign before the parenthesis* belongs to the whole quantity inside, and not to its first term. For example, $-(a - b) = -(+a - b)$; or, writing the same expression with a vinculum, $-\overline{a - b} = -\overline{+a - b}$. The expression $-(7 - 3 + 4 - 12) = -(-4) = +4$. Here the first term, 7, is positive The beginner should not forget this.

SECTION III.

EXERCISE IN REMOVING THE SYMBOLS.

101. Exercise in Removing the Symbols of Aggregation from Simple Expressions.—The results are to be simplified (**88.** 4).

1. $3x^2 - 2y + (2x - 1)$.
2. $3x + 2y - (x - y)$.

3. $x^2 + 2xy + y^2 - (x^2 - 2xy + y^2)$.

4. $2b + (b - 2c) - \{b + 2c\}$.

5. $-[a - b] - [b - c] - [c - a]$.

6. $-\overline{3m + 2n} - \overline{3m - 2n} + 9m$.

7. $ab - (m - 3ab + 2ax) - 7ab$.

8. $\overline{3a - b + 7c} - \overline{2a + 3b} - \overline{5b - 4c} + \overline{3c - a}$.

9. $1 - (1 - a) + (1 - a + a^2) - (1 - a + a^2 - a^3)$.

10. $(a - x - y) + (x - y - b) + (c + 2y)$.

102. Exercise in removing the Symbols from Complex Expressions. — The results are to be simplified.

1. $3a - [7a + 2 + \{4a - 5 - (-2a + 9 - \overline{a - 6})\}] - 10$.

Let us first remove the vinculum by changing the signs of a and 6, rule 4.

$3a - [7a + 2 + \{4a - 5 - (-2a + 9 - a + 6)\}] - 10$;

Next we remove the parenthesis, rule 4.

$3a - [7a + 2 + \{4a - 5 + 2a - 9 + a - 6\}] - 10$;

Next, the brace, rule 3.

$$3a - [7a + 2 + 4a - 5 + 2a - 9 + a - 6] - 10;$$

Finally, the bracket, rule 4.

$$3a - 7a - 2 - 4a + 5 - 2a + 9 - a + 6 - 10.$$

By simplifying, **88**, 4, we have

$8 - 11a$. *Ans.*

2. $2a - \{2b - (3c + 2b) - a\}$.

3. $3a + x - [a + 5x - (3a - 2x)]$.

4. $1 - \{2 - (1 - x + x^2)\}$.

Note. — This method of removing the symbols, viz., the innermost first; then the next, and so on, is the best at first. The expert algebraist writes the result directly.

5. $5a - [a + 5x - \{a - x - \overline{3a - 2x}\}]$.

6. $a + 2b - \{6a - [3b + (8x - \overline{2 + 6y} - x) + 4a] - 3b\}$.

7. $2x - [3y - \{4x - (5y - 6x)\}]$.

8. $ax + 4cx - (mx + cx - y) + [mx - (cx + y)]$.

9. $m + 4x - [-4y + 2x + (ay - x) + p]$.

10. $(a^2bc + 3c^2) + 3a^2bc - (m - c) - \{-(4a^2bc + c) - (-3c^2 - m)\}$.

11. $(6a - 3b - 2c) - \{5a - (10b + 5c)\} + \{2b - (2c - 2a)\}$.

12. $\{a - [b + (c + 2x) - (y - z)]\}$.

SECTION IV.

Uses of the Symbols.

103. General Exercise. The Uses of the Symbols of Aggregation.

1. Write the polynomial $a + b + c + d$ as a binomial so that the sums of a and b and c and d may be regarded as single quantities.

2. Write $a^2 + 2ab + b^2 - c^2 - 2cd - d^2$ as a residual binomial, regarding the first three and last three terms as single quantities.

3. Write $a + b + c$ as a binomial so as to show the sum of a and b as one quantity.

4. Write $a^2 + 2ab + b^2 - 2ac - c^2 + 2bc - a + b - c$ as a trinomial of which the sum of the squares forms one term, the sum of the products a second, and the sum of the factors a third.

5. Write $12ax + 12ay + 4by - 12bz - 15cx + 6cy + 3cz$ so that the terms which contain x may appear as one quantity, the terms which contain y as one quantity, and

the terms which contain z as one quantity, but subtracted from the other two.

6. A man who owns a mill worth \$$a$, has personal property worth \$$b$ and has \$$d$ in the bank. But he owes \$$e$ for his engine and owes his millwright \$$f$. However, the millwright is indebted to him for \$$g$ worth of flour. How shall we represent what he is worth, indicating the worth of the mill as one part and all the rest as another; also, keeping his personal property separate from his accounts, and indicating the difference he owes the millwright as one sum?

7. Remove the parentheses from the preceding result, obtaining the man's fortune indicated as assets and liabilities.

CHAPTER VII.

MULTIPLICATION.

104. The definition of multiplication was given in Art. **34**. Consult again that article.

We have,

$$+3 \times +4 = +12;\ 3 \times -9 = -27;\ -4 \times 8 = -32;$$
$$-12 \times -15 = +180.$$
$$+x \times +y = +xy;\ x \times -y = -xy;\ -x \times y = -xy;$$
$$-x \times -y = +xy.$$

a. In Art. 35 there was given an investigation of the rule of signs in multiplication. It will be helpful to give similar proofs at this point, using letters, which may stand for any numbers, instead of figures.

1. To multiply $+b$ by $+a$.

Once $+b$ is $+b$; twice $+b$ is $+2b$; three times $+b$ is $+3b$; and so on. Then a times $+b$ is $+ab$.

2. To multiply $-d$ by c.

Once $-d$ is $-d$; twice $-d$ is $-2d$; three times $-d$ is $-3d$. Then c times $-d$ is $-cd$.

3. To multiply n by $-m$.

From what was learned in Chapter II., this means that n is to be taken $0-m$ times; i.e., no times less m times. But, $0 \times n = 0$, and $m \times n = mn$. Subtracting the latter product from the former, we have,

$$0 - mn = -mn.$$

4. To multiply $-n$ by $-m$.

By the same reasoning as before, multiplying $-n$ by $0-m$, we would have,

$$0-(-mn) = +mn.$$

105. Rule for Signs in Multiplication. — Like signs give plus, and unlike minus. This rule applies to all other expressions as well as to single numbers.

106. Multiplication of Monomials. — In the multiplication of monomials, as in the product $-5a^2x \times +8ay$, there are three things to be considered; viz., the signs, the numerical coefficients, and the literal part.

1. *The Signs.* — The fundamental rule of signs is that just given in **105**. Extended, it includes the rules of **39** and **40**.

2. *The Numerical Coefficient.* — The numerical coefficients can be multiplied as in arithmetic, giving the numerical coefficient of the product.

Thus $5a \times 6b = 5 \times 6 \times ab = 30ab$.

3. *The Literal Part.* — We learned in **56** and **57** that a^4 is the shorthand way of writing $a \times a \times a \times a$; a^3 of $a \times a \times a$; a^2 of $a \times a$; or, dropping the multiplication sign, $a^2 = aa$, $a^3 = aaa$, $a^4 = aaaa$, etc.

(1. Let us seek the product of, say, a^2c, a^3b, and cx^2. Writing out the factors, we get,

$$a \times a \times c \times a \times a \times a \times b \times c \times x \times x.$$

Or, by **38**, 2, changing the order so as to bring the same letters together, we have,

$$a \times a \times a \times a \times a \times b \times c \times c \times x \times x = a^5bc^2x^2.$$

For, $a \times a \times a \times a \times a = a^5$, $c \times c = c^2$, and $x \times x = x^2$.

And just so it will be in every example of such multiplication. A letter that appears in two or more factors is repeated as many times as indicated by the sum of its exponents. A letter that appears in only one term is repeated in the product merely as many times as there are units in its exponent.

(2. In the same manner let us multiply x^m by x^n.

$x^m = x \cdot x \cdot x \cdot x \cdot x \ldots x$, m factors being multiplied together.

$x^n = x \cdot x \cdot x \cdot x \ldots x$, n factors being multiplied together.

$x^m \cdot x^n = x \cdot x \cdot x \cdot x \cdot x \cdot x \ldots x$, $m + n$ factors being multiplied together.

$\therefore x^m x^n = x^{m+n}$;

i.e., the exponents of the factors are *added* for the exponent of the same letter in the product.

(3. Multiply together the following compound expressions.

$(a + b)^2 \, mx$ and $(a + b)(m + n) \, y$.

Writing the product in full, we have

$(a + b)(a + b)(a + b)(m + n) \, mxy = (a + b)^3 (m + n) \, mxy.$

4. *Examples of the Multiplication of Monomials.*

(1. $4 \, a^2 \times 6 \, abc = 24 \, a^3bc.$

REASONS. $+$ by $+$ gives $+$; $4 \times 6 = 24$; $a^2 \times a = a^3$, or adding the exponents $2 + 1 = 3$; b and c appearing in only one factor are brought down.

(2. $-3 \, xy \times xy^2 = -3 \, x^2y^3.$

REASONS. $-$ by $+$ gives $-$; $3 \times 1 = 3$; $x \times x = x^2$; $y \times y^2 = y^3$.

(3. $7 \, m^2n^sp^t \times - mn^u \times - 2 \, n^2p^2 \times - 3 \, p^{r-2} = - 42 \, m^3n^{s+u+2} \, p^{t+r}.$

REASONS.—An odd number of $-$ signs, viz., three, gives $-$ in the product; $7 \times 1 \times 2 \times 3 = 42$; $m^2 \times m = m^3$, or, $2 + 1 = 3$; $n^s \times n^u \times n^2 = n^{s+u+2}$, for $s + u + 2$ is the sum of the exponents of n: $p^t \times p^2 \times p^{r-2} = p^{t+r}$, for the sum of the exponents is $t + 2 + r - 2 = p + r$.

107. Rule for the Multiplication of Monomials.

1. An odd number of minus signs gives minus in the product; otherwise the product is positive (**39**).

2. Multiply together the coefficients which are expressed in figures.

3. In the literal part, add the exponents in the factors for the exponent of the corresponding letter in the product, and bring down single factors.

a. For convenience it is customary to arrange the literal part in the order in which the letters come in the alphabet. Thus, ax would be written, and not xa, unless for some special reason; x^2am^2y would be arranged into am^2x^2y; z^2yx^2 into x^2yz^2; and so on.

108. Exercise in the Multiplication of Monomials.

1. $(+8)(-2)$.
2. $(+6a)(-2a)$.
3. $(+5mn)(+9mn)$.
4. $(+8ab)(-3c)$.
5. $(2ab^2x)(3a^2b^3x^2)$.
6. $-x^2y \times 2ax$.
7. $7a^5c \times 2ab$.
8. $-2a^4 \times -3a^2b$.
9. $a^2b^c \times a^2b^c$.
10. $2xy^2 \times 2x^2y$.
11. $-4ab^2x \times 5ax^2y$.
12. $-x^2y^3z \times -x^4y^5z^2$.
13. $2x^m \times x$.
14. $y^a \times -y^b$.
15. $-3(a+b)^3 \times -9(a+b)^2$.
16. $2(x+y) \times 4a^2(x+y)$.
17. $2m \times n \times -a \times -2b$.
18. $8ab^3 \times 2a^2c \times -5c^2$.
19. $-6ab^2y^3 \times 2by^3 \times 4a^2y$.
20. $-3abxy \times -2a^4bx^2 \times -8cx^2 \times -5y^2 \times -a$.
21. $-5a^2b \times ab^3 \times -3a^2c \times -5abc$.
22. $-3c^2dm \times -2c^2dm \times -5c^2dm$.
23. $2x^2y \times 3xy^2 \times z^2 \times -xz \times -x^3z$.
24. $-7am^2 \times -3b^2n^2 \times -4a^2b \times -a^2bn \times -2b^2n \times -mn^2$.

25. $-e^2x \times -3x \times -b^2c \times -ay^4 \times -5c^2y \times -cxy \times y^2$.

26. $\frac{1}{4}ax \times 3cx \times -\frac{1}{2}mx \times -4y^2 \times 6m$.

27. $x^a \times x^b \times x^c \times x^d$.

109. Multiplication of Polynomials.

1. To multiply a polynomial by a monomial, as,

$$a - b - c + d \text{ by } m.$$

To multiply $a - b - c + d$ by m is, of course, the same as to add m quantities each equal to $a - b - c + d$. In this sum a would appear m times; $-b$, m times; $-c$, m times; and $+d$, m times. Hence, $m(a - b - c + d) = ma - mb - mc + md$.

2. Since by the commutative law the *order* of multiplication is indifferent,

$$(a - b - c + d)\,m = am - bm - cm + dm.$$

And so it is in general. Whatever be the number of terms within the parenthesis, each term of the quantity inside must be multiplied by the quantity outside.

3. To multiply two polynomial sums, as,

$$a + b + c + d \text{ by } m + n + p + q.$$

Here it is plain that $a + b + c + d$ is to be taken the sum of m, n, p, and q times. This gives,

$$m(a + b + c + d) + n(a + b + c + d) + p(a + b + c + d) + q(a + b + c + d) = ma + mb + mc + md + na + nb + nc + nd + pa + pb + pc + pd + qa + qb + qc + qd.$$

Thus every term of the multiplicand is multiplied by each term of the multiplier, as in multiplication in Arithmetic. Indeed, the explanation just given is the justification of the arithmetical rule.

4. Finally, let us multiply two differences, say,

$$x - y \text{ by } m - n.$$

Now, $x - y$ taken $m - n$ times means $x - y$ taken m times *less* $x - y$ taken n times, or,

$$m(x - y) - n(x - y) = mx - my - (nx - ny).$$

Therefore, removing the parenthesis in the last,

$$(m - n)(x - y) = mx - my - nx + ny.$$

By inspection one sees that to obtain the product both terms of the multiplicand have been multiplied by each term of the multiplier, according to the rule of signs **(105)**.

Thus, $-my$ comes from multiplying $-y$ by m, and $+ny$ from multiplying $-y$ by $-n$. This result may be considered as a generalized proof of the rule of signs **(36)**.

By uniting sets of positive terms into one, and negative terms into one with parentheses, any example of the multiplication of polynomials can be reduced to the multiplication of two residual binomials. Thus,

$$\begin{aligned}&(a - b + c - d - e)(l - o - p - q - r)\\ &= [(a + c) - (b + d + e)][l - (o + p + q + r)]\\ &= (a + c)l - (a + c)(o + p + q + r) - (b + d\\ &+ e)l + (b + d + e)(o + p + q + r).\end{aligned}$$

Now, remembering 3, above, it follows that in the final product every term of the multiplicand would be multiplied by each term of the multiplier.

5. The Distributive Law. — The multiplication of every term of the multiplicand by each term of the multiplier is termed the *distributive law*.

110. Examples of the Multiplication of Polynomials. — The multiplication of polynomials is indicated by enclosing them in parentheses.

a. The operation of multiplication is similar to that in arithmetic, but with this modification, that we usually let the work extend to the right instead of to the left. To this end the first term of the multiplier is set under the first term of the multiplicand, the second term under the second term, and so on. Moreover, we begin at the left to multiply. In the following examples the multiplication is first indicated, and then the work is performed beneath. As in arithmetic, the simpler of the two factors is generally taken as the multiplier.

1. $4a(3a - 2b) = ?$

$$\begin{array}{l} 3a - 2b \\ 4a \\ \hline 12a^2 - 8ab \end{array}$$

2. $(3y^2 + y - 2)xy = ?$

$$\begin{array}{l} 3y^2 + y - 2 \\ xy \\ \hline 3xy^3 + xy^2 - 2xy \end{array}$$

3. $-7a^3cx^2(4ay - 3x + 2c) = ?$

$$\begin{array}{l} 4ay - 3x + 2c \\ -7a^3cx^2 \\ \hline -28a^4cx^2y + 21a^3cx^3 - 14a^3c^2x^2 \end{array}$$

4. $(3a^2 - 4b^2 - 2c^2) \times -a^2c = ?$

$$\begin{array}{l} 3a^2 - 4b^2 - 2c^2 \\ -a^2c \\ \hline -3a^4c + 4a^2b^2c + 2a^2c^3 \end{array}$$

$$\begin{array}{llll} 2x^2 - 3x - 4 & & & \\ 2x^2 - 3x - 4 & & & \\ \hline 4x^4 - 6x^3 - 8x^2 & & & \text{multiplying by } 2x^2 \\ \quad -6x^3 + 9x^2 + 12x & & & \text{multiplying by } -3x \\ \qquad -8x^2 + 12x + 16 & & & \text{multiplying by } -4 \\ \hline 4x^4 - 12x^3 - 7x^2 + 24x + 16 & & & \textit{Ans.} \end{array}$$

6. $(3a^3b - 2ab^3 + b^4)(2ab + b^2) = ?$

$$\begin{array}{ll} 3a^3b - 2ab^3 + b^4 & \\ 2ab + b^2 & \\ \hline 6a^4b^2 - 4a^2b^4 + 2ab^5 & \text{multiplying by } 2ab \\ \qquad -2ab^5 + 3a^3b^3 + b^6 & \text{multiplying by } b^2 \\ \hline 6a^4b^2 - 4a^2b^4 + 3a^3b^3 + b^6 & \textit{Ans.} \end{array}$$

Explanation. — The first term of the second product is $3a^3b^3$, but it is not similar to any term in the first product, and so we set it to the right and in a line beneath the first product. The second term of the second product is $-2ab^5$, and we set it under the term to which it is similar; and so generally, as soon as dissimilar terms appear, they are placed to the right.

7. $(x^4 - 3x^2 + 2x + 1)(x^3 - 2x - 2) = ?$

$$
\begin{array}{lllllll}
x^4 - 3x^2 + 2x + 1 & & & & & & \\
x^3 - 2x - 2 & & & & & & \\
\hline
x^7 & -3x^5 & +2x^4 & +x^3 & & & \\
 & -2x^5 & & +6x^3 & -4x^2 & -2x & \\
 & & -2x^4 & & +6x^2 & -4x & -2 \\
\hline
x^7 & -5x^5 & & +7x^3 & +2x^2 & -6x & -2 \quad \textit{Ans.}
\end{array}
$$

8. $(a^2 - ab - ac + b^2 - bc + c^2)(a + b + c)$

$$
\begin{array}{llllllllll}
a^2 - ab - ac + b^2 - bc + c^2 & & & & & & & & & \\
a + b + c & & & & & & & & & \\
\hline
a^3 & -a^2b & -a^2c & +ab^2 & -abc & +ac^2 & & & & \\
 & +a^2b & & -ab^2 & -abc & & +b^3 & -b^2c & +bc^2 & \\
 & & +a^2c & & -abc & -ac^2 & & +b^2c & -bc^2 & +c^3 \\
\hline
a^3 & & & & -3abc & & +b^3 & & & +c^3
\end{array}
$$

Ans.

9. $(a^2 + ab + b^2)(a^2 - ab + b^2)(a^2 - b^2) = ?$

$$
\begin{array}{llllll}
a^2 + ab + b^2 & & & & & \\
a^2 - ab + b^2 & & & & & \\
\hline
a^4 & +a^3b & +a^2b^2 & & & \\
 & -a^3b & -a^2b^2 & -ab^3 & & \\
 & & +a^2b^2 & +ab^3 & +b^4 & \\
\hline
a^4 & & +a^2b^2 & & +b^4 & \\
a^2 - b^2 & & & & & \\
\hline
a^6 & +a^4b^2 & +a^2b^4 & & & \\
 & -a^4b^2 & -a^2b^4 & -b^6 & & \\
\hline
a^6 & & & -b^6 & \quad \textit{Ans.} &
\end{array}
$$

111. General Rule for Multiplication.

1. Monomials. See **107.**

2. Polynomial and Monomial.

Multiply every term of the polynomial by the monomial, according to the rule for monomials, and unite the products.

3. Polynomials.

(1. Two Polynomials. — Multiply every term of the multiplicand by each term of the multiplier and add the partial products.

(2. Three or more Polynomials. — Multiply any two together and their product by a third, and so on.

a. If one factor of a product is 0, of course the product is 0.

112. Exercise in Multiplication.

1. Multiply together $8a$, $-6a$, $2a^2$, $-b$, $-3ab$, and b^3.
2. $(2a + 3b) \cdot \times 2 \times 2 \times c$.

REMARK. — Set down this and the following exercises as in 110, always combining monomial factors — when there are more than one — into a single monomial product, before using the polynomial. Thus the monomial factor here is $4c$.

3. $-\frac{3}{7}ax(by - 2cx + 5)$.
4. $(13m^4n^2 - 11a^2b^6 + 5x^4y^2) \times 6a^3m^4y$.
5. Multiply $-m - n - a - c$ by $-m$.
6. $(a - 2b - 3c)(-14c)$.
7. $-(a - b - c)(-2abc)$.
8. $4ab(a^2 - 3ab - 5b^2)(-2ab^2)$.
9. Multiply $x + 3$ by $x + 2$.
10. Multiply $x + 8$ by $x - 2$.
11. Multiply $p^2 + q^2$ by $x^2 - y^2$.
12. Multiply $x^2 + y^2$ by $x^2 - y^2$.
13. $(x^2 - 4x + 16)(x + 5)$.
14. $(x - 9)(x - 5)$.
15. $x + a$ and $x + a$.
16. $m + a$ and $m + b$.

17. $5x + 4y$ and $3x - 2y$.

18. $a + b - c$ and $m - n$.

19. $3x + y - 3z$ and $x + y + z$.

20. $x^2 + y$ and $x + y^2$.

21. $x^2 - x + 1$ and $x + 1$.

22. $(2ac^2 - 3by)(2c^3 - 3y^2)$.

23. $(x^2 + xy + y^2)(x^2 - xy + y^2)$.

24. $(a^4 + a^2c^2 + c^4)(a^2 + c^2)$.

25. $c^7 - c^5 + c^3 - c$ and $c^2 + 1$.

26. $x + y + w$ and $y + z + w$.

27. $a^3 + a^2b + ab^2 + b^3$ by $a + b$.

28. $a + b - c$ and $a - b + c$.

29. $a^3 - a^2 + a - 1$ by $a^2 - a + 1$.

30. $(x^4 + 2x^3 + 3x^2 + 2x + 1)(-3x^2)(-2x)$.

31. $\{ac - (a - b)(b + c)\} - b\{b - (a - c)\} = ?$

SUGGESTION.—First perform the multiplication indicated in $(a - b)(b + c)$. The product enclosed in a parenthesis takes the place of the factors. Then remove the two inner parentheses. Lastly b is multiplied into each term of the second bracket quantity, according to the distributive law, and this product is subtracted from the first brace quantity.

32. $(x - 1)(x - 2) - 3x(x + 3) + 2\{(x + 2)(x + 1) - 3\}$.

33. $4[4a - (4a + 1)(a - 3) - 9] - a\{(12a - 2)2 - (a - 3)(a - 4)\}$.

34. $a^m + b^n$ by $a^m + b^n$.

35. $(1 + y + y^2 + y^3 + y^4)(1 - y)$.

36. $a^x + a^y + a^z$ by $a + 1$.

37. $(x - 5)(x - 6)(x - 7)$.

113. Use of one Letter to stand for a Quantity expressed by any number of letters.

In **109**, 4, we had x, y, m, and n, each representing sums of other letters. Likewise, in any example, one letter may take the place of any combination of letters. For, when the combination is reduced to its numerical value, it gives one number, and the letter used to take the place of the combination has this number for its value.

Thus, A may stand for $\frac{a^2 + 3\,bc}{m}$

or, d may replace $(3\,c^3 + 7\,xy)^2$.

For, if $a = 5$, $b = 4$, $c = 3$, $m = 11$,

$$\frac{a^2 + 3\,bc}{m} = \frac{5^2 + 3 \cdot 4 \cdot 3}{11} = \frac{61}{11}.$$

Thus A must have the value $\frac{61}{11}$

In like manner, if $c = 2$, $x = 3$, and $y = 1$.

$$d = (3 \cdot 2^3 + 7 \cdot 3 \cdot 1)^2 = 45^2 = 2025.$$

a. A rule or theorem expressed by means of letters is called a *formula*.

114. Theorems in Multiplication.

There are certain problems in multiplication which occur so many times that a great saving is made if the learner be able at once to write down the product instead of going through the operation of multiplication. They are called *theorems* because they are proofs of general truths (**201**). The theorems of multiplication are of a very simple character.

1. The square of the sum of two quantities.

(1. Examples. The student will write down and multiply the following as in **112**.

(1) $(a + b)^2 = (a + b)\,(a + b)$ (**57**). *Ans.*, $a^2 + 2\,ab + b^2$.

(2) $(m^2 + n^2)^2$. (4) $(3\,x + 4)^2$.

(3) $(5\,a^2b + 2)^2$. (5) $(3\,a^2b + 7\,c^3)^2$.

(2. In order to generalize this problem we say,

Let A denote any quantity (**113**),

and B " " second quantity.

$$\begin{array}{l} A + B \\ A + B \\ \hline A^2 + AB \\ \quad + AB + B^2 \\ \hline A^2 + 2\,AB + B^2 \end{array}$$

The square is found by simple multiplication. The result, since it is true for any quantities, may be translated into a general theorem as follows: —

(3. THEOREM I. *The square of the sum of two quantities is equal to the square of the first, plus twice the product of the first by the second, plus the square of the second.*

(4. EXERCISE. — The result is to be written down without multiplying.

(1) $(3+4)^2 = 3^2 + 2\ 3 \times 4 + 4^2 = 9 + 24 + 16 = 49 = 7^2$.
(2) $(4\,ax + 9\,y)^2 = 16\,a^2x^2 + 72\,axy + 81\,y^2$.
(3) $(2\,a + 4\,b)^2$. (4) $(5\,m + 2\,np)^2$.

2. The square of the difference of two quantities.

(1. Examples. — To be set down and the multiplication performed as in **112**.

(1) $(11 - 6)^2$. (4) $(4\,m^2 - 6\,n^2)^2$.
(2) $(a - b)^2$. (5) $(5\,x^3 - 1)^2$.
(3) $(a^2 - 2\,b)^2$. (6) $(4\,xy - 3\,z)^2$.

(2. Generalization.

$$\begin{array}{l} A - B \\ A - B \\ \hline A^2 - \ AB \\ \quad - \ AB + B^2 \\ \hline A^2 - 2\,AB + B^2 \end{array}$$

Let A denote any quantity, and B denote any second quantity. Then tho square of $A - B$ is $A^2 - 2\,AB + B^2$. Hence,

(3. THEOREM II. *The square of the difference of two quantities is equal to the square of the first minus twice the product of the first by the second, plus the square of the second.*

(4. EXERCISE. — The answers must be *written down* by the theorem.

(1) $(a - 3ab)^2$.
(2) $(2x - 3y)^2$.
(3) $(4y^2 - 3)^2$.
(4) $(6y - 7z)^2$.
(5) $(3y^3 - z^2)^2$.
(6) $(10 - 9mn)^2$.

3. The product of the sum and difference of two quantities.

(1. Examples. — Set down and multiply as in **112**.

(1) $(5a + b)(5a - b)$.
(2) $(3m^2 + 5p)(3m^2 - 5p)$.
(3) $(11m^2 - 6n^3)(11m^2 + 6n^3)$.
(4) $(15a^4 - 1)(15a^4 + 1)$.

(2. Generalization.

$$\begin{array}{l} A + B \\ A - B \\ \hline A^2 + AB \\ \quad - AB - B^2 \\ \hline A^2 - B^2 \end{array}$$

Let A denote any quantity, and B denote any second quantity. Then the product of A + B and A − B is $A^2 - B^2$. Hence,

(3. Theorem III. *The product of the sum and difference of two quantities is equal to the difference of their squares.*

(4. Exercise. — The answers must be written down by the theorem.

(1) $(8 + 5)(8 - 5)$.
(2) $(6ab + 9)(6ab - 9)$.
(3) $(7xyz - 1)(7xyz + 1)$.
(4) $(a^n + b^n)(a^n - b^n)$.

115. Exercise in the Use of the Theorems of Multiplication.

1. $(a + b)(a + b)$.
2. $(a + 4)(a + 4)$.
3. $(7 + 5)(7 + 5)$.
4. $(a - x)(a - x)$.
5. $(m + n)(m - n)$.
6. $(x + y)^2$.
7. $(3a + 2b)^2$.
8. $(5c^2d - 4cd^2)^2$.
9. $(\frac{1}{2}m^2 + \frac{1}{3}n^2)^2$.
10. $(4g^2h^3 - \frac{1}{4})^2$.

11. $(6+3)(6-3)$.

12. $(4a^2+3b^2)(4a^2-3b^2)$.

13. $(8abc-9ab^2c^3)^2$.

14. $(\frac{1}{2}m^3n-1)^2$.

15. $(a^5-b^6)^2$.

16. $(a^x-b^y)^2$.

17. $(a^m+b^n)^2$.

116. Other Noteworthy Cases in Multiplication.

1. Prove by Theorem I. that

$$(a+b+c)^2=a^2+b^2+c^2+2ab+2ac+2bc.$$

We have, $[(a+b)+c]^2=(a+b)^2+2(a+b)c+c^2=a^2+2ab+b^2+2ac+2bc+c^2.$

2. Prove by the same theorem that

$$(a+b+c+d)^2=a^2+b^2+c^2+d^2+2ab+2ac+2ad+2bc+2bd+2cd.$$

$$[(a+b)+(c+d)]^2=(a+b)^2+2(a+b)(c+d)+(c+d)^2=a^2+2ab+b^2+2ac+2ad+2bc+2bd+c^2+2cd+d^2.$$

The formulæ of this and the preceding example generalized give the following theorem.

3. THEOREM IV. *The square of the sum of any number of quantities is equal to the sum of their squares, plus twice the product of each quantity by all those that follow it.*

4. Prove by actual multiplication that

$$(a+b)^3=(a+b)(a+b)(a+b)=a^3+3a^2b+3ab^2+b^3.$$

From this formula we have,

5. THEOREM V. *The cube of the sum of two quantities is equal to the cube of the first, plus three times the square of the first times the second, plus three times the first times the square of the second, plus the cube of the second.*

6. $(a+b)^4=a^4+4a^3b+6a^2b^2+4ab^3+b^4.$

Prove this equality by actual multiplication of the former result $(a+b)^3$ by $a+b$. State a theorem for this case.

7. Perform the multiplications indicated in the following:

(1. $(x+3)(x+4)$.	(7. $(x+2)(x-6)$.
(2. $(x+6)(x+9)$.	(8. $(x-3)(x-4)$.
(3. $(x+13)(x+17)$.	(9. $(x+9)(x-12)$.
(4. $(x+1)(x-5)$.	(10. $(x-15)(x+16)$.
(5. $(x+18)(x-4)$.	(11. $(x-12)(x-8)$.
(6. $(x+20)(x-4)$.	(12. $(x-25)(x-30)$.

8. THEOREM VI. *The product of two binomials having a term common is equal to the square of the common term, plus the* SUM *of the other terms times the common term, plus the* PRODUCT *of the other terms.*

9. Exercise in the use of the theorems of this article.

(1. $(x+11)(x+13)$.	(6. $(a-b+3)^2$.
(2. $(x-6)(x-1)$.	(7. $(4ax+5b+6c)^2$.
(3. $(x-20)(x+8)$.	(8. $(15x-9y-12z)^2$.
(4. $(x+100)(x-2)$.	(9. $(3m+n+2p+1)^2$.
(5. $(a+b+1)^2$.	(10. $(a+2b-(3c-d))^2$.

(11. $(x+3)(x+4)(x-3)(x-4)$.

SUGGESTION. — Multiply the first and third factors together, and the second and fourth, and then these products.

(12. $(1+c)(1+c)(1-c)(1+c^2)$.

(13. $(x+2y)^3$.

117. Simple Powers. — See definition in **57** and treatment of powers of simple algebraic numbers in **42**.

1. Powers of Monomials.

(1. Find the fourth power of $3a^2b^3c$.

By the rules of multiplication for monomials,

$$3a^2b^3c \times 3a^2b^3c \times 3a^2b^3c \times 3a^2b^3c = 81a^8b^{12}c^4.$$

(2. Rule.

(1) SIGNS. — The rules of **42,** 1 apply equally here.

(2) The coefficient expressed in figures is raised to the power indicated.

(3) By the rule for multiplication (**107**) the exponents of the factors are added; but here the numbers to be added are equal, and we can *multiply* each exponent by the number denoting the power to which the quantity is raised.

(3. EXERCISE.

(1) Multiply $3\,(2\,a^2b)^2$ by $4\,(-\,ab^2c)^3$.

$$(2\,a^2b)^2 = 4\,a^4b^2;\ \text{and}\ (-\,ab^2c)^3 = -\,a^3b^6c^3.$$

$\therefore$ we have $3 \times 4\,a^4b^2 \times -\,4\,a^3b^6c^3 = -\,48\,a^7b^8c^3$.

Notice that the coefficients outside of the parentheses are not affected by the exponents.

(2) Square the following:

$3\,xy^2,\ bx^4,\ -\,3\,cx^8,\ -\,2\,abc^2,\ 4\,a^4b^5x^2,\ a^m,\ x^{2p}$.

(3) Cube

$2\,ab^2n^3,\ -\,3\,a^3b,\ -\,2\,ab^2mn^3,\ 7\,pq^2n^3,\ a^mb^n$.

(4) Develop the following:

$3\,(4\,x^4)^4,\ (-2x^3y)^5,\ (-\,a^2x)^6,\ (-\,pq)^6,\ 2\,(-\,b)^7,\ \{(a^2)^3\}^4$.

(5) Develop and multiply,

$7\,(-\,2\,xy)^4 \times (-\,2\,am)^3;\ (-\,1)\,(-\,1)^2\,(-\,1)^3\,(-\,)^4$.

REMARK. — Since $1 = 1^2 = 1^3 = 1^4 = 1^5$, etc., we may regard 1 as having any exponent desired.

2. Powers of Polynomials.

(1. Square of Polynomials. — See Theorem IV. of the last article. Being true for any number of terms, it includes Theorems I. and II. of Art. **114**.

(2. **Cube and Fourth Power of Polynomials.** — See **116**, 5 and 6 for cube and fourth power of a *binomial*. The further investigation of the cases already given and of the development of higher powers is remanded to a subsequent chapter on Involution.

CHAPTER IX.

DIVISION.

118. The definition of division was given in Art. **43**. All the rules for signs were explained there.

119. Division of Monomials.

1. The signs. — See **43**.

2. The Coefficients. — Divide the coefficient of the dividend by the coefficient of the divisor to obtain the coefficient of the quotient.

3. The Literal Part. — In multiplication we *added* the exponents of the same letter in the factors. Thus, $a^2 \times a^5 = a^7$. Hence, by the definition of division (**43**), we must *subtract* the exponent in the divisor from that in the dividend for the exponent in the quotient.

Thus, $a^7 \div a^5 = a^2$; $x^m \div x^n = x^{m-n}$.

Three special cases deserve attention.

(1. *When a letter* is found in the dividend and not in the divisor, it must appear in the quotient, else when the divisor and quotient are multiplied they will not give the dividend.

(2. When a letter contains the same exponent in both dividend and divisor it gives rise to the factor 1 in the quotient, and is not written; for, any factor is contained in itself once.

Thus, $6\,a^2b \div 2\,a^2 = 3 \cdot 1 \cdot b = 3\,b$.

(3. When a letter is found having a greater exponent in the divisor than in the dividend, or appearing in the divisor and not in the dividend, the quotient is usually written as a fraction. A factor in the dividend cancels an equal factor in the divisor, as in the previous case. Hence, we subtract the exponent in the dividend from that in the divisor, and write the letter with this exponent in the *denominator*. When a letter is found in the divisor and not in the dividend, it is written in the *denominator* of the quotient.

A fraction as here used is to be regarded as an expressed division, the numerator being the dividend and the denominator the divisor.

4. Examples in the Division of Monomials.

(1. $6\,a^4b^2 \div 2\,a^2b = 3\,a^2b.$

REASONS. $+ \div +$ gives $+$; $6 \div 2 = 3$; $4 - 2 = 2$; $2 - 1 = 1$.

(2. $9\,m^3\,n^2p \div 3\,n^2 = 3\,m^3\,p.$

(3. $-7\,r^5s^2 \div -4\,rs^3 = \dfrac{7\,r^4}{4\,s}.$

(4. $-45\,a^4b^3c \div 15\,a^5b^3 = -\dfrac{3\,c}{a}.$

120. Rule for the Division of Monomials.

1. The Signs. — Like signs in the dividend and divisor give plus in the quotient, unlike signs give minus.

2. Numerical Coefficients. — Divide the coefficient of the dividend by that of the divisor for the coefficient of the quotient.

3. The Literal Part. — Subtract the exponent of any quantity in the divisor from its exponent in the dividend for its exponent in the quotient. When the remainder is negative the factor is written in the *denominator* of the

quotient. Cancel quantities having equal exponents, and bring down the other quantities, if any, in the dividend (numerator) and divisor (denominator) of the quotient.

121. Exercise in the Division of Monomials.

1. $+81 \div 3$.
2. $-64 \div 16$.
3. $48\,a^2b \div 2$.
4. $a^4x^3 \div a^2x$.
5. $12\,a^5b^3 \div -3\,a^2b$.
6. $\dfrac{-10\,e^2x^5y^5v}{-2\,evy^4}$
7. $\dfrac{-2\,acx^8y^4v^2}{-14\,ax^5y^4}$.
8. $15\,a^4x^3y \div -10\,a^2x^4y^2$.
9. $\dfrac{-12\,a^6c^6}{3\,a^3c^4}$.
10. $\dfrac{a^nb^2}{ab}$.
11. $\dfrac{-21\,b^3cx}{17\,bcx^3}$.
12. $\dfrac{-6\,(a-b)^3}{-2\,(a-b)}$
13. $\dfrac{6\,(m+n)^5}{3\,(m+n)^2}$.
14. $\dfrac{x^2-y^2}{x-y}$.
15. $\dfrac{(x-y)^2}{(x-y)^3}$.
16. $\dfrac{(a+b)^p}{(a+b)^q}$.
17. $14a^mb^nc^p \div -2\,a^2b^3c^q$.
18. $-16\,x^ay^b \div -8\,x^b$.
19. $\dfrac{(x-y)^3\,(m-n)^4}{(x-y)^2\,(m-n)^2}$.

122. Leading Letter in a Polynomial.

In most polynomial expressions in division there is what is called a *leading letter*.

Thus, in $ax^6+bx^5+cx^4+dx^3+ex^2+fx+g$, x is the leading letter. Usually it is quite easy to distinguish it. Sometimes, however, one letter seems as prominent as another, as in $a^2+2\,ab+b^2+2\,ac+2\,bc+c^2$. In such cases it is convenient to regard that letter which comes first in the alphabet as the leading one, and to let the others come after it as nearly as may be in their alphabetical order. Thus, a would be the leading letter in the above expression. In the terms in which a does not appear b would be the leading letter, and so on.

123. A **Polynomial** is said to be **arranged** with reference to the powers of the leading letter when the exponents of this letter either increase or decrease in regular order.

For example,

$$2a^4 + 2a^3b + 5a^2b^2 - 6ab^3 + 4b^4$$

is arranged with reference to the descending powers of a.

Also, $$1 + 2x - 6x^3 + 8x^5$$

is arranged with reference to the ascending powers of x.

124. Division of Polynomials. — Explanations.

There are two cases corresponding to short and long division in arithmetic.

1. Division of a Polynomial by a Monomial.

As in multiplication we multiply every term of the polynomial by the monomial; so in division, for similar reasons, we divide every term of the polynomial by the monomial. Such examples may be set down as in short division in arithmetic.

$$\begin{array}{l} 3a^2b)\ \underline{9a^3b^2 + 15a^2b^3c - 3a^2bc^3} \\ \qquad\quad 3ab + 5b^2c - c^3 \end{array}$$

2. Division of a Polynomial by a Polynomial.

(1. Investigation of the Process of Division. — Since division is the converse of multiplication, let us multiply two polynomials; then, using the product as a dividend, and one factor as the divisor, let us try to obtain the other factor as quotient by retracing our steps in multiplication.

$$\begin{array}{l}
2\,x^2 - 3\,ax + b^2 \\
\underline{3\,x^2 + 4\,bx - a^2} \\
6\,x^4 - 9\,ax^3 + 3\,b^2x^2 \\
\qquad\qquad\qquad + 8\,bx^3 - 12\,abx^2 + 4\,b^3x \\
\qquad\qquad\qquad\qquad\qquad\qquad\qquad\qquad - 2\,a^2x^2 + 3\,a^3x - a^2b^2 \\
\hline
6\,x^4 - 9\,ax^3 + 3\,b^2x^2 + 8\,bx^3 - 12\,abx^2 + 4\,b^3x - 2\,a^2x^2 + 3\,a^3x - a^2b^2
\end{array}$$

For reasons given farther on, we arrange the dividend with reference to x.

$$\begin{array}{lll}
\text{DIVISOR.} & & \text{QUOTIENT.} \\
2\,x^2 - 3\,ax + b^2) & \text{DIVIDEND.} & (3\,x^2 + 4\,bx - a^2 \\
\hline
\multicolumn{3}{l}{6\,x^4 - 9\,ax^3 + 8\,bx^3 - 2\,a^2x^2 - 12\,abx^2 + 3\,b^2x^2 + 3\,a^3x + 4\,b^3x - a^2b^2} \\
\multicolumn{3}{l}{6\,x^4 - 9\,ax^3 \qquad\qquad\qquad\qquad\qquad\qquad + 3\,b^2x^2} \\
\hline
\multicolumn{3}{l}{\qquad\qquad 8\,bx^3 - 2\,a^2x^2 - 12\,abx^2 \qquad\qquad + 3\,a^3x + 4\,b^3x - a^2b^2} \\
\multicolumn{3}{l}{\qquad\qquad 8\,bx^3 \qquad\qquad - 12\,abx^2 \qquad\qquad\qquad\quad + 4\,b^3x} \\
\hline
\multicolumn{3}{l}{\qquad\qquad\qquad - 2\,a^2x^2 \qquad\qquad\qquad + 3\,a^3x \qquad\qquad - a^2b^2} \\
\multicolumn{3}{l}{\qquad\qquad\qquad - 2\,a^2x^2 \qquad\qquad\qquad + 3\,a^3x \qquad\qquad - a^2b^2} \\
\hline
\end{array}$$

EXPLANATION. — We know *in advance* that the quotient is the former multiplier, $3\,x^2 + 4\,bx - a^2$.

In order to obtain the first term of the quotient, $3\,x^2$, the first term of the dividend, $6\,x^4$, is divided by $2\,x^2$, the first term of the divisor. Then, as in arithmetic, the divisor is multiplied by the first term of the quotient, giving $6\,x^4 - 9\,ax^3 + 3\,b^2x^2$, and the product is subtracted from the dividend. In order to obtain the second term of the quotient, $4\,bx$, we divide the first term of the remainder, $8\,bx^3$, by the first term of the divisor, $2\,x^2$. Then we multiply the whole divisor by this quotient term, and subtract the product from the first remainder.

To get the third term of the quotient the first term of the last remainder, $-2\,a^2x^2$ is divided by the first term of the divisor. After multiplying the whole divisor by the quotient term just found, and subtracting, there is no remainder, and the division is completed.

a. Remark on the Process of Long Division. — If we look closely at the example given above, we see that the process of division succeeds in separating the dividend into three trinomial parts, into each of which the divisor is contained exactly.

The same separation of the dividend is seen in long division in arithmetic.

$$\begin{array}{l}
25)\ 675\ (27 \\
\quad\ \ \underline{50(0)} \\
\quad\ \ 175 \\
\quad\ \ \underline{175}
\end{array}
\qquad \text{i.e.} \qquad
\begin{array}{r}
25)\ \underline{500 + 175 = 675} \\
20 + \quad 7 = 27.
\end{array}$$

(2. Divide $4x^5 - x^3 + 4x$ by $2x^2 + 3x + 2$.

DIVIDEND.

$$
\begin{array}{l|l}
4x^5 - x^3 + 4x & 2x^2 + 3x + 2 \quad \text{DIVISOR.} \\
\underline{4x^5 + 6x^4 + 4x^3} & \underline{2x^3 - 3x^2 + 2x} \quad \text{QUOTIENT.} \\
\quad -6x^4 - 5x^3 & \\
\quad \underline{-6x^4 - 9x^3 - 6x^2} & \\
\qquad\qquad 4x^3 + 6x^2 + 4x & \\
\qquad\qquad \underline{4x^3 + 6x^2 + 4x} &
\end{array}
$$

Here the dividend is separated into

$$(4x^5 + 6x^4 + 4x^3) + (-6x^4 - 9x^3 - 6x^2) + (4x^3 + 6x^2 + 4x).$$

By adding these we may verify that their sum is the dividend. Writing the dividend in this form and dividing as in short division

$$
\begin{array}{l}
2x^2 + 3x + 2)\ \underline{(4x^5 + 6x^4 + 4x^3) + (-6x^4 - 9x^3 - 6x^2) + (4x^3 + 6x^2 + 4x)} \\
\qquad\qquad\qquad 2x^3 \qquad\quad - \qquad\quad 3x^2 \qquad\quad + \qquad\quad 2x
\end{array}
$$

b. Necessity of dividing the first term of the remainder by the first term of the divisor.

From all that has been said it is plain that the *first term* of the dividend and of each remainder *correspond* to the *first term* of the divisor, and when divided by it give the different terms of the quotient. Now not only the dividend but each successive remainder must be kept arranged with reference to the leading letter. Had we, by a mistake in the arrangement of the first remainder above, brought $-12abx^2$ in advance of $8bx^3$, we should have obtained $-6ab$ as a term of the quotient!

125. Examples in the Division of Polynomials.

1. Divide $3a^mb^3 + 6a^4b^n - 3ab^3$ by $-3ab^2$.

$$
\begin{array}{l}
-3ab^2)\ \underline{3a^mb^3 + 6a^4b^n - 3ab^3} \\
\qquad\qquad -a^{m-1}b - 2a^3b^{n-2} + b \quad \textit{Ans.}
\end{array}
$$

2. Divide $4y^4 - 5y^2 + 1$ by $1 - y^2 - y$.

Arranging both dividend and divisor according to the ascending powers of y, we have

$$
\begin{array}{l|l}
1 - 5y^2 + 4y^4 & 1 + y - 2y^2 \quad \text{Divisor.} \\
\underline{1 + y - 2y^2} & \overline{1 - y - 2y^2} \quad \text{Quotient.} \\
\quad - y - 3y^2 & \\
\quad \underline{- y - y^2 + 2y^3} & \\
\qquad - 2y^2 - 2y^3 + 4y^4 & \\
\qquad \underline{- 2y^2 - 2y^3 + 4y^4} &
\end{array}
$$

a. Had both dividend and divisor been arranged according to the descending powers of y, it would merely *have reversed the order of the terms in the quotient.* Sometimes one arrangement is preferable, sometimes another.

b. In writing the partial products it is often difficult to have terms come under those to which they are similar. When dissimilar terms are written in the same column, care is necessary on the part of the beginner to keep from subtracting their coefficients.

c. The divisor and quotient of the first example of the preceding article were written respectively at the left and right of the dividend as in arithmetic. For convenience of multiplying in Algebra both are written at the right.

d. When there is no term in the minuend similar to a term in the subtrahend, the latter is subtracted from zero, and according to the rule for subtraction, is *brought down with its sign changed.*

3. Divide $d^4 + 1$ by $d + 1$.

$$
\begin{array}{l|l}
d^4 + 1 & d + 1 \\
\underline{d^4 + d^3} & \overline{d^3 - d^2 + d - 1} + \dfrac{2}{d+1} \\
\quad - d^3 & \\
\quad \underline{- d^3 - d^2} & \\
\qquad + d^2 & \\
\qquad \underline{+ d^2 + d} & \\
\qquad\quad - d + 1 & \\
\qquad\quad \underline{- d - 1} & \\
\qquad\qquad 2 &
\end{array}
$$

As in arithmetic, the remainder is written over the divisor and added to the other terms of the quotient. In this addition a + sign must be used. See **52** *b*.

126. Rule for Division.

1. Monomials. See **120**.

2. Polynomial by a Monomial. — Divide each term of the Polynomial by the Monomial.

3. Polynomial by a Polynomial.

(1. Arrange both dividend and divisor according to the descending or ascending powers of the same leading letter.

(2. Divide the first term of the dividend by the first term of the divisor and place the result in the quotient; multiply the divisor by this quotient term and subtract the product from the dividend, and arrange the remainder with reference to the leading letter.

(3. Divide the first term of the remainder by the first term of the divisor for the second term of the quotient; multiply the divisor by this term and subtract as before; and so continue until all the terms are brought down.

(4. If there is a remainder write it over the divisor, and add this fraction to the other terms of the quotient.

127. Exercise in Division.

1. $30\,a^4b \div -5\,a^2b = ?$

2. $27\,m^3n^4\,(a+c)^5 \div 9\,mn^4\,(a+c)^3$.

3. $3\,x^{m+n} \div x^{m-n}$.

4. $a^s \div a^t$.

5. $(35\,m^3y + 28\,m^2y^2 - 14\,my^3) \div -7\,my$.

6. $(12\,a^4y^6 - 16\,a^5y^5) \div 4\,a^4y^3$.

7. Divide $36\,p^3q^2rs^3 - 72\,p^2q^2rt^3 - 84\,pq^2rx^2$ by $-12\,pq^2$.

8. $\dfrac{4\,a^2bc + 8\,abc^2 + 12\,ab^2c}{2\,abc}$.

9. $\dfrac{27\,(a-b)^5 - 18\,(a-b)^3 + 9\,(a-b)^2}{9\,(a-b)}$.

10. Divide $2x^2 + 7xy + 6y^2$ by $x + 2y$.

11. Divide $ac + bc - ad - bd$ by $a + b$.

12. $(12 - 4a - 3a^2 + a^3) \div (4 - a^2)$.

13. $(4a^4 - 5a^2b^2 + b^4) \div (2a^2 - 3ab + b^2)$.

14. Divide $5x^2y + y^3 + x^3 + 5xy^2$ by $4xy + y^2 + x^2$.

15. Divide $\frac{1}{3}x^3 + \frac{17}{6}x^2 - \frac{5}{4}x + \frac{9}{4}$ by $\frac{1}{3}x + 3$.

16. Find the multiplicand when the product is $3x^4 + 14x^3 + 9x + 2$ and the multiplier is $x^2 + 5x + 1$.

17 $(a^4 - x^4) \div (a - x)$.

18. $(a^5 - 243) \div (a - 3)$.

19. Divide $a^4 - 3x^4$ by $a + x$.

20. Divide $6a^6x^3 - 14a^5x^6 + 12a^4x^9 - a^7$ by $2a^2x^3 - a^3$.

21. Divide $ax^2 - ab^2 + b^2x - x^3$ by $(x + b)(a - x)$.

22. $(a^3 + b^3 + c^3 - 3abc) \div (a^2 + b^2 + c^2 - bc - ac - ab)$.

23. $\left(\frac{1}{2}p^2 + \frac{1}{2}pq + pr - q^2 + \frac{7}{2}qr - \frac{3}{2}r^2\right)$

$$\div \left(\frac{p}{2} - \frac{q}{2} + \frac{3r}{2}\right).$$

24. Divide $x^{4n} + x^{2n}y^{2n} + y^{4n}$ by $x^{2n} - x^ny^n + y^{2n}$.

25. Divide $x^6 - 2x^3 + 1$ by $x^2 - 2x + 1$.

26. Divide $x^6 - a^6$ by $x^3 + 2ax^2 + 2a^2x + a^3$.

27. Divide $a^8 + a^6b^2 + a^4b^4 + a^2b^6 + b^8$ by $a^4 + a^3b + a^2b^2 + ab^3 + b^4$.

128. Zero and Negative Exponents. — Theorems of Notation.

Zero exponents and negative exponents arise when the exponent in the divisor is equal to or greater than that in the dividend.

Thus, $a^5 \div a^5 = a^{5-5} = a^0$, and $b^6 \div b^{10} = b^{6-10} = b^{-4}$.

Their use was obviated in **119**, 3, as explained in (2 and (3 of that Article.

1. Zero Exponents.

Let A be any quantity and n be any exponen

Then, $\frac{A^n}{A^n} = A^{n-n} = A^0$; but $\frac{A^n}{A^n} = 1$,

$\therefore A^0 = 1$.

Thus, $5^0 = 1$, $12^0 = 1$, $1000^0 = 1$.

THEOREM I. *Any quantity whose exponent is 0 is equal to unity.*

NOTE. — There is a sharp distinction between zero used as a coefficient and zero used as an exponent.

For, $0 \times a = 0$, while $a^0 = 1$.

The zero exponent shows that the given factor cancels out of the quotient, and is therefore used *no times* as a factor in it.

For example, $\frac{5\,a^2b^3c}{3\,a^2d} = \frac{5\,b^3c}{3\,d} \times \frac{a^2}{a^2} = \frac{5\,b^3c}{3\,d} \times 1 = \frac{5\,b^3c}{3\,d}$.

2. Negative Exponents.

In Article **119** it was noted that when the exponent of a letter in the divisor is greater than that of the same letter in the dividend, the dividend quantity is cancelled out, and the letter with the excess exponent remains in the divisor.

Thus, $\frac{6\,a^5b^2}{3\,a^2b^4} = \frac{2\,a^3}{b^2}$.

But, now, if the rule to subtract the exponent of the divisor from that of the dividend were followed strictly, a negative exponent would result.

Thus, $\frac{6\,a^5b^2}{3\,a^2b^4} = 2\,a^3b^{-2}$.

This suggests the idea that changing the sign of the exponent removes the factor itself to the opposite term of the division.

To test this a means must be found of transferring a factor from one term of the division to the other. If, now, we seek to remove such a factor as a^6 in $\frac{5a^6b^2}{3c^2d}$ from the numerator, it may be done (using a principle well known in arithmetic) by dividing both terms of the fraction by a^6. (See **156.**)

$$\frac{5a^6b^2 \div a^6}{3a^0c^2d \div a^6} = \frac{5a^{6-6}b^2}{3a^{0-6}c^2d} = \frac{5a^0b^2}{3a^{-6}c^2d} = \frac{5b^2}{3a^{-6}c^2d}.$$

Similarly, C^{-n} in $\frac{A^m}{BC^{-n}}$ may be transferred by multiplying both terms by C^n.

$$\frac{A^m}{BC^{-n}} = \frac{A^m \times C^n}{BC^{-n} \times C^n} = \frac{A^mC^n}{BC^{n-n}} = \frac{A^mC^n}{BC^0} = \frac{A^mC^n}{B}.$$

In like manner, any other factor could be transferred from the denominator to the numerator or from the numerator to the denominator, by changing the sign of its exponent.

Theorem II. *Any factor may be transferred from one term of a division to the other, providing the sign of its exponent is changed.*

Remark. — Here we see the opposite nature of the signs when prefixed to exponents. If a *positive* exponent denotes that its quantity is to be *multiplied* into another quantity, a *negative* shows that its quantity is to be used as a divisor of the other quantity.

The student is asked to note that we have come quite naturally to the meaning of a negative exponent. In subtracting exponents we have simply followed the rule for division (which in turn came from the rule for multiplication), and have merely subtracted a greater exponent from a less.

a. The following definition may prove helpful to the student.

A $\left\{\begin{matrix}\text{positive}\\\text{negative}\end{matrix}\right\}$ integral power of a number is the continued $\left\{\begin{matrix}\text{product}\\\text{quotient}\end{matrix}\right\}$ of 1 by that number.

b. The **Reciprocal** of a quantity is 1 divided by that quantity. The reciprocal of a is $\frac{1}{a}$. But by this article $\frac{1}{a} = \frac{a^{-1}}{1} = a^{-1}$. Hence the exponent -1 denotes the reciprocal of a quantity.

3. Exercise in the use of Zero and negative Exponents.

(1. What will x^{n-2} equal when $n = 2$?

(2. What does $a^2 \times a^3 \times a^{-6}$ equal?

(3. Reduce the expression $\frac{5\,a^2b^{-1}c^{-4}}{6\,a^2m^{-3}d}$ to an equivalent one having positive exponents.

By cancelling the *a's* and transposing those letters which have negative exponents to the opposite term of the division, we derive

$$\frac{5\,a^2b^{-1}c^{-4}}{6\,a^2m^{-3}d} = \frac{5\,m^3}{6\,bc^4d}$$

(4. $3^{+2} = ?$ 3^{-2}, 3^0, 3^{-1}, etc. Write all the values in a series from the exponent $+4$ to -4.

(5. Reduce to positive exponents $6\,a^{-1}b^3c \div 2\,abc^{-3}$; $2^{-1}\,x^{-3} \div y^{-1}$; $(3\,am^{-4}n \times 2^{-2}\,mn) \div \frac{3}{4}am^{-3}n^2$.

(6. Reduce to positive exponents,

$$p^{-2}q^{-3};\quad \frac{1}{r^{-3}\,5^{-2}};\quad \frac{a^{-8}b^4c^2}{a^{-6}b^{-4}d}.$$

(7. Simplify

$$b^2 + \frac{1}{b^{-2}} + 3\,x - \frac{1}{x^{-1}};\quad \left(\frac{1}{m^{-1}} + \frac{1}{n^{-1}}\right);\quad \left(m^{-1} + n^{-1}\right);\quad (a^{-1})^3.$$

129. Simple Roots. — Dividing a number by one of its roots two or more times will give the quotient unity, since a root of a number is one of its equal factors (**60**).

1. Roots of Monomials.

(1. EXAMPLE. — Having a root of a quantity, say the fourth root of $81\,a^8b^{12}c^4$, to find, we see that if we extract the fourth root of the coefficient and then divide each exponent by 4, we get a quantity ($3\,a^2b^3c$), which being divided into $81\,a^8b^{12}c^4$ and then into the quotient, and so on four times, gives a quotient 1. Therefore by the definition, $3\,a^2b^3c$ is the fourth root of $81\,a^8b^{12}c^4$. This process, it is plain, applies generally.

(2. Rule for extracting the root of a monomial.

(1) Extract the required root of the coefficient as in arithmetic.

(2) Divide each exponent in the literal part by the index of the root required to obtain the exponents of the respective letters in the root.

(3) Find the sign of the root by the rules in **45**.

(3. Exercise in the extraction of the roots of monomials.

(1) $\sqrt{16\,a^2b^4}$.

(2) $\sqrt{9\,a^2b^6}$.

(3) $\sqrt[3]{8\,x^3y^9}$.

(4) $\sqrt[4]{256}$.

(5) $\sqrt[3]{-27\,m^3}$.

(6) $\sqrt[5]{-32\,a^5c^{10}}$.

(7) $\sqrt{625\,p^8q^{12}}$.

(8) $\sqrt[4]{81\,x^{16}}$.

(4. Different ways of regarding the same power.

(1) Thus $8\,a^8 = (a^2)^4 = (a^4)^2$; $a^{12} = (a^6)^2 = (a^4)^3 = (a^3)^4$.

(2) $a^6 - b^6 =$ the difference of two sixth powers of a and b,
also $=$ " " " the cubes of a^2 and b^2,
also $=$ " " " " squares of a^3 and b^3,

(3) $a^{12} + b^8 = (a^3)^4 + (b^2)^4 = (a^6)^2 + (b^4)^2$.

(4) $a^{16} - b^{16} = (a^2)^8 - (b^2)^8 = (a^4)^4 - (b^4)^4 = (a^8)^2 - (b^8)^2$.

(5) $a^{18} - b^{12} = ?$

2. Rules for finding the square, cube, etc., roots of polynomials will be given in Chapter XVIII.

130. Theorems in Division: Divisibility of Binomials. —

1. THEOREM I. *The difference of the same powers of two quantities is divisible by the difference of the quantities.*

(1. Let the student verify the following by performing the divisions indicated (**124**).

$(A - B) \div (A - B) = 1$; $(A^2 - B^2) \div (A - B) = A + B$.
$(A^3 - B^3) \div (A - B) = A^2 + AB + B^2$.
$(A^4 - B^4) \div (A - B) = A^3 + A^2B + AB^2 + B^3$.
$(A^5 - B^5) \div (A - B) - A^4 + A^3B + A^2B^2 + AB^3 + B^4$.
$(A^6 - B^6) \div (A - B) = A^5 + A^4B + A^3B^2 + A^2B^3 + AB^4 + B^5$.

From these examples we are led to suppose[1] that the theorem as given is true.

(2. Rule for writing down the quotient.

We derive this rule from an inspection of the examples given.

(1) The first term of the quotient is the first term of the dividend divided by the first term of the divisor.

(2) Disregarding signs, the second term of the quotient is found by dividing the first term of the quotient by the first term of the divisor, and then multiplying this result by the second term of the divisor.

(3) The third term of the quotient is obtained from the second in the same way as the second was derived from the first, and so on. All the signs in the quotient are plus.

2. THEOREM II. *The difference of the same* EVEN *powers of two quantities is also divisible by the sum of the quantities.*

(1. Verify the following divisions.

$(A^2 - B^2) \div (A + B) = A - B$; $(A^4 - B^4) \div (A + B) = A^3 - A^2B + AB^2 - B^3$; $(A^6 - B^6) \div (A + B) = A^5 - A^4B + A^3B^2 - A^2B^3 + AB^4 - B^5$.

[1] The truth of the theorems of division for any powers whatever is commonly shown by mathematical induction in advanced algebra.

From these examples we are led to suppose the theorem true.

(2. The rule for writing down the quotient is the same as for theorem I, with this exception, *the signs are alternately plus and minus.*

3. THEOREM III. *The sum of the same* ODD *powers of two quantities is always divisible by the sum of the quantities.*

(1. Verify the following divisions.

$(A + B) \div (A + B) = 1$; $(A^3 + B^3) \div (A + B) = A^2 - AB + B^2$.

$(A^5 + B^5) \div (A + B) = A^4 - A^3B + A^2B^2 - AB^3 + B^4$.

$(A^7 + B^7) \div (A + B) = A^6 - A^5B + A^4B^2 - A^3B^3 + A^2B^4 - AB^5 + B^6$.

These examples suggest the theorem.

(2. For the quotient use the rule given for theorem II.

REMARK.—When the second term of the divisor is *minus*, the process of division in these examples makes every successive dividend plus, so that every term of the quotient is plus. When the second term of the divisor is plus, only every other successive dividend is plus, and the signs of the quotient are thus made to alternate. The process of division shows also the reason for the rule given above for writing down the quotient. In actual division we first multiply a quotient term by the second term of the divisor and *afterwards* divide by the first term of the divisor. By the rule we *divide* first and *multiply* afterwards, which is easier.

4. THEOREM IV. *The sum of the same even powers of two quantities is* NOT EXACTLY *divisible by either the sum or the difference of the quantities.*

(1. Perform the following indicated operations:

$(A^2 + B^2) \div (A + B) = A - B + \frac{2B^2}{A + B}$ i.e., is not exactly divisible.

$(A^4 + B^4) \div (A + B) = ?$ $(A^6 + B^6) \div (A + B) = ?$
$(A^8 + B^8) \div (A + B) = ?$ $(A^8 + B^8) \div (A^2 + B^2) = ?$
$(A^2 + B^2) \div (A - B) = ?$ $(A^4 + B^4) \div (A - B) = ?$
$(A^6 + B^6) \div (A - B) = ?$
$(A^6 + B^6) \div (A^2 + B^2) = ?$

The last is exactly divisible, because it really comes under Theorem III., although it seems to belong here.

(2. The quotients may be written by the previous rules, if the proper remainders are also found and added.

5. Examples in the theorems of Divisibility. — (With particular reference to powers which are themselves simple powers.)

(1. What is $x^3 - 27\,y^3$ divisible by? *Ans.* $x - 3\,y$; for it is the difference of the third powers of x and $3\,y$, Theorem I. The quotient is $x^2 + 3\,x + 9$. For $x^3 \div x = x^2$; then to find the second term of the quotient $(x^2 \div x) \times 3 = 3\,x$; to find the third term, $(3\,x \div x) \times 3 = 9$. The signs of the quotient are all plus.

(2. $16\,y^4 - 81$ is divisible by what binomial? By theorem I. it is divisible by $2\,y - 3$. For, $\sqrt[4]{16\,y^4} = 2\,y$ and $\sqrt[4]{81} = 3$. To find the quotient: $16\,y^4 \div 2\,y = 8\,y^3$; $(8\,y^3 \div 2\,y) \times 3 = 12\,y^2$; $(12\,y^2 \div 2\,y) \times 3 = 18\,y$; $(18\,y \div 2\,y) \times 3 = 27$. The signs are all $+$. Hence the quotient is $8\,y^3 + 12\,y^2 + 18\,y + 27$.

(3. By what is $x^5 + 32\,y^5$ divisible? *Ans.* $x + 2\,y$. Theorem III., since $\sqrt[5]{x^5} = x$ and $\sqrt[5]{32\,y^5} = 2\,y$. To find the quotient: —

$x^5 \div x = x^4$; $(x^4 \div x) \times 2\,y = 2\,x^3y$; $(2\,x^3y \div x) \times 2\,y = 4\,x^2y^2$; $(4\,x^2y^2 \div x) \times 2\,y = 8\,xy^3$; $(8\,xy^3 \div x) \times 2\,y = 16\,y^4$.
Quotient, $x^4 - 2\,x^3y + 4\,x^2y^2 - 8\,xy^3 + 16\,y^4$.

(4. What is $m^9 - n^6$ divisible by? (Cf. **129**, 1, (4, for this and the following examples.)

This quantity is to be looked upon as the difference of two cubes, viz., of m^3 and n^2, and by theorem I. is divisible by $m^3 - n^2$.

$$(m^9 - n^6) \div (m^3 - n^2) = m^6 + m^3n^2 + n^4 \text{ by the rule.}$$

(5. What is $x^8 - y^4$ divisible by? By Theorem II

$(x^8 - y^4) \div (x^2 + y) = x^6 - x^4y + x^2y^2 - y^3$ (fourth roots.)
$(x^8 - y^4) \div (x^4 + y^2) = x^4 - y^2$ (square roots.)
$(x^8 - y^4) \div (x^2 - y) = ?$ (Theorem I)

(6. What is $x^9 - y^9$ divisible by? By Theorem I

$$(x^9 - y^9) \div (x - y) = x^8 + x^7y + x^6y^2 + x^5y^3 + x^4y^4 + x^3y^5 + x^2y^6 + xy^7 + y^8.$$
$$(x^9 - y^9) \div (x^3 - y^3) = x^6 + x^3y^3 + y^6.$$

(7. What factors has $x^8 - y^8$?

By Theorem I., $(x^8 - y^8) \div (x - y) = ?$ $(x^8 - y^8) \div (x^2 - y^2) = ?$ $(x^8 - y^8) \div (x^4 - y^4) = ?$

By Theorem II., $(x^8 - y^8) \div (x + y) = ?$ $(x^8 - y^8) \div (x^2 + y^2) = ?$ $(x^8 - y^8) \div (x^4 + y^4) = ?$

(8. What are the factors of $a^6 - 1$? (See Remark in **117**.)

By Theorem I., $(a^6 - 1^6) \div (a - 1) = a^5 + a^4 + a^3 + a^2 + a + 1.$
$(a^6 - 1^6) \div (a^2 - 1^2) = ?$ $(a^6 - 1) \div (a^3 - 1) = ?$

By Theorem II., $(a^6 - 1) \div (a + 1) = ?$ $(a^6 - 1) \div (a^3 + 1) = ?$

(9. What are the factors of $x^9 + x^{12}$? One is $x^3 + x^4$; what is the other?

131. Exercise in the Use of the Theorems of Divisibility.— The quotient factors are to be written down directly, by reference to the appropriate rule, and not to be found by division.

1. $(a^2 - b^2) \div (a - b) = ?$
2. $(m^3 - n^3) \div (m - n) = ?$
3. $(5^3 - 2^3) \div (5 - 2) = ?$
4. $(1 - y^3) \div (1 - y) = ?$
5. $(8\,a^3 + 27\,b^3)$.
6. $(x^{18} + a^{36}) \div (x^6 + a^{12})$.
7. $(x^8 - y^2 z^6)$.
8. $(a^{12} - b^9)$.
9. $(a^{21} + b^{15})$.
10. $(a^{24} - b^{18})$.
11. $64\,a^3 + 1000\,b^3$.
12. $64\,a^6 - b^6$.
13. $16x^4 - 1$.
14. $m^6 + n^6$.
15. $(81\,y^4 - 16\,z^4) \div (3\,y - 2\,z)$.
16. $a^6 + 64\,z^6$.
17. $z^8 + t^8$. By $z + t$; also by $z^2 + t^2$, also by $z^4 + t^4$.
18. $(x^8 + 1)$.
19. $256\,y^4 + x^8$.
20. $x^8 + x^{12}$.

CHAPTER X.

FACTORING.

132. Factoring in algebra is the process of separating a quantity into others, which, multiplied together, will produce it.

The terms used in arithmetic to describe factors are used in precisely the same way in algebra.

1. A *divisor* of a quantity is any quantity exactly contained in it.

2. A *multiple* of a quantity is some number of times the quantity, and therefore will contain the quantity itself exactly.

3. A *prime quantity* is divisible only by itself and unity.

4. A *composite quantity* is divisible by other quantities besides itself and unity.

SECTION I.

MONOMIALS.

133. Monomial Factors.

1. Monomials standing alone.

(1. Factor $30\,ax^2y^3$. *Ans.* $2 \cdot 3 \cdot 5 \cdot a \cdot x \cdot x \cdot y \cdot y \cdot y$.

(2. $14\,m^3n$. (3. $91\,a^4bc^3$. (4. $9\,(a+b)^3$. (5. $21\,(m+n)^2\,(p+q)^3$.

2. Monomials appearing in polynomials.

(1. Separate $2m^2n + 6mn^2$ into a monomial and a polynomial. $2mn$ is the monomial factor, being contained in each term of the polynomial (**124, 1**). The other factor is $m + 3n$.

(2. Separate $20x^3 - 45xy^2$ into its factors.
(3. $12ax^2 - 3bx^2 + x^2$.
(4. $x^4y^2 - x^3y^3 + x^2y^2$.
(5. $14bc^4x - 21b^2c^3x + 7b^3c^2x$.
(6. $6bc^2x - 15bc^3 - 3b^2c^2$.

a. In factoring the first thing to be done with any given polynomial is to seek for a monomial factor in it, and if one is found to remove it.

SECTION II.

BINOMIALS.

134. The Factoring of **Binomials**.

1. *The difference of two squares is factored into the sum and difference of their roots.*

This is the converse operation of the third theorem of mlutiplication

(1. $9m^2 - 16n^2$.

By the first theorem of division $9m^2 - 16n^2$ is divisible by $3m - 4n$, and by the second it is divisible by $3m + 4n$. Or, more simply, reversing the theorem of multiplication the difference of the two squares is factored into the product of the sum and difference of the square roots.

$$9m^2 - 16n^2 = (3m + 4n)(3m - 4n).$$

(2. $a^4 - b^4$.

(3. $a^2b^2 - c^2d^2$.

(4. $a^2 - \frac{1}{49}$.

(5. $9a^2 - 4m^2$.

(6. $4 - x^2$.

(7. $a^2 - \frac{b^2}{4}$

(8. $16y^2 - 1$.

(9. $1 - 81x^4$.

(10. $4a^2 - 25$.

(11. $36x^2 - 49y^2$.

(12. $m^4x - 36n^4x$.

(13. $a^8 - b^6$.

(14. $a^3 - ab^2$.

(15. $a^{2m} - b^{4n}$.

(16. $9x^{2m} - 25y^{2m}$.

2. *The difference of the same powers.* See Theorem I., **130**.

(1. $a^4 - 16$.

By the theorem, $a^4 - 16$ is divisible by $a - 2$ and the quotient is $a^3 + 2a^2 + 4a + 8$. (See **130**, 1, (2.)

(2. $x^3 - y^3$.

(3. $x^3 - 8$.

(4. $x^5 - y^5$.

(5. $y^3 - 216$.

(6. $8a^3 - 27b^3$.

(7. $m^3n^3 - p^3$.

(8. $a^6 - b^3$.

(9. $64y^3 - 1000z^3$.

(10. $a^{12} - c^9$.

(11. $y^3 - 343z^3$.

3. *The difference of the same even powers.* Theorem II., **130**.

By the preceding case the difference of *any* powers is divisible by the difference of the quantities, and by the present theorem the difference of *even* powers is divisible by the *sum* of the quantities. Hence the difference of two even powers can be factored in two ways, and two sets of answers should be given to the following. See **130**, 2, (2.

(1. $m^4 - n^4$.

By this theorem $m^4 - n^4$ is divisible by $m + n$, and the quotient is $m^3 - m^2n + mn^2 - n^3$. By the preceding theorem it is divisible by $m - n$, and the quotient is $m^3 + m^2n + mn^2 + n^3$.

(2. $x^6 - y^6$.

(3. $a^6 - 64$.

(4. $y^4 - 1$.

(5. $81\,a^4 - 16\,b^4$.

(6. $m^4 - 81$.

(7. $x^8 - x^2y^6$.

(8. $x^{12} - y^6$.

(9. $16\,x^4 - 625$.

(10. $a^4b^4 - c^4d^4$.

(11. $a^6 - b^6c^6$.

(12. $a^8 - 1$.

(13. $729 - r^6$.

(14. $a^{12} - b^8$.

(15. $96\,x^4 - 486\,y^4$.

(16. $72\,x^4 - 1152$.

4. *The sum of the same two odd powers.* Theorem III., **130**.

(1. $x^3 + 27$.

By the theorem $x^3 + 27$ is divisible by $x + 3$, and the quotient is $x^2 - 3\,x + 9$. See **130**, 3, (2.

(2. $x^3 + 27\,y^3$.

(3. $x^5 + z^5$.

(4. $27\,m^3 + 8\,n^3$.

(5. $z^3 + 1$.

(6. $1 + 64\,a^6$.

(7. $m^6 + 8\,n^3$.

(8. $r^5 + 32\,s^5$.

(9. $x^3 + 64\,z^3$.

(10. $x^8 + 128\,xz^7$.

(11. $x^9 + y^6$.

(12. $a^{15} + b^{12}$.

(13. $8\,a^3b^3c^3 + 1$.

(14. $8\,x^3 + 64\,y^3$.

(15. $4\,x^6 + 108\,z^6$.

(16. $243\,z^5 + 1$.

5. *Binomials separable into two trinomials.*

(1. $a^4 + 4\,b^4$.

This binomial is not divisible by a binomial. See theorem IV in division. Nor can it be factored as it stands. It will be shown in **136**, 2, however, that a quadrinomial which is the difference of two squares can be factored. Now it is possible to make such a quadrinomial of the given expression. For, to make $a^4 + 4b^4$ a square, $4a^2b^2$ must be added; and if this term be also subtracted the expression will be unaltered. Thus,

$$a^4 + 4b^4 = a^4 + 4a^2b^2 + 4b^4 - 4a^2b^2,$$

which is a quadrinomial the difference of two squares. By the article referred to,

$$(a^4 + 4a^2b^2 + 4b^4) - 4a^2b^2 = (a^2 + 2ab + 2b^2)(a^2 - 2ab + 2b^2).$$

Exercise in factoring such binomials will be given in **136,** 2.

SECTION III.

TRINOMIALS.

135. Trinomials.

1. *Trinomial Squares.* — See theorems I. and II. in multiplication. Reversing the process of squaring in theorems I. and II., **114,** we derive this rule for factoring trinomial squares: Extract the square root of the first term and of the last; if the middle term is twice the product of these two, the trinomial is a perfect square, and the sum or difference of the square roots is one of the two equal factors according as the middle term is plus or minus.

(1. $x^2 + 16x + 64.$

$\sqrt{x^2} = x$: $\sqrt{64} = 8$. $2 \times 8 \cdot x = 16x$, $\therefore x^2 + 16x + 64 = (x + 8)(x + 8) = (x + 8)^2.$

(2. $x^2 + 10x + 25.$

(3. $x^2 + 4x + 4.$

(4. $x^2 + 12xz + 36z^2.$

(5. $9x^2 + 30x + 25.$

(6. $x^2y^2 + 22xyz + 121z^2.$

(7. $a^2b^2 + 2abcd + c^2d^2.$

(8. $4x^4 + 12x^2y + 9y^2.$

(9. $36x^4 + 84x^2y^2 + 49y$

(10. $9y^6 + 30y^3 + 25$.

(11. $x^2 - 6x + 9$.

$\sqrt{x^2} = x$. $\sqrt{9} = 3$. $2 \times x \times 3 = 6x$, $\therefore x^2 - 6x + 9 = (x - 3)(x - 3) = (x - 3)^2$.

(12. $x^2 - 26y + 169$.

(13. $x^2 - x + \frac{1}{4}$.

(14. $4x^2 - \frac{4}{3}xy + \frac{y^2}{9}$.

(15. $\frac{9}{16}y^2 - \frac{15}{2}y + 25$.

(16. $x^2 - 2x + 1$.

(17. $5a^4 - 10a^2b + 5b^2$.

(18. $24x^3b - 72x^2b^2 + 54xb^3$.

(19. $a^{2n} - 2a^n + 1$.

(20. $m^x - 4m^{\frac{x}{2}} + 4$.

2. ***Simple trinomials not squares.***

Reversing the theorem of **116,** 8, we learn that if one term of a trinomial contain as a factor the square root of another term, it may be factored into two binomials having this square root as one term in each. The other terms must be so chosen that their sum is the coefficient of the square root and their product the third term.

(1. $x^2 + 7x + 12$.

The second term contains x, which is the square root of the first term. Also the coefficient of x in the second term equals $4 + 3$, and the third term $= 4 \times 3$.

Therefore $x^2 + 7x + 12 = (x + 4)(x + 3)$.

Verify this by actual multiplication, and the reasons will become more clear. Of course the particular numbers for each separate example have to be found by *trial.*

(2. $x^2 + 5x + 6$.
(3. $z^2 + 11z + 30$.
(4. $x^2 + 7x + 6$.
(5. $z^2 + 29qz + 100q^2$.
(6. $x^2 + 21x + 110$.
(7. $b^2 + 13b + 12$.
(8. $x^8y^4 + 7x^4y^2 + 12$.
(9. $r^2s^2 + 23rsz + 90z^2$.
(10. $m^2 + 17mnp + 52n^2p^2$.

(11. $a^2 - 11a + 24$. Here the SUM must be -11, and the PRODUCT $+24$.

$$a^2 - 11a + 24 = (a - 8)(a - 3).$$

(12. $x^2 - 9x + 20$.
(13. $x^2 - 11x + 30$.
(14. $48 - 14x + x^2$.
(15. $x^8 - 15x^4 + 54$.
(16. $x^{2m} - 19x^m + 90$.
(17. $a^2b^2 - 13abc + 40c^2$.

(18. $x^2 + 9x - 22$. Here the SUM must be $+9$, and the PRODUCT -22.

$x^2 + 9x - 22 = (x + 11)(x - 2)$, since $11 - 2 = 9$, and $11 \times -2 = -22$.

(19. $a^2 + 2a - 15$.
(20. $x^4 + x^2 - 30$.
(21. $y^2 + \frac{1}{2}y - \frac{3}{16}$.
(22. $m^2 + \frac{3m}{4} - \frac{1}{4}$.
(23. $c^2d^2 + \frac{2}{3}cde - \frac{35}{9}e^2$.
(24. $x^4y^2 + 7x^2y - 8$.

(25. $x^2 - 4x - 21$. Here the SUM must be -4, and the PRODUCT -21.

$x^2 - 4x - 21 = (x - 7)(x + 3)$, as $-7 + 3 = -4$, and $-7 \times 3 = -21$.

(26. $x^2 - x - 30$.
(27. $y^2 - 4yz - 45z^2$.
(28. $a^2b^2c^2 - 6abc - 40$.
(29. $a^2 - 2a - \frac{45}{4}$.
(30. $m^{4p} - 3m^{2p} - 70$.
(31. $x^{6s} - 5x^{3s}y^{2s} - 104y^{4s}$.

3. *Trinomials the product of any two binomials.*

(1. Let us first take any two binomials which will give a trinomial product, say $2x + 3$ and $3x + 4$, and form their product.

$$\begin{array}{l} 2x + 3 \\ 3x + 4 \\ \hline 6x^2 + 9x \\ \quad\ + 8x + 12 \\ \hline 6x^2 + 17x + 12 \end{array}$$

If, now, $6x^2 + 17x + 12$ be given to find its factors, we know just three things concerning them as the multiplication in the margin shows.

First, that the product of the two coefficients is 6.

Second, that the product of the last two numbers is 12.

Third, that the sum of the *cross products* is $17x$.

Now, if the product is given to find the factors, we do not know whether the 6 is produced by multiplying 2 by 3, or 6 by 1; or, whether the 12 results from the multiplication of 3 by 4, or 2 by 6, or 12 by 1; for, we do not know in advance what arrangement of the coefficients will, upon cross multiplication and addition, give the 17.

What has been said suggests the following rule.

(2. Rule.

(1). Take a set of four coefficients, the product of the first terms being equal to the first coefficient, and the product of the last terms being equal to the last coefficient, and examine whether upon cross multiplication they will give the middle coefficient of the trinomial.

(2). If the first set chosen does not give the middle coefficient, try another arrangement, and so on until a set is found satisfying the last condition, or it is shown that none will answer, which would indicate that the trinomial is a prime quantity.

(3. Exercise.

(1). Required the factors of $6x^2 - 25xy + 4y^2$.

Using 2 and 3 as the factors of 6, we write the following sets,

$2 - 2$	$2 - 4$	$2 - 1$
$3 - 2$	$3 - 1$	$3 - 4$
middle coefficient $= -10$	middle coefficient $= -14$	middle coefficient $= -11$

Next using 6 and 1 as the factors of 6

$6 - 2$	$6 - 4$	$6 - 1$
$1 - 2$	$1 - 1$	$1 - 4$
middle coefficient $= -14$	middle coefficient $= -10$	middle coefficient $= -25$

The last arrangement gives the middle term as desired.

Hence, $6x^2 - 25xy + 4y^2 = (6x - y)(x - 4y)$. *Ans.*

(2) $9x^2 + 9x + 2$.

(3) $3x^2 + 13x + 14$.

(4) $4x^2 + 11x - 3$.

(5) $9x^2 + 64x + 7$.

(6) $3x^2 + 10xy - 8y^2$.

(7) $2x^2 - x - 1$.

(8) $3x^2 - 19x - 14$.

(9) $2c^2 - 13cd + 6d^2$.

(10) $2m^2 - 3my - 2y^2$.

(11) $12x^2 - 31x - 15$.

(12) $15z^2 - 224z - 15$.

(13) $24x^2 - 29xy - 4y^2$.

(14) $3 + 11x^3 - 4x^6$.

(15) $20 - 9x - 20x^2$.

(16) $2x^4 + x^2 - 28$.

(17) $24a^2b^2c^2 - 37abc - 72$.

(18) $6 + 32x - 21x^2$.

(19) $5x^2 - \frac{1}{2}xz - \frac{1}{10}z^2$.

4. *Trinomials of the form* $9a^4 - 4a^2b^2 + 4b^4$.

By adding $16a^2b^2$ to the second term and subtracting it again in a fourth term, as was done in **134**, 5, the trinomial becomes a quadrinomial which is the difference of two squares. Exercise in factoring such trinomials will be given in **2** of the next article.

5. *Trinomials the Product of a Binomial and Trinomial.*

If we multiply $3x^2 + 4x + 5$ by $3x - 4$, the coefficient of x^2 in the product is zero, and the product reduces to a trinomial. In the same way $5x^2 + 3x - 4$ and $3x + 4$ multiplied together give zero as the coefficient of the first power of x in the product. Such trinomials, having their coefficients related in one or other of these two ways may be factored.

(1. $4x^3 - 43x - 21$.

To find the factors assume that 2 is the first coefficient in each factor. The second term of the binomial is one or other of the factors of 21. Trying 7 as the second coefficient in each and 3 as the third term of the trinomial, the factors assumed are $2x^2 + 7x + 3$ and $2x - 7$, which, upon cross-multiplication, give $-43x$. Hence these quantities are the two factors required.

(2. $25x^3 - 61x - 12$.

(3. $8x^3 - 24x^2 + 25$.

(4. $21x^3 + 26x^2 + 25$.

(5. $9x^3 + 5x + 50$.

SECTION IV.

Quadrinomials.

136. Quadrinomials.

1. *The cube of a binomial.*

(1. The product obtained in **116, 4,** is the *form* of a binomial cube, as $a^2 + 2ab + b^2$ is the form of a binomial square. From the cube product there given, viz., $(a + b)^3 = a^3 + 3a^2b + 3ab^2 + b^3$, we derive the rule for obtaining the cube root.

(2. Rule.

(1) Extract the cube root of the two leading terms, the first and last as usually arranged.

(2) With these roots see if the middle terms are respectively three times the square of the first into the second, and three times the first into the square of the second.

(3) The sum of the roots or their difference (depending on whether all the terms of the quadrinomial are positive, or the second and fourth minus) is one of the three equal factors of the quantity.

(3. Exercise.

(1) $8a^3 - 36a^2b + 54ab^2 - 27b^3$.

$\sqrt[3]{8a^3} = 2a$; $\sqrt[3]{27b^3} = 3b$; and $3(2a)^2(3b) = 36a^2b$, the second term; $3(2a)(3b)^2 = 54ab^2$. Since the second and fourth terms are negative the cube root is $2a - 3b$.

(2) $a^3 + 3a^2 + 3a + 1$.
(3) $64b^9 + 48b^6 + 12b^3 + 1$.
(4) $8x^3 - 60x^2y + 150xy^2 - 125y^3$.
(5) $216x^3 - 108x^2y + 18xy^2 - y^3$.

2. *Quadrinomials the difference of two Squares.*

(1. Factor $x^2 + 2xy + y^2 - a^2$.

Writing the quadrinomial as the difference of two squares, we have $(x+y)^2 - a^2$ which by **134**, 1, is factorable into the *sum* and *difference* of its square roots.

$(x+y)^2 - a^2 = ((x+y)+a)((x+y)-a) = (x+y+a)(x+y-a)$.

(2. $4a^2 - (9b^2 - 6bc + c^2)$.
(3. $4a^2 - (9b^2 - 12bc + 4c^2)$.
(4. $4a^2 - 9b^2 + c^2 + 4ac$.

In this and some of the following examples, the order of the terms is disarranged, and to unite the terms properly is the first thing to be done. Here the first, third, and last terms go together to make the square of $2a + c$. Hence we have $(2a+c)^2 - 9b^2$, the factors of which are $(2a+c+3b)(2a+c-3b)$.

(5. $l^2 - m^2 - n^2 + 2mn$
(6. $2ab + a^2 - x^2 + b^2$.
(7. $b^2 - 1 - 2ab + a^2$.

(8. $(3a^2 - b^2)^2 - (a^2 - 3b^2).^2$
(9. $16 - (3a - 2b)^2$.
(10. $x^4 - x^2 + 12x - 36$.
(11. $-4y^2 + x^2 + 2yz - 9z^2$.
(12. $42ab + 1 - 49a^2 - 9b^2$.
(13. $-12ab - 4a^2 + 9x^2 - 9b^2$.
(14. $2ab - a^2 - b^2 + 1$.
(15. $a^2 + 2bc - b^2 - c^2$.
(16. $x^{2m} - (y - z)^{2m}$.
(17. $a^4 + a^2b^2 + b^4$. (See **135**, 4.)

SUGGESTION. — Add and subtract a^2b^2. Thus we get

$$(a^4 + 2a^2b^2 + b^4) - a^2b^2 = (a^2 + b^2 + ab)(a^2 + b^2 - ab).$$

(18. $25x^4 - 36x^2y^2 + 4y^4$.

SUGGESTION. — To have a trinomial square, the middle term ought to be $+20x^2y^2$ or $-20x^2y^2$. If $-16x^2y^2$ be taken out of $-36x^2y^2$ and a fourth term be made of it, the expression can be factored by this case.

$(25x^4 - 20x^2y^2 + 4y^4) - 16x^2y^2 = (5x^2 - 2y^2 + 4xy)(5x^2 - 2y^2 - 4xy)$.

(19. $16x^4 - 17x^2y^2 + y^4$.
(20. $9x^4 + 38x^2y^2 + 49y^4$.
(21. $9a^4 + 21a^2c^2 + 25c^4$.
(22. $25x^4 - 41x^2y^2 + 16y^4$.
(23. $x^4 + 64$. (See **134**, 5.)

SUGGESTION. — Add and subtract $16x^2$. Thus, $x^4 + 16x^2 + 64 - 16x^2$.

(24. $4x^4 + 81z^4$.
(25. $64x^4y^4 + 81z^4$.
(26. $4m^4 + 625$.

3. *Quadrinomials the product of binomials.*

(1. Examples.

(1) Factor $ac + ad + bc + bd$.

Here a can be taken out of the first two terms and b out of the last two, giving

$$a(c+d)+b(c+d)$$

which is plainly divisible by the binomial $c+d$,

$$c+d)\underline{\;a(c+d)+b(c+d)\;}$$
$$a \quad +b$$

Therefore the factors of $ac+ad+bc+bd$ are $c+d$ and $a+b$.

(2. Factor $6ax-2by+3bx-4ay$.

Taking $3x$ out of the first and third terms, and $2y$ out of the second and fourth

$$6ax-2by+3bx-4ay=3x(2a+b)-2y(b+2a)$$

$$2a+b)\underline{\;3x(2a+b)-2y(b+2a)\;}$$
$$3x-2y$$

$$\therefore\; 6ax-2by+3bx-4ay=(2a+b)(3x-2y).$$

(2. Rule.

(1) Take a monomial factor out of two terms, and a second monomial factor (if necessary) out of the two remaining terms. If this is done properly, a binomial factor is seen directly (i.e., if the quadrinomial can be factored at all).

(2) The other factor is found by division.

(3. Exercise.

(1) $ac-bd+bc-ad$.

(2) a^3+a^2+a+1.

(3) $xy-3y+2x-6$.

(4) $6ax-4ay-21xy+14y^2$.

(5) $9am-4bm-27an+12bn$.

(6) $m^3-mn+m^2n^2-n^3$.

(7) $x^5-xy^4-x^4y+y^5$.

(8) $cdx^2-cxy+dxy-y^2$.

(9) $x^3-x^2y-xy^2+y^3$.

(10) $abcy - b^2dy - acdx + bd^2x$.

(11) $12x^3 - 8xy - 9x^2y^2 + 6y^3$.

(12) $15a^3 + 12a^2 + 10a + 8$.

4. *Quadrinomials the product of binomials and trinomials.*

(1) $4x^2 - 9y^2 - 2xz + 3yz$.

Factoring the first two terms and taking z out of the last two

$$\underline{2x - 3y)}\ \underline{(2x + 3y)\,(2x - 3y) - z\,(2x - 3y)}$$
$$2x + 3y - z.$$

(2) $x^3 - 6x^2 + 6x - 1$.

$$\underline{x - 1)}\ \underline{x^3 - 1 - 6x\,(x - 1)}$$
$$x^2 + x + 1 - 6x = x^2 - 5x + 1.$$

$\therefore$ the factors are $x - 1$ and $x^2 - 5x + 1$.

(3) $m^3 + 5m^2 + 5m + 1$.

(4) $4a^2b^2 - 169c^2 + 6abd + 39cd$.

(5) ** $10x^3 + x^2 - x - 28$.

Assuming that this can be factored into a binomial and trinomial, let us set down trial coefficients, leaving the middle term of the trinomial blank.

$$\begin{array}{lr}
5x^2 - (\ )x - 7 & \\
2x + 4 & \\
\hline
10x^3 + (\ )x^2 - 14x & \\
\quad 20x^2 & -28 \\
\hline
10x^3 & -28
\end{array}$$

Choosing as coefficients, 5, -7, 2, and 4, as indicated, and multiplying $5x^2$ by 4 the product is $20x^2$. Since the coefficient of x^2 in the product is 1, the product of 2 by the blank coefficient of x must be -10. But this value makes the coefficient of x, -52 instead of -1. Hence this arrangement of the coefficients fails.

$$\begin{array}{llll}
2x^2 + (\,3)x + 4 & & & \\
5x - 7 & & & \\
\hline
10x^3 + (15)x^2 + 20x & & & \\
\quad - 14x^2 & -(21)x & -28 & \\
\hline
10x^3 + x^2 & -x & -28 &
\end{array}$$

This arrangement gives $-14x^2$, which requires $+(15)x^2$, i.e., the coefficient $+3$ in the trinomial. Now $+3$ in the trinomial gives $-x$ in the product as desired. Hence, $2x^2 + 3x + 4$, and $5x - 7$ are the factors sought.

(6) $4x^3 - 21x^2 + 44x - 30$.

(7) $6x^3 - 3x^2 - 33x - 6$.

SECTION V.

Polynomials of more than Four Terms.

137. Factoring Polynomials of more than Four Terms.

1. *Expressions which are powers.*

(1. $A^4 + 4A^3B + 6A^2B^2 + 4AB^3 + B^4$.

If an expression of five terms have two fourth powers and three other terms formed out of the two roots as the middle terms above are formed from A and B, it is a perfect fourth power, and can be factored into an expression of the form, $(A + B)^4$ or $(A - B)^4$.

(2. $16x^4 - 96x^3y + 216x^2y^2 - 216xy^3 + 81y^4$.
$\sqrt[4]{16x^4} = 2x$; $\sqrt[4]{81y^4} = \pm 3y$; now, $4 \times (2x)^3 \times -3y = -96x^3y$; $6 \times (2x)^2 \times (-3y)^2 = 216x^2y^2$; $4 \times (2x)(-3y)^3 = -216xy^3$.
Therefore the factors are $(2x - 3y)^4$.

Note.—Higher powers than the fourth contain more terms, and involve a greater number of conditions, but are solved in the same way as the third and fourth powers.

2. *Polynomials the difference of two squares.*

(1. Factor $a^2 + 2ab + b^2 - c^2 + 2cd - d^2 = (a + b)^2 - (c - d)^2$.

The factors are the sum and difference of the roots $a + b$ and $c - d$. (Cf. **134**, 1, and **136**, 2).

$\overline{a + b} + \overline{c - d}$ and $\overline{a + b} - \overline{c - d} =$

or, $a + b + c - d$ and $a + b - c + d$.

(2. $(a + b + c)^2 - d^2$.

(3. $4m^2 - 12mn + 9n^2 - p^2 + 4pq - 4q^2$.

(4 $(a + b + c)^2 - (e + f + g)^2$.

(5. $(a + b + c + d)^2 - (e - f)^2$.

(6. $(3x^2 - 4x - 2)^2 + (3x^2 + 4x - 2)^2$.

(7. $4(ab + cd)^2 - (a^2 + b^2 - c^2 - d^2)^2$.

3. *Polynomial the square of a trinomial.*

(1. $a^2 + 4b^2 + 9c^2 + 4ab + 6ac + 12bc$.

Extracting the roots of the square terms, we have a, $2b$, and $3c$. The other terms are double the products of these, and therefore (See **116**, 3), the factors are $(a + 2b + 3c)^2$.

(2. $a^2 + 9b^2 + 25c^2 - 6ab + 10ac - 30bc$.

4. *Other Polynomials.*

(1. $x^2 + 2xy + y^2 + 6x + 6y$ or, $(x + y)^2 + 6(x + y)$ which is evidently divisible by $(x + y)$.

$$x + y)\ (x + y)^2 + 6\ (x + y)$$
$$x + y + 6$$

(2. * $6x^2 - 11xy + 3y^2 - xz - 7yz - 2z^2$.

Factoring *the first three terms* as in **135**, 3, and setting down the factors in the customary way, we have,

$$2x - 3y,$$
$$3x - y.$$

We see now that if $+ z$ be annexed to the first factor and $-2z$ to the second, we shall obtain the additional terms of the expression.

Therefore $2x - 3y + z$ and $3x - y - 2z$ are the factors sought.

(3. $x^2 - 2xy + y^2 + 5x - 5y$.

(4. $2x^2 - xy - 3y^2 - 5yz - 2z^2$.

(5. $2a^2 + 6ax - 18a + 4x^2 - 24x + 36$.

(6. $2a^2 - 4ab - 4ac + 2b^2 + 4bc + 2c^2$.

(7. ** $a^3 + b^3 + c^3 + 3(a^2b + ab^2 + a^2c + ac^2 + b^2c + bc^2) + 6abc$.

This is the *form* of the cube of a trinomial, as may be verified by forming and arranging the product $(a + b + c)(a + b + c)(a + b + c)$.

(8. $a^3 - b^3 - c^3 - 3(a^2b - ab^2 + a^2c - ac^2 + b^2c + bc^2) + 6abc$.

General Remark. — When the number of terms in a polynomial exceeds four, it is usually difficult to factor it. And, as a rule, the greater the number of terms the greater the difficulty of factoring.

In the foregoing classification of algebraic expressions with respect to the number of their terms *it is intended that the expressions are to be in their expanded forms.* Otherwise there will be more or less confusion. Thus $(a+b)^2 - c^2$ is a quadrinomial, not a trinomial or binomial.

138. Promiscuous Exercise in Factoring.

1. $a^6 + 2a^3b^3 + b^6$. (2. $9x^2 - 49a^2y^2$.
3. $18x^2 + 33axy + 14a^2y^2$.
4. $16a^2b^2c^2 + 24ab^2c^3 + 9b^2c^4$.
5. $8c^2 - 6cd - 5d^2$.
6. $a^9 - x^6$.
7. $a^3b^3 + 512$.
8. $a^2 - y^2 - 2yz - z^2$.
9. $(a+b)^2 - c^2$.
10. $8a^3 + 6a^2y^2 - 9ay^4 - 27y^6$.
11. $54a^2mx + 12a^2m^2 + 18am$.
12. $(7x+4y)^2 - (2x+3y)^2$.
13. $x^3 - 6x^2y + 12xy^2 - 8y^3$.
14. $6x^2 - 5xy - 6y^2$.
15. $8a^2 + 2a - 3$.
16. $m^3x + m^3y - n^3x - n^3y$.
17. $1 - a^2x^2 - b^2y^2 + 2abxy$.
18. $a^2y^3 - b^2yx^2 - a^2dy^2 + b^2dx^2$.
19. $x^3 + bx^2 + ax + ab$.
20. $a^2b^2 - a^2 - b^2 + 1$.
21. $x^4 - 7x^2 - 18$.

22. $x^4 - 4x^2y^2z^2 + 4y^4z^4$.

23. $x(x + z) - y(y + z)$.

24. $x^3 + y^3 + 3xy(x + y)$.

25. $a^2bc - ac^2d - ab^2d + bcd^2$.

26. $m^2 + n^2 + p^2 + 2mn + 2mp + 2np$.

27. $c^5d^3 - c^2 - a^2c^3d^3 + a^2$.

28. $(a + b)^2 + (a + b)$.

29. $a^{12} - b^8$.

30. $a^3b^3c^3 - 1$.

31. $18x^3yz + 9x^2y^2z + 6x^2yz^2 + 3xy^2z^2$.

32. $-a^2x^2 - b^2y^2 + 2abxy + (ax + 1)^2$.

33. $6am + 4an + 9bm + 6bn$.

34. $1 - (x - y)^3$.

35. $a^4 - (b - c)^4$.

36. $(a - 3x)^2 - 16y^2$.

37. $8(x + y)^3 - (2x - y)^3$.

38. $a^6 - b^6c^6$.

39. $a^4 + a^2b^2 - b^2c^2 - c^4$.

40. $x^4 - 7x^2y^2 + y^4$.

41. $31x - 35 - 6x^2$.

42. $ax(y^3 + b^3) + by(bx^2 + a^2y)$.

43. $1 + (b - a^2)x^2 - abx^3$.

44. $3a^2 - 6ab + 3b^2 + 6ac - 6bc$.

45. $(a^2 + b^2 + 2ab)^2 - (a^2 + b^2 - 2ab)^2$.

46. $(x^2 + y^2 + z^2 - 2xy + 2xz - 2yz) - (y + z)^2$.

47. $z^2 + (x - y)^2 - 2z(x - y)$.

48. $m^3 - n^3 - m(m^2 - n^2) + n(m - n)^2$.

49. $(a^2 + a - 4)^2 - 4$.

50. $x^2 - (x - 6)^2$.

CHAPTER XI.

COMMON FACTORS.

139. Common Factors. — Highest Common Factor.

1. A Common Factor of two or more quantities is a factor that appears in each of them.

2. The Highest Common Factor of two or more quantities is the product of all the prime factors common to each.

a. A common factor is the same as a common divisor. The highest common factor (abbreviated into h. c. f.) in algebra corresponds to the greatest common divisor in arithmetic. Indeed, by many authors the latter is the term used in algebra.

SECTION I. First Method.

THE HIGHEST COMMON FACTOR BY FACTORING.

140. Principle involved in finding the h. c. f. by Factoring. — If we have a quantity expressed as the product of its factors, each factor is a divisor of it, and furthermore, *the product of any set of its prime factors is also a divisor.*

Thus, the prime factors of 2310 are 2, 3, 5, 7, and 11. Now any one of these, of course, is a divisor of 2310; the product of any two of them, as 33 ($= 11 \times 3$), is likewise a divisor; so, also, the product of any three of them as 70 ($= 2 \times 5 \times 7$); and so on. And finally the product of *all* of the factors is a divisor, being the number itself.

Similarly the product of any set of the prime factors of $x^8 - y^8$, $(1, x - y, x + y, x^2 + y^2$, and $x^4 + y^4)$ is a divisor of $x^8 - y^8$.

It is evident from what has just been said that the product of all factors *common* to two or more quantities, (i.e., found in each of them) is a divisor of each of them. Moreover, this is the h. c. f., since if an additional factor were introduced *all of the quantities* would not contain the result.

If, then, we have the different quantities expressed as the product of their factors, we can pick out, merely by inspection, all those which are common, and so have the factors of the h. c. f.

To illustrate. Suppose it is required to find the h. c. f. of $a^2 - 2ab + b^2$, $a^2 - b^2$, and $ac - bc$.

Factoring each quantity,

$$a^2 - 2ab + b^2 = (a - b)(a - b),$$
$$a^2 - b^2 = (a + b)(a - b).$$
$$ac - bc = c(a - b).$$

Here $a - b$ is a divisor of each product, and therefore by definition the common factor.

To illustrate still further, let us find the h. c. f. of $12a^2b$ $(a - x)^2(a + x)^3$, $3a^2b(a - x)^2(a + x)^3$, and $9a^3b^2(a - x)^3$ $(a + x)$.

$$12a^2b(a - x)^2(a + x)^3 = 12a^2b(a - x)(a - x)(a + x)(a + x)(a + x).$$
$$3a^2b(a - x)^2(a + x)^3 = 3a^2b(a - x)(a - x)(a + x)(a + x)(a + x).$$
$$9a^3b^2(a - x)^3(a + x) = 9a^3b^2(a - x)(a - x)(a - x)(a + x).$$

By inspection, we see that $3a^2b$ is the product of all the monomial factors common; $(a - x)^2$ is the highest power of $a - x$, and $(a + x)$ the highest power of $a + x$, which are contained in *each* of the quantities. Therefore, by definition, the h. c. f. is $3a^2b(a - x)^2(a + x)$

141. Rule to find the h. c. f. of two or more Quantities by Factoring.

1. Separate each quantity into its *prime* factors.

2. Choose out the greatest coefficient and the highest power of every other factor that will still be contained in each quantity.

3. The product of these will be the h. c. f.

142. Exercise. — Represent the h. c. f. as the product of all the prime common factors.

1. $6a^2b$, $9a^3b$, $24a^4xy$. 2. $284a$, $126a$, $210a$.

3. $12ab^2c$, and $25b^4c^4$.

4. $48x^3y^3z^4$, $12x^3y^3z^3$, $24x^4y^4z^4$, $20x^4y^3z^2$.

5. $2a^3 - 2ab^2$, and $4b(a+b)^2$.

6. $35a^2pxy$, and $42b^2qxy$.

7. $12a^3x^2y - 4a^3xy^2$, and $30a^2x^3y^2 - 10a^2x^2y^3$.

8. $x^2 - y^2$, $x^3 - y^3$, and $x^2 - 7xy + 6y^2$.

9. $x^2 - 2x - 3$, and $x^2 - x + 12$.

10. $3x^3 + 6x^2 - 24x$, and $6x^3 - 96x$.

11. $3x^2 - 6x + 3$, $6x^2 + 6x - 12$, and $12x^2 - 12$.

12. $x^6 - x^3 - 30$, $x^6 - 13x^3 + 42$, and $x^6 + x^3 - 42$.

13. $x^{2m} + x^m - 30$, and $x^{2m} - x^m - 42$.

14. $ac(a-b)(a-c)$, and $bc(b-a)(b-c)$.

NOTE. $b - a = -(a-b)$, and $a - b$ is contained in $(b-a)$, -1 times.

15. $a^6b^2 - 4a^4b^4$, and $a^6b^2 - 16a^2b^6$.

16. $a^3 + 3a^2b + 2ab^2$, and $a^4 + 6a^3b + 8a^2b^2$.

17. $a^3 - a^2x$, $a^3 - ax^2$, $a^4 - ax^3$.

18. $x^5 - xy^2$, and $x^3 + x^2y + xy + y^2$.

19. $x^3 + 3x^2y + 2xy^2$, $x^3 + 6x^2y + 5xy^2$.

20. $3a^2 - 4ab + b^2$, $4a^4 - 5a^3b + a^2b^2$.

21. $2b^2 - 5b + 2$, $12b^3 - 8b^2 - 3b + 2$.

22. $x^2 - 1$, $x^2 - 2x - 3$, $6x^2 - x - 20$.

23. $6a^2 + 7ax - 3x^2$, $6a^2 + 11ax + 3x^2$.

24. $3x^2 + 16x - 35$, $5x^2 + 33x - 14$.

25. $c^2x^2 - d^2$, $acx^2 - bcx + adx - bd$.

26. $2x^2 + 9x + 4$, $2x^2 + 11x + 5$, $2x^2 - 3x - 2$.

27. $3x^4 + 8x^3 + 4x^2$, $3x^5 + 11x^4 + 6x^3$, $3x^4 - 16x^3 - 12x^2$

28. $8a^3 + 1$, $16a^4 + 4a^2 + 1$.

29. $8x^3 - 27$, $16x^4 + 36x^2 + 81$.

Remark. — The method by factoring is not adapted to the solution of problems whose expressions are difficult to resolve into their factors; and recourse is then had to the method by continued division.

SECTION II.

Finding the H. C. F. by Continued Division.

143. Principles involved in finding the h. c. f. by Continued Division.

1. The definitions and principles of the first method.

2. A divisor of a number (or quantity) is a divisor of any number of times that number (or quantity.)

If, for example, 5 is contained in 20, it is contained in 60 ($= 3 \times 20$) three times as often as in 20. It is contained in 20 four times, and in 60 three times four times, or twelve times.

If a is contained in A, b times exactly, it is contained in $2A$, $2b$ times exactly, and in mA, mb times exactly.

3. A common divisor of two numbers (or quantities) is a divisor of their sum or difference.

This follows from **124, 1**.

Thus, $6)\frac{24 \quad + \quad 66 \quad = 90}{4 \text{ times} + 11 \text{ times} = 15 \text{ times}}.$

$6)\frac{66 \quad - \quad 12 \quad = 54}{11 \text{ times} - 2 \text{ times} = 9 \text{ times}}.$

144. Application of the Principles to justify the Method of finding the h. c. f. by Continued Division.

1. Application to an arithmetical example. — To find the g. c. d. of 258 and 731.

EXPLANATION. — 731 is divided by 258 and contains it twice with a remainder 215. Then 258 is divided by this remainder, and so on. 43 is contained in 215 exactly and is the g. c. d. *It is required to prove that this process gives the* g. c. d. *of the two numbers.*

```
258) 731 (2
     516
     215) 258 (1
          215
           43) 215 (5
               215
```

(1. Is any divisor of 258 a divisor of 516? By which principle? Is any common divisor of 516 and 731 contained in 215? By which principle? Does it follow that any common divisor of 258 and 731 is a divisor of 215? Why? Can it be inferred from this reasoning that the g. c. d. of 258 and 731 *cannot be greater* than 215?

(2. Furthermore, is any common divisor of 215 and 258 (and, of course, of twice 258) also contained in 731? State the principle. Would it be allowable then to drop 731 entirely and proceed as before with 215 and 258? Also, after a second similar operation with 43 and 215?

(3. Is 43 the g. c. d. of itself and 215? By the foregoing reasoning is 43 contained in 258 and 731, and can their g. c. d. be greater than 43? Does it follow then that 43 is the g. c. d. sought?

By the preceding method we always have two numbers before us from which the h. c. f. required is to be found. First, the two numbers themselves are taken; then the divisor and a third number derived from the first two; then the last derived and a fourth, and so on. Now, every derived number used was obtained by principles *2* and *3*. But these principles will give other numbers. Thus, a divisor of 258 and 731 is a divisor of their sum 989, or their difference, 473; and 258 and 989, or, 258 and 473, may be used instead of 258 and 731.

2. Algebraic demonstration of the method of finding the h. c. f. by continued division.[1]

To find the h. c. f. of A and B.

Dividing as in the margin, just as in the arithmetical example, letting the Q's and R's stand for the successive quotients and remainders, suppose R_2 is contained in R_1, Q_3 times exactly; then R_2 is the h. c. f. sought. The proof follows the lines of that for the arithmetical problem.

$$
\begin{array}{l}
A)\ B\ (Q \\
\quad \underline{QA} \\
\quad B - QA = R)\ A\ (Q_1 \\
\qquad\qquad\qquad \underline{Q_1R} \\
\qquad\qquad\qquad A - Q_1R = R_1)\ R\ (Q_2 \\
\qquad\qquad\qquad\qquad\qquad\qquad \underline{Q_2R_1} \\
\qquad\qquad\qquad\qquad\qquad\qquad R - Q_2R_1 = R_2)\ R_1\ (Q_3 \\
\qquad\qquad\qquad\qquad\qquad\qquad\qquad\qquad\qquad \underline{Q_3R_2} \\
\qquad\qquad\qquad\qquad\qquad\qquad\qquad\qquad\qquad R_1 - Q_3R_2 = 0
\end{array}
$$

(1. Every divisor of A is also a divisor of QA (prin. 2); every common divisor of B and QA is a divisor of their difference, R (prin. 3); hence, any common divisor of B and A is a divisor of R, and so, of course, *cannot be higher than R.* Furthermore, any common divisor of R and A is also a divisor of B (prin. 3); therefore B can be dropped, and

[1] This demonstration should be omitted by beginners.

the process can be continued with A and R; in the next operation, the common divisor of A and R cannot *be higher than R_1*, and so on, as long as it is necessary to continue the operation. Hence, the h. c. f. cannot be of a higher degree than the last divisor.

(2. The last divisor is contained in the given quantities, as also in each of the remainders, as is made evident by the factored values, (**140**).

By hypothesis R_1 contains R_2, Q_3 times. Therefore

$$R_1 = Q_3R_2$$

$$R = Q_2R_1 + R_2 = Q_2(Q_3R_2) + R_2 = (Q_2Q_3 + 1)R_2 \qquad \textbf{(43)}$$

$$A = Q_1R + R_1 = Q_1(Q_2Q_3 + 1)R_2 + Q_3R_2 = (Q_1Q_2Q_3 + Q_1 + Q_3)R_2$$

$$B = QA + R = Q(Q_1Q_2Q_3 + Q_1 + Q_3)R_2 + (Q_2Q_3 + 1)R_2$$
$$= (QQ_1Q_2Q_3 + QQ_1 + QQ_3 + Q_2Q_3 + 1)R_2$$

145. Method and Principles in the Solution of Algebraic Examples. — Modifications. The argument by which we have proved that the process of continued division gives the g. c. d. of two arithmetical numbers applies equally well (as we have just seen) in algebra. But there are a number of very *important modifications in algebra*. To show how these arise, and how they are dealt with, an algebraic example is given.

It is required to find the h. c. f. of

$$8x^5 - 16x^4 - 30x^3 + 18x^2, \text{ and } 18x^4 - 60x^3 - 6x^2 + 72x.$$

Taking out the monomial factors

$$8x^5 - 16x^4 - 30x^3 + 18x^2 = 2x^2(4x^3 - 8x^2 - 15x + 9)$$
$$18x^4 - 60x^3 - 6x^2 + 72x = 6x(3x^3 - 10x^2 - x + 12).$$

In this form we see by inspection that $2x$ is the monomial factor *common*, and therefore a factor of the h. c. f. Since we have no ready means of factoring the polynomials to see if they have a common factor, we resort to division as in **144** to find their h. c. f.

But we are confronted at the outset with the dilemma that whichever polynomial is made the dividend, the first term of the other will not be contained in it exactly. Thus, $4x^3$ is not contained in $3x^3$, nor *vice versa.* To obviate this difficulty, and others of a similar character, certain courses are pursued in algebra which were not necessary in arithmetic. They depend upon the principles already given.

1. *We may, if we choose, multiply one of the two quantities in hand by a factor not found in the other.*

2. *We may discard a factor found in one quantity and not in the other.*

For *no common factor* is either *introduced into*, or *taken away from* the h. c. f. However, we *must not* multiply one quantity by a factor found in the other; for in so doing we make this factor common to the two, and introduce an irrelevant factor into the h. c. f.

3. *It is immaterial at any time whether a quantity itself is used, or its opposite.* The h. c. f. when found may have one value or its opposite. For, finding a h. c. f. is simply a question of *divisibility.*

To give a simple illustration, if 6 is a divisor of $+ 30$, it is contained in $- 30$ also; and if $- 3$ is contained in 18, so also is $+ 3$.

a. In order to change the sign of a quantity, we must change the sign of each term (100, 2, 4). To change the sign of a polynomial in the present case, it is customary to divide through by a negative factor if any is to come out, otherwise by $- 1$. It is preferable to retain the first term of polynomial quantities positive, changing signs when necessary.

Let us proceed now with the example begun above by first multiplying $3x^3 - 10x^2 - x + 12$ by 4 so as to make it divisible by $4x^3 - 8x^2 - 15x + 9$.

$$3x^3 - 10x^2 - x + 12$$
$$4$$
$$12x^3 - 40x^2 - 4x + 48 \quad (4x^3 - 8x^2 - 15x + 9$$
$$12x^3 - 24x^2 - 45x + 27 \quad (3 \text{ 1st quo.}$$
$$-1\,(-16x^2 + 41x + 21 \quad (4)$$
$$\text{2d divis. } 16x^2 - 41x - 21)\ 16x^3 - 32x^2 - 60x + 36\ (x \text{ 2d quo.}$$
$$16x^3 - 41x^2 - 21x$$
$$3\,(9x^2 - 39x + 36$$
$$\text{3d divid. } 16x^2 - 41x - 21\ (3x^2 - 13x + 12 \text{ 3d divis.}$$
$$15x^2 - 65x + 60 \quad (5 \text{ 3d quo.}$$
$$x^2 + 24x - 81$$
$$3x^2 - 13x + 12\ (x^2 + 24x - 81$$
$$3x^2 + 72x - 243 \quad (3. \text{ 4th quo.}$$
$$-85\,(-85x + 255$$
$$x - 3)\ x^2 + 24x - 81\ (x + 27$$
$$x^2 - 3x$$
$$27x - 81$$
$$27x - 81$$

Hence $x - 3$ is the h. c. f. of the two polynomials, and $2x(x - 3)$ the h. c. f. of the original quantities.

EXPLANATION.—The first divisor is multiplied by 4 in order to contain the first remainder. The second remainder is divided by 3 following the directions in 2. of this article. There is danger if this were not done of introducing the factor 3 into the other polynomial, and 3 would thus become a *common* factor, which would be manifestly wrong, as neither of the polynomials we began with contains 3. Instead of multiplying $3x^2 - 13x + 12$ by 16, or $16x^2 - 41x - 21$ by 3 to make the first term of the dividend contain the first term of the divisor, and in so doing obtaining large coefficients, we divide $16x^2 - 41x - 21$ by $3x^2 - 13x + 12$ obtaining a remainder of the same degree as the divisor. This amounts to multiplying $3x^2 - 13x + 12$ by 5 (prin. 2), and subtracting the product from $16x^2 - 41x - 21$ (prin. 3). The remainder is used with $3x^2 - 13x + 12$ to continue the operation (**144**). The divisor $x - 3$ is contained in its dividend, and is therefore the h. c. f. of the polynomials.

146. Rule for finding the h. c. f. by Continued Division.

This method may be used to solve every kind of problem, but it should only be used in the case of polynomials too difficult to factor. It is assumed that all monomial factors have been taken out, and either retained for the h. c. f. or suppressed as no longer needed.

1. According to the rule for division **126**, 3 divide the polynomial of higher degree by the other (or if of the same degree either may be divided by the other) until the remainder is of as low or of a lower degree than the divisor.

2. Divide the divisor by the remainder just obtained in the same way, and so continue.

3. That divisor which is contained in its dividend without a remainder is the h. c. f. sought.

4. If there are three or more quantities, first use the two whose h. c. f. is most easily obtainable; then take the h. c. f. of these along with a third quantity, and obtain their h. c. f.; and so on. The last one found will be the one required, for it will be the highest factor contained in each polynomial.

a. A factor found in one quantity and not in the other should be discarded at any time.

b. We may introduce at any time into one a factor which is not found in the other quantity.

c. We may at any time multiply one polynomial by one factor and the other by another factor (by 1 in **145**) and add or subtract the products (prin. 3), and use this result with either of the given polynomials, or with a second result similarly obtained to continue the operation. This plan may be followed when the regular process is about to give large numerical coefficients.

d. When a remainder is obtained which does not contain the letter of arrangement, there is no common divisor.

147. Examples in finding the h. c. f. by Continued Division.

1. Given $x^2 - 6x + 8$, and $4x^3 - 21x^2 + 15x + 20$.

$$\begin{array}{l|l} 4x^3 - 21x^2 + 15x + 20 & x^2 - 6x + 8 \\ \underline{4x^3 - 24x^2 + 32x} & \overline{4x + 3} \\ \quad 3x^2 - 17x + 20 & \\ \quad \underline{3x^2 - 18x + 24} & \end{array}$$

$$\text{h. c. f.} \quad x - 4)\, x^2 - 6x + 8\, (x - 2$$
$$\begin{array}{r} \underline{x^2 - 4x} \\ -2x + 8 \\ \underline{-2x + 8} \end{array}$$

2. Given $7x^2 + 25x - 12$, and $17x^2 + 67x - 4$.

The solution of the problem in this form promises to give large numerical coefficients. We prefer therefore to derive two other expressions (**146**, c).

$$\begin{array}{ll} 7x^2 + 25x - 12 & 17x^2 + 67x - 4 \\ \underline{2} & \underline{14x^2 + 50x - 24} \\ 14x^2 + 50x - 24 & 3x^2 + 17x + 20 \quad \text{1st derived expression.} \end{array}$$

$$\begin{array}{ll} 7x^2 + 25x - 12 & 17x^2 + 67x - 4 \\ \underline{5} & \underline{2} \\ 35x^2 + 125x - 60 & 34x^2 + 134x - 8 \\ \underline{34x^2 + 134x - 8} & \\ x^2 - 9x - 52 \quad \text{2d derived expression.} & \end{array}$$

We proceed with the two derived expressions as if they were the quantities given. The following arrangement of the work of finding the h. c. f. avoids the necessity of re-writing divisors.

2d quotient.	1st divisor and 2d dividend.	1st dividend.	1st quotient.
$x - 13$	$x^2 - 9x - 52$	$3x^2 + 17x + 20$	3
	$\underline{x^2 + 4x}$	$\underline{3x^2 - 27x - 156}$	
	$-13x - 52$	$44(44x + 176$	
	$\underline{-13x - 52}$	h. c. f. $x + 4$	

148. Exercise in finding Highest Common Factors. — The choice of method is left to the student. (See remark in **142**.)

1. 2945 and 3441. 2. 630, 1134, 1386.

3. 3094, 4420, 2652, 4662.

4. $3x^2+16x-35$, and $5x^2+33x-14$.

5. $2x^2-xy-6y^2$, and $3x^2-8xy+4y^2$.

6. $6a^2+7ax-3x^2$, and $3x^2+11ax+6a^2$.

7. $a^5c-4a^3cm+3acm^2$, and $a^4c^2-6a^2c^2m+5c^2m^2$.

8. $6x^3-7ax^2-20a^2x$, and $3x^2+7ax+4a^2$.

9. $2x^4-7x^3+5x^2$, and x^3+3x^2-4x.

10. $12x^5-51x^3+12x$, and $2x^5-4x^4-2x^3+4x^2$.

11. $3y^3-13y^2+23y-21$, and $6y^3+y^2-44y+21$.

12. $a^3x^3-a^2bx^2y+ab^2xy^2-b^3y^3$; $2a^2bx^2y-ab^2xy^2-b^3y^3$.

13. $2x^2+9x+4$, $2x^2+11x+5$, and $2x^2-3x-2$.

14. $3x^4+8x^3+4x^2$, $3x^5+11x^4+6x^3$, and $3x^4-16x^3-12x^2$.

15. $x^3-9x^2+26x-24$, $x^3-10x^2+31x-30$, and $x^3-11x^2+36x-36$.

16. $x^4-2a^2x^2+a^4$, and $x^4+2ax^3+a^2x^2-x^2-2ax-a^2$ (**137**, 2).

17. $4a^2-5ab+b^2$, $3a^3-3a^2b+ab^2-b^3$, and a^4-b^4.

18. x^3+7x^2-x-7, x^3+5x^2-x-5, and $(x^2-2x+1)^2$ (**136**, 3).

19. $6x^2-25x+14$, $4x^2-20x+21$, $2x^2-15x+28$, and $2x^2+5x-42$.

20. $2x^4-4x^3+8x^2-12x+6$, and $3x^4-3x^3-6x^2+9x-3$.

21. $x^4+ax^3-9a^2x^2+11a^3x-4a^4$, and $x^4-ax^3-3a^2x^2+5a^3x-2a^4$.

22. $3x^2+17x+20$, and $x^2-9x-52$.

The following example and exercises should be passed over until the subject is reviewed.

23. Given $6a^3+13a^2b-73ab^2+60b^3$, and $8a^3-26a^2b+27ab^2-9b^3$.

We shall use the first, and the difference of the first and second instead of the two quantities themselves.

	1st Divisor.	1st Dividend.	
	$2a^3-39a^2b+100ab^2-69b^3$	$6a^3+13a^2b-73ab^2+60b^3$	3.
	65 2d. Dividend	$6a^3-117a^2b+300ab^2-207b^3$	1st
a	$130a^3-2535a^2b+6500ab^2-4485b^3$	$b)\ 130a^2b-373ab^2+267b^3$	quo.
2d	$130a^3-373a^2b+267ab^2$	$130a^2-373ab+267b^2$	
quo.	$-23b)\ -2162a^2b+6233ab^2-4485b^3$	2	
	$94a^2-271ab+195b^2$	$260a^2-746ab+534b^2$	
	3		
	$282a^2-813ab+585b^2$		
	$260a^2-746ab+534b^2$		
	$22a^2-67ab+51b^2$		
	6		
	$132a^2-402ab+306b^2$		
	$130a^2-373ab+267b^2$		
$a-13$	4th Dividend. $2a^2-29ab+39b^2$	$22a^2-67ab+51b^2$	
4th	$2a^2-3ab$	$22a^2-319ab+429b^2$	3d
quo.	$-26ab+39b^2$	$126b)\ 252ab-378b^2$	quo.
	$-26ab+39b^2$	$2a-3b$ 4th Divisor.	
		h. c. f.	

EXPLANATION. — Having found the second remainder, $-2162a^2b+6233ab^2-4485b^3$, and the numbers being very large, by the method of g. c. d. in arithmetic their greatest common divisor was found to be 23. Using $94a^2-271ab+195b^2$ and $130a^2-373ab+267b^2$, we derive $22a^2-67ab+51b^2$ by **146**, c. Next, using $22a^2-67ab+51b^2$ in a similar manner $2a^2-29ab+39b^2$ is derived. Finally, using $2a^2-29ab+39b^2$ and $22a^2-67ab+51b^2$ the h. c. f. is obtained in the usual manner.

24. $6x^3+13x^2+15x-25$, and $2x^3+4x^2+4x-10$.

25. $6x^4-x^3-3x^2-4x-4$, and $8x^4-2x^3-19x^2+3x+10$.

26. $y^3-5y^2+11y-15$, y^3-y^2+3y+5, $2y^3-7y^2+16y-15$.

27. $2y^6+2y^5-3y^4-5y^3-14y^2-7y$, and $2y^5+2y^4-5y^3-5y^2-7y-7$.

28. $x^6+4x^5-3x^4-16x^3+11x^2+12x-9$ and $6x^5+20x^4-12x^3-48x^2+22x+12$.

CHAPTER XII.

COMMON MULTIPLES.

149. Common Multiples — Lowest Common Multiple.

1. A multiple of a quantity is literally the quantity multiplied by something.

2. A common multiple of two or more quantities is a quantity that is some number of times each one.

3. The lowest common multiple of two or more quantities is that one which is of the lowest degree.

a. A multiple of a quantity is often defined as one that will contain it; and a common multiple of two or more quantities as a quantity that will contain each. The lowest common multiple (abbreviated into l. c. m.) contains no factors except those necessary to make it a multiple of the several quantities.

Of common multiples there can be any number, but there can be only *one* lowest common multiple. To illustrate this, if we take any quantity as $a + m$, then $3(a + m)$, $6a + 6m$, $a^2 + am$, etc., are all multiples; and there may be any number of such multiples. For, anything at all by which we multiply $a + m$ gives a multiple of it. The expression $3am^2$ is a multiple of 3, a, $3a$, m, am, $3am$, m^2, etc., and so a common multiple of any two or more of them. It is not a multiple of m^3, nor of x, which is a factor not found in it. Now $3am^2$ is the l. c. m, of the quantities enumerated above. It is not, however, the l. c. m. of a, $3a$, and $3m$, for $3am$ will contain each of them, and is of a lower degree than $3am^2$.

b. The l. c. m. is used in adding and subtracting fractions, in clearing equations of fractions, etc. And it is highly desirable that the student should be thoroughly conversant with the examples and exercises about to be given.

150. Principles applicable in finding Common Multiples.

1. Any number of times a given multiple of a quantity is also a multiple of that quantity. This is self-evident.

2. The product of any number of quantities will always be one of their common multiples, since it will contain each of them.

3. The lowest common multiple of two or more quantities must contain all the prime factors of each of them, and no others, each factor appearing with the highest exponent which it has in any one quantity.

For, the product of all such factors is necessary and sufficient to contain every quantity.

151. Rules for finding the l. c. m. of two or more Quantities.

1. After factoring each quantity, by inspection write down in a product all the prime factors, using no factor oftener than it occurs in any one quantity.

2. Or, set down the factored quantities on a horizontal line, and divide by any prime factor that will divide two or more of them, and bring down the undivided quantities to the line beneath. Divide this new line of quantities by any prime factor that will divide two or more of them without a remainder, and so continue to divide until the last line of quotients and undivided numbers are prime to each other.

Multiply the divisors and last line of undivided quantities together for the l. c. m.

A little reflection will show that this accomplishes what was required in 3. of the last article.

3. Method when the quantities cannot be factored.

(1. When there are but two quantities find their h. c. f.; then divide each by the h. c. f; and last of all multiply the two quotients and the h. c. f. together. This process fulfils the requirements in principle 3 of **150**.

(2. When there are more than two quantities find the l. c. m. of two of them, then of this result and a third, and so on. The last l. c. m. is the one required.

a. There is usually an advantage in retaining all expressions in their factored form.

b. When quantities have no common factor their product is their l. c. m.

152. Examples in finding Lowest Common Multiples.

1. Find four common multiples and the l. c. m. of a^2, ab^3c^2, and bc^4.

(1. Their product $a^3b^4c^6$ is a common multiple.

(2. $a^4b^5c^7$, or any higher exponents.

(3. $a^3b^3c^6$ still contains each.

(4. $a^2b^3c^4$ the l. c. m.

2. Find the l. c. m. of $3a(x^2-1)$, $6ab(x^2-x)$, and $4a^3b(x-1)$. Factoring the quantities and writing the results side by side,

$$3a(x^2-1) = 3a(x+1)(x-1)$$
$$6ab(x^2-x) = 2\cdot3abx(x-1)$$
$$4a^3b(x^3-1) = 2\cdot2a^3b(x-1)(x^2+x+1).$$

$\therefore$ the l. c. m, by rule 1, is $12a^3bx(x+1)(x-1)(x^2+x+1)$.

3. Find the l. c. m. of $x^3-6x^2+11x-6$, $x^3-9x^2+26x-24$, and $x^3-8x^2+19x-12$.

$$x^3-9x^2+26x-24\ (x^3-6x^2+11x-6$$
$$\underline{x^3-6x^2+11x-6}\qquad(1$$
$$-3\ (-3x^2+15x-18$$
$$x^2-5x+6)\ x^3-6x^2+11x-6\ (x-1$$
$$\underline{x^3-5x^2+6x}$$
$$-x^2+5x-6$$
$$\underline{-x^2+5x-6}$$

$\therefore x^2-5x+6\ (=(x-2)(x-3))$ is the l. c. m. of these two quantities. Dividing the third quantity, $x^3-8x^2+19x-12$ by x^2-5x+6, their h. c. f. is soon found to be

$x - 3$. We are now in position to factor the three quantities, since we know two of the factors of the first two quantities, and one factor of the third. We find

$$x^3 - 6x^2 + 11x - 6 = (x - 1)(x - 2)(x - 3)$$
$$x^3 - 9x^2 + 26x - 24 = (x - 2)(x - 3)(x - 4)$$
$$x^3 - 8x^2 + 19x - 12 = (x - 1)(x - 3)(x - 4).$$

By inspection, we have for the l. c. m. $(x - 1)(x - 2)(x - 3)(x - 4)$.

153. Exercise in finding the l. c. m. of Sets of Quantities.

1. 54, 81, 24, 27.
2. 60, 12, 120, 48, 36.
3. 432, 270.
4. $18ax^2$, $7ay^2$, $12xy$.
5. x^2 and $ax + x^2$.
6. $x^2 - 1$ and $x^2 - x$.
7. $3a^2b$, $4b^2c$, $2c^2d$, $9ad^2$.
8. $x^2 - y^2$, $x + y$, and $x - y$.
9. $ab^2c^3x^4$, $a^4bc^2x^2$, and a^3b^3cx.
10. $3x^2yz^3$, $15xy^3z^2$, $10x^2y^2z^2$.
11. $3a^2bc$, $27a^3b^2c^2$, and 6.
12. $9a^4b^2$, $12b^3c$, a^2c^3, 36, 8.
13. $14a^2b^2$, $7b^2x^3$, 3, 2, 5.
14. $2x(x - y)$, $4xy(x^2 - y^2)$, $6xy^2(x + y)$.
15. $4(1 + x)$, $8(x - 1)$, $1 - x^2$.
16. $a^2 + ab$, $ab + b^2$.
17. $3x^2$, $4x^2 + 8$.
18. $4x^2y - y$, $2x^2 + x$.
19. $x^3(x - y)^2$, $y^3(x + y)^2$, $xy(x^2 - y^2)$
20. $8a^2b + 8ab^2$, $6a - 6b$.

21. $a(x-b)(x-c)$, $b(c-x)(x-a)$, and $c(a-x)(b-x)$. (See Sug. Ex. 14, **142**.)

22. x^2+2x, x^2+3x+2.

23. $3(x+y)$, $3(x-y)$, $3(x^3+y^3)$.

24. $a+b$, $a^2+2ab+b^2$, and a^4-b^4.

25. x^2+5x+6, and x^2+6x+8.

26. $x^2+11x+30$, and $x^2+12x+35$.

27. a^2-x^2, $a^2-2ax+x^2$, $a^2+2ax+x^2$.

28. x^3-1, x^2+x+1.

29. $6x^2-x-1$, $2x^2-3x-2$.

30. $(x-x^2)^2$, x^2-1, and $4x(1+x)$.

31. x^2-4a^2, $(x+2a)^3$, and $(x-2a)^3$.

32. x^2-1, x^3+1, x^3-1, and x^6+1.

33. $3(a^3-b^3)$, $4(a-b)^3$, $5(a^4-b^4)$, $6(a^2-b^2)$, and $(a^2-b^2)^3$.

34. x^4-2x^2+1, $x^4+4x^3+6x^2+4x+1$.

35. $3x^4+26x^3+35x^2$, $6x^2+38x-28$, and $27x^3+27x^2-30x$.

36. $12x^2-23xy+10y^2$, $4x^2-9xy+5y^2$, and $3x^2-5xy+2y^2$.

37. $x^4+ax^3+a^3x+a^4$, $x^4+a^2x^2+a^4$.

38. $15x^3-14x^2y+24xy^2-7y^3$, and $27x^3+33x^2y-20xy^2+2y^3$.

39. $x^2-3xy-10y^2$, $x^2+2xy-35y^2$, $x^2-8xy+15y^2$, and $x^2+4xy-21y^2$.

40. x^2+7x+9, x^2-3x+7, and $x^2-2x+11$.

CHAPTER XIII.

FRACTIONS.

154. A Fraction in Algebra is any expression written in the fractional form, that is, with a numerator and denominator, and used to indicate a division.

a. Thus, if a and b represent whole numbers $\frac{a}{b}$ represents a simple fraction in algebra. The unit is divided into b parts and a parts are taken. Or, if we look upon the fraction as an indicated division, a is divided by b, just as 3 is divided by 5 in $\frac{3}{5}$.

But letters are supposed to have any values, integral or fractional. Now, if, for example, $a = \frac{1}{2}$ and $b = \frac{4}{7}$, we cannot any longer say that the unit is divided into $\frac{4}{7}$ parts and $\frac{1}{2}$ part is taken, for such language has no meaning; but we can say that $\frac{1}{2}$ is divided by $\frac{4}{7}$ giving an arithmetical complex fraction. For this reason it is customary to look upon all algebraic fractions as indicated divisions (Cf. **53**, *a*), the numerator being the dividend, the denominator the divisor, and the value of the fraction the quotient.[1]

As in arithmetic, the numerator and denominator are called the terms of the fraction. Besides such expressions as the one given, $\frac{a}{b}$, the fraction may have any complex quantity for either or both of its terms, e.g., $\frac{6\,ab + 9\,ac + 15\,b^2c}{16\,abc + 9\,c^2}$ has a trinomial for its numerator and a binomial for its denominator.

b. It is the *form* of an indicated division, and not any numerical value it may have, which makes an expression fractional. Indeed, what we should call an integral quantity, as $3\,ab$, may have a fractional value; thus, putting $a = \frac{1}{2}$, $b = \frac{3}{4}$, $3\,ab = \frac{9}{8}$. While on the other hand $\frac{3a}{6}$ which is an algebraic fraction becomes equal to 2, an integer.

[1] For a thoroughgoing treatment of all the fundamental questions of algebra the teacher should consult treatises on the subject like Peacock's or Chrystal's.

SECTION I.

Classification and Principles.

155. Classification of Fractions.

1. With respect to their origin.

(1. A Simple Fraction, the original form of the fraction, contains entire quantities for its numerator and denominator.

$$\text{Thus, } \frac{6\,ab^2c}{5\,m^3n},\ \frac{4\,a^2+3\,b^2+c}{2\,my+3\,nz-t^2}.$$

(2. A Complex Fraction arises upon dividing one fraction by another. A complex fraction has a fractional expression in one or both of its terms.

$$\text{Thus, } \frac{\frac{5\,ab}{m}}{n},\ \frac{\frac{m^2-n^2}{m+q}}{\frac{rs}{p+q+r}}.$$

(3. A Mixed Quantity is one which is partly integral and partly fractional.

$$\text{E. g., } 5\,a+\frac{c}{d}.$$

2. With respect to their capability of reduction to a mixed quantity.

(1. When the numerator does not contain the denominator an entire number of times, the expression may be called by analogy a *proper fraction.*

$$\text{E. g., } \frac{6\,abc}{mn},\ \frac{3\,a+2\,b}{4+7\,c}.$$

(2. When the numerator does contain the denominator an entire number of times, it may be called an *improper fraction.*

$$\text{E. g., } \frac{3\,a+4\,b}{a+b+c}=3+\frac{b-c}{a+b+c} \qquad \textbf{(126)}.$$

156. Fundamental Principle in Fractions. — If a fraction is regarded as an expressed division, then multiplying or dividing both terms by the same number will not change the quotient, i.e., the value of the fraction.

A proof of this principle may be given as follows: —

Let $\frac{a}{b}$ denote any fraction, and x its value; then $x = \frac{a}{b}$.

Whence, $a = bx$ (def., **43**)

Let m be any number; then from the equation just written, it follows self-evidently that

$$ma = mbx \qquad \text{(207, 3)}$$

$$\text{or, } ma = mb \cdot x \qquad \text{(38, 2)}$$

Therefore, $\frac{ma}{mb} = x$, (def. **43**)

i. e., $\frac{ma}{mb} = \frac{a}{b}$ *Q. E. D.*

It follows, conversely, that both terms of a fraction may be *divided* by the same quantity without altering its value.

157. The Three Signs connected with every fraction. There are three signs expressed or understood belonging to every fraction, viz., those of the numerator, denominator, and fraction itself.

$$\text{E. g., } \frac{a}{b} = + \frac{+a}{+b}; \quad - \frac{+(m+p)}{-(n+q)}.$$

The same is true of a polynomial numerator or denominator.

$$\text{To illustrate, } \frac{a^2 - b^2 - c^2}{-ab + ac - bc} = + \frac{+(a^2 - b^2 - c^2)}{+(-ab + ac - bc)}$$

$$= + \frac{+(a^2 - b^2 - c^2)}{-(ab - ac + bc)},$$

parentheses being used to constitute the polynomial one quantity.

The sign before the fraction applies to the *value of the fraction*, i.e., the *quotient* of the numerator divided by the denominator.

1. Effect upon the fraction produced by changing these signs.

(1. Evidently, changing the sign before a fraction changes its *value* from plus to minus, or from minus to plus.

(2. Changing the sign of either numerator or denominator changes the sign of the quotient, (**43**) which is the value of the fraction.

Thus, $\frac{20}{4} = +5$, while $\frac{-20}{4} = -5$, and $\frac{20}{-4} = -5$.

(3. If the sign of either numerator or denominator and at the same time the sign of the fraction be changed, the fraction is changed back to its former sign and *remains unaltered.*

(4. If the signs of both terms of a fraction be changed, the value of the fraction remains unaltered.

Thus, $\frac{15}{3} = 5$, and $\frac{-15}{-3} = 5$; $\frac{-18}{6} = \frac{18}{-6} = -3$.

(5. Finally, since changing the signs of the two terms leaves the fraction the same, changing all three signs changes the sign of the fraction.

2. Rules for changing the signs of a fraction.

(1. Changing one or all three, i.e., an odd number of the signs of a fraction, changes the sign of the fraction.

(2. Changing *any two* of the signs of a fraction does not alter its value.

SECTION II.

REDUCTION.

158. Reduction of Fractions. — Reduction in all mathematics is the process of changing the *form* of a quantity without altering its value.

a. There are five cases of reduction of fractions commonly given: reduction to lowest terms; reduction of an improper fraction to a mixed quantity; reduction of a mixed quantity to an improper fraction; reduction of an entire quantity or a fraction to the form of a fraction having a given denominator; reduction of two or more fractions to equivalent fractions having a least common, or any common denominator.

I.—TO LOWEST TERMS.

159. Reduction of a Fraction to its Lowest Terms. Principle and Rule. — A fraction is in its lowest terms when its numerator and denominator are prime to each other.

By **156**, dividing both terms of a fraction by the same quantity changes its form, but does not alter its value. Therefore we may,

1. Factor the numerator and denominator into their prime factors, and then cancel out of both terms the factors common.

2. Or, find by continued division the h. c. f. of the terms, and divide both terms by it. The resulting fraction is in its lowest terms.

160. Exercise in reducing Fractions to their Lowest Terms.

1. $\frac{6a^2b^3c}{9ab^4c}$. Dividing both terms by $3ab^3c$ we get $\frac{2a}{3b}$. *Ans.*

2. $\frac{15}{45}$, $\frac{81}{144}$, $\frac{243}{162}$

3. $\frac{1274}{2002}$, $\frac{18607}{24587}$

4. $\frac{3x^2y^3}{8xy^3}$

5. $\frac{105b^2y^3}{15by^2}$

6. $\frac{129a^3b^3x^2}{27a^4b^4x^4}$.

7. $\frac{a^n}{a^{n+1}}$

8. $\frac{mx^a}{nx^{a-1}}$

9. $\frac{4(a+b)^2}{5(a^2-b^2)}$

10. $\frac{x^2-1}{2xy+2y}$

11. $\frac{ax+x^2}{ac^2+c^2x}$

12. $\frac{abx+bx^2}{acx+c^3}$

13. $\frac{a^5-a^3b^2}{a^4-b^4}$

14. $\frac{x^3-b^2x}{x^2+2bx+b^2}$

15. $\frac{12m^3x^4+2m^2x^5}{18mn^2x+3n^2x^2}$

16. $\frac{(a+b)^2+(a-b)^2}{a^4-b^4}$

17. $\frac{a^2-a-20}{a^2+a-12}$

18. $\frac{a^2x^2-16a^2}{ax^2+9ax+20a}$

19. $\frac{27a+a^4}{18a-6a^2+2a^3}$

20. $\frac{x^2+2xy+y^2}{x^3-xy^2}$

21. $\frac{2bx^3-2b^3x}{2bx^2+4b^2x+2b^3}$

22. $\frac{6a^2+7ax-3x^2}{6a^2+11ax+3x^2}$

23. $\frac{3ax^4+9ab^3+6a^2b^2}{a^4+a^3b-2a^2b^2}$

24. $\frac{6ac+10bc+9ad+15bd}{6c^2+9cd-2c-3d}$

25. $\frac{a^3-3a^2+3a-2}{a^3-4a^2+6a-2}$.

REMARK. — In most instances hereafter it will be desirable to use fractions in their lowest terms, and in some cases it is necessary.

II. — REDUCTION OF FRACTIONS TO MIXED OR ENTIRE QUANTITIES.

161. To Reduce an Improper Fraction to an Entire or Mixed Quantity. Principle and Rule.

Since a fraction is an expressed division, divide the numerator by the denominator, and if there is a remainder place it over the denominator. This amounts to the same as dividing both terms by the denominator, which is justifiable by **156**.

162. Exercise in reducing Fractions to Entire or Mixed Quantities.

1. $\frac{ax+2x^2}{a+x}$ Operation. $ax+2x^2\,(a+x$
$$\begin{array}{l} ax+x^2 \\ \hline x^2 \end{array} \quad x+\frac{x^2}{a+x}\ \textit{Ans.}$$

2. $\frac{25}{8}$, $\frac{147}{19}$, $\frac{75}{13}$, $\frac{1425}{111}$

3. $\frac{ab+b^2}{a}$

4. $\frac{x^2-y^2}{x-y}$

5. $\frac{x^2+y^2}{x}$

6. $\frac{4x^2-2x}{2x^2-x+1}$

7. $\frac{a^6x-3ax^6}{a^3-ax^2}$

8. $\frac{x^2+3x+2}{x+3}$

9. $\frac{2a^2-2ab+4b^2}{a-b}$.

10. $\frac{22a^3b^3-33a^4b^4+7ax}{11a^2b}$

11. $\frac{x^4+1}{x-1}$

12. $\frac{a^3+x^3-x^4}{a+x}$

13. $\frac{1-a-ab+a^2b}{ab-b}$

14. $\frac{10a^2-13ax-3x^2}{2a-3x}$

15. $\frac{x^5+y^5}{x+y}$

16. $\frac{12c^3+8ac^2x^2-3acx-2a^2x^3}{3c+2ax^2}$

163 Dissection of a Fraction. — Principle and Rule. — Divide each term of the numerator by the denominator (**126**, 2), writing the quotients in the fractional form connected by their proper signs. Reduce each fraction to its lowest terms.

164. Exercise in Dissecting Fractions.

1. $$\frac{adn+bcn-bdm}{bdn}=\frac{adn}{bdn}+\frac{bcn}{bdn}-\frac{bdm}{bdn}$$
$$=\frac{a}{b}+\frac{c}{d}-\frac{m}{n}\quad \textbf{(159)}.$$

2. $\dfrac{6a^2 - 3b^2 + 10c^2}{30abc}$

3. $\dfrac{abc + bcd + acd + abd}{abcd}$

4. $\dfrac{(m-n)(n+q) - (m+n)(p-q)}{(m-n)(p-q)}$

5. $\dfrac{(a+b)(m-n) - (a-b)(m+n)}{a^2 - b^2}$

165. Fractions written as Entire Quantities.

By **128**, 2, we can transfer factors from the denominator of a fraction to the numerator by changing the signs of their exponents. This enables us to write any expression in the integral form.

166. Exercise.

1. $\dfrac{x^3b}{ay^2}$ By changing the signs of the exponents of a and y^2, we have $a^{-1}bx^3y^{-2}$

2. $\dfrac{5ab^2}{b^3c^2}$

3. $\dfrac{x^3y^2}{a^{-1}b}$

4. $\dfrac{5a^{-4}c}{3a^2d^3}$

5. $\dfrac{x(a-b)^2}{4(a-b)^3}$

6. $\dfrac{x^3(m+n)}{(m-n)^{-2}}$

7. $\dfrac{(a+b)^{-2}x^5y^6}{(a-b)^{-4}y^3}$

III.—MIXED QUANTITIES TO IMPROPER FRACTIONS.

167. To Reduce a Mixed Quantity to the Form of an Improper Fraction. Principle and Rule.

This case is the reverse of the preceding. There the dividend and divisor were given to find the quotient. Here the quotient, remainder, and divisor are given to find the dividend, which being found is written over the divisor for the equivalent fraction. Therefore, the entire quantity is to be multiplied by the denominator and the numerator added to the product (or subtracted if the sign before the fraction is minus), and the result placed over the denominator.

168. Exercise in reducing a Mixed Quantity to the Form of a Fraction.

1. $3a + y + \dfrac{2ax - xy}{x - y}$.

Operation. $(3a + y)(x - y) = 3ax - 3ay + xy - y^2$

Adding the Numerator, $\quad 2ax \qquad - xy$

$\qquad\qquad 5ax - 3ay \qquad - y^2$

$$\therefore 3a + y + \frac{2ax - xy}{x - y} = \frac{5ax - 3ay - y^2}{x - y} \quad \textit{Ans.}$$

2. $2 + 3y - \dfrac{y - 5}{4y}$.

Operation. $\dfrac{(2 + 3y)4y - (y - 5)}{4y} = \dfrac{12y^2 + 7y + 5}{4y}$.

a. The minus sign before a fraction changes the sign of every term of the numerator according to the rule for subtraction. (**92**, 1). Thus, $-\dfrac{y - 5}{4y}$ is the same as $\dfrac{-y + 5}{4y}$.

3. $4\frac{19}{23}$, $7\frac{3}{4}$, $4\frac{6}{23}$.

Remark.—In arithmetic we write, e.g., $4\frac{7}{8}$ and not $4 + \frac{7}{8}$, while in algebra, $a + \frac{b}{c}$ must be written so, and not $a\frac{b}{c}$, since this last would mean *a times* $\frac{b}{c}$. This illustrates a difference in one particular between the arithmetical and algebraical notations. It may be said that when no sign is written in arithmetic + is understood, while in algebra the sign of multiplication is understood. In arithmetic the numerator is always added to the product of the denominator and entire quantity; in algebra it is added or subtracted according as the sign before the fraction is + or −. (Cf. **52**, *b*).

4. $2a - 2b + \frac{a - x}{3}$

5. $a^2 + \frac{b^3}{a}$

6. $5 + \frac{2x - 7}{3x}$

7. $1 + \frac{a - b}{a + b}$

8. $3x - \frac{4x^2 - 5}{5x}$

9. $a + b - \frac{a^2 - 2ab + b^2}{a + b}$

10. $a^2 - ax + x^2 - \frac{x^3}{a + x}$

11. $a^2 - a + b\ b^2 - \frac{b^3 - a^3}{a + b}$

12. $15 - \frac{3}{7}c - \frac{a^2}{d}$

13. $1 - \frac{a^2 - 2ab + b^2}{a^2 + b^2}$

14. $x^3 + x^2 + x + 1 + \frac{2}{x - 1}$

15. $1 + u - \frac{(1 - u)^2}{u + 1}$.

IV.—REDUCTION OF FRACTIONS OR INTEGERS TO EQUIVALENT FRACTIONS HAVING GIVEN DENOMINATORS.

169. To Reduce a Fraction or an Integer to an Equivalent Fraction having a Given Denominator. Principle and Rule.

Multiply both terms of the given fraction by the quotient obtained by dividing the required denominator by the denominator of the fraction (**156**). For an integer the denominator 1 is understood, and both terms are multiplied by the required denominator.

170. Exercise in reducing Fractions to Equivalent Expressions having Given Denominators.

1. Reduce a to the denominator d.

   Operation. $\frac{a}{1} \times \frac{d}{d} = \frac{ad}{d}$. *Ans.*

2. Reduce $\frac{m}{n}$ to the denominator $3n^2$.

   Operation. $3n^2 \div n = 3n$; $\frac{m}{n} \times \frac{3n}{3n} = \frac{3mn}{3n^2}$.

3. Reduce $5a^2b$ to the denominator ab^2d.

4. Reduce $\frac{7}{8}$ to the denominator 56.

5. Reduce $\frac{3a+b}{a+b}$ to the denominator $a^2 - b^2$.

6. Reduce $5(a+b)^2$ to the denominator $(a-b)^2$.

7. Reduce $ac + bd + ad$ to the denominator $ab + cd$.

8. Reduce $f + 2x$ to the denominator $a + b$.

9. Reduce $\frac{x^2+xy+y^2}{x+y}$ to the denominator $x^3 + y^3$.

10. Reduce $\frac{a+b+c}{a-b+c}$ to the denominator $a^2 - b^2 + 2bc - c^2$.

171. To Reduce Two or More Fractions to Equivalent Fractions having a Common Denominator, (usually the lowest Common Denominator). Principle and Rule.

Of the common denominators which deserve especial consideration, there are two: the *product* of all the denominators of the fractions, and their lowest common multiple. Taking these as required denominators and remembering the process of **169**, we get two distinct rules.

1. Find the l. c. m. of the denominators of the fractions, (called the *lowest common denominator*), and multiply both terms of each fraction (**156**) by the quotient of the l. c. d. divided by its denominator. The results are the equivalent values of the respective fractions, having, it is plain, the l. c. m. for a common denominator.

2. Multiply both terms of each fraction by the product of all the denominators except its own. (**156**.)

Evidently the product of all the denominators divided by any one will give the product of all except that one.

a. The l. c. d. and each denominator should be written *as the product of their factors*. Then the quotient of the l. c. d., divided by any denominator, is the product of all the factors not in that denominator.

In addition and subtraction of fractions, as we shall see, the multiplications have to be performed in full in obtaining the numerators;

but not so the denominators. Let the student follow this rule, *never to multiply factors together until it is seen to be necessary*, and a great saving in time and labor will thus be made.

172. Exercise in the Reduction of Fractions to Equivalent Fractions having a Common Denominator.

1. $\frac{1+x}{1-x}, \frac{1+x^2}{1-x^2}, \frac{1+x^3}{1-x^3}$

Operation. — Writing the denominators in the factored form, we have,

$$\frac{1+x}{1-x}, \quad \frac{1+x^2}{(1-x)(1+x)}, \quad \frac{1+x^3}{(1-x)(1+x+x^2)},$$

and by inspection we see that $(1+x)(1-x)(1+x+x^2)$ is the l. c. d.

Now, $\frac{(1+x)(1-x)(1+x+x^2)}{(1-x)} = (1+x)(1+x^2)$,

and multiplying the numerator by this quotient, and placing the product over the l. c. d., we have,

$\frac{(1+x)^2(1+x+x^2)}{(1+x)(1-x)(1+x+x^2)}$ for the equivalent value of $\frac{1+x}{1-x}$, the first fraction. (169)

In like manner,

$$\frac{1+x^2}{1-x^2} = \frac{(1+x^2)(1+x+x^2)}{(1+x)(1-x)(1+x+x^2)}; \quad \frac{1+x^3}{1-x^3} = \frac{(1+x^3)(1+x)}{(1+x)(1-x)(1+x+x^2)}$$

2. $\frac{6}{9}, \frac{9}{12}, \frac{12}{20}, \frac{7}{10}$

3. $\frac{a}{x}, \frac{b}{x^2}, \frac{1}{x^3}$

4. $\frac{3a}{7b}, \frac{5b}{21c}$

5. $\frac{a}{b}, \frac{c}{d}, \frac{e}{f}$

6. $\frac{a}{5bc}, \frac{b}{10ac}, \frac{c}{3ab}$

7. $\frac{3x}{2a}, \frac{2x}{5a^2}, \frac{m}{n}$

8. $\frac{9a}{8x}$, $\frac{7b}{36x}$, $\frac{11a}{28}$, $\frac{7(a+b)}{4x}$

9. $\frac{a-b}{ab}$, $\frac{a-c}{ac}$, $\frac{b-c}{bc}$

10. $\frac{ab}{cd}$, $\frac{cd}{ab}$, $\frac{ef}{gh}$, $\frac{gh}{ef}$

11. $\frac{a+b}{a-b}$, $\frac{a-b}{a+b}$

12. $\frac{1}{x+1}$, $\frac{3}{4x+4}$, $\frac{x}{x^2-1}$

13. $\frac{a^2}{a^2-x^2}$, $\frac{1}{a-x}$, $\frac{a}{a+x}$

14. $\frac{x}{x+y}$, $\frac{y}{x+y}$, $\frac{x}{x-y}$, $\frac{y}{x-y}$

15. $\frac{3}{2x^3-2xy^2}$, $\frac{5}{6(x+y)^2}$

16. $\frac{1}{(a+b)^2}$, $\frac{1}{a(a-b)^2}$, $\frac{1}{b(a^2-b^2)}$

SECTION III.

The Fundamental Operations in Fractions.

173. Addition, Subtraction, Multiplication, and Division of Fractions. The Rules for these operations are the same as in Arithmetic.

I. — ADDITION.

174. Addition of Fractions. Principle and Rule.

Let $\frac{a}{c}=x$, and $\frac{b}{c}=y$, be two fractions having a common denominator.

Then, $\left.\begin{matrix} a = xc \\ b = yc \end{matrix}\right\}$ (Def. of division, **43**)

Adding these equal quantities (Ax. 1, **207**)

$$a+b=(x+y)c \qquad \textbf{(97)}$$

$$\therefore \frac{a+b}{c}=x+y \qquad \text{(Def. of division, } \textbf{43}\text{)}$$

Thus, when two fractions have a common denominator, the numerator of the sum is the sum of their numerators. This demonstration may be extended to include any number of fractions united by both + and − signs.

RULE. Reduce the fractions to equivalent fractions having a lowest common denominator, add their numerators and place the sum over the common denominator.

175. Exercise in Adding Fractions.

1. $\frac{x}{x-1}+\frac{x}{x+1}+3.$

Operation. — We write 1 for the denominator of the integer. $1(x-1)(x+1)$ is the l. c. d. The equivalent fractions are,

$$\frac{x(x+1)}{(x+1)(x-1)}+\frac{x(x-1)}{(x+1)(x-1)}+\frac{3(x-1)(x+1)}{(x+1)(x-1)}$$

Expanding the numerators (171, *a*) and adding them, we have,

$$x^2+x+x^2-x+3x^2-3=5x^2-3,$$

which, placed over the l. c. d., gives,

$$\frac{5x^2-3}{x^2-1}. \quad \textit{Ans.}$$

2. $\frac{3}{4}+\frac{7}{8}+\frac{17}{24}$

3. $\frac{2-x}{5}+\frac{3x-1}{2}$

4. $\frac{7x}{8}+\frac{x}{12}+\frac{x}{4}$

5. $\frac{a}{b}+\frac{c}{d}+\frac{m}{n}$

6. $x+\frac{3x-5}{2}+\frac{2x-4}{3}$

7. $\frac{3x}{a}+\frac{2y}{b}$

8. $\frac{a-b}{ab}+\frac{c-a}{ac}+\frac{b-c}{bc}$

9. $c+\frac{c+d}{2}+\frac{c-d}{2}$

10. $5x-\frac{2x}{7}+\frac{5x}{9}+x^2$

11. Add $3x$, $x+\frac{3a}{4}$, $4x-\frac{6a}{5}$

SUGGESTION. — Add the entire parts and the fractions separately.

12. $4a^2 + \frac{x-2}{3} + \frac{x+2}{3} + 5a^2 + \frac{3x-1}{7}$

13. $\frac{1}{x+2} + \frac{1}{x+3}$

14. $\frac{m}{m+p} + \frac{p}{m-p}$

15. $\frac{ma+x}{m+n} + \frac{ma-x}{m+n}$

16. $\frac{n}{n-1} + \frac{1-2n}{n^2-1}$

17. $\frac{1-x^2}{1+x^2} + \frac{1+x^2}{1-x^2}$

18. $\frac{1}{x(x-y)} + \frac{1}{y(x+y)}$

19. Add $\frac{a}{a-b}, \frac{b}{b-c}, \frac{c}{c-d}$

20. Add $\frac{a}{1-a}, \frac{a^2}{(1-a)^2}, \frac{a^3}{(1-a)^3}$

21. $\frac{1}{x^2-7x+12} + \frac{1}{x^2-5x+6}$

22. $\frac{5}{5+x-18x^2} + \frac{2}{2+5x+2x^2}$

23. Add $\frac{1}{x^2-7x+12}, \frac{2}{x^2-4x+3}, \frac{3}{x^2-5x+4}$

24. $\frac{1}{8-8x} + \frac{1}{8+8x} + \frac{x}{4+4x^2} + \frac{x}{4-4x^4}$

II. – SUBTRACTION.

176. Subtraction of Fractions. Rule.

Reduce the fractions to equivalent fractions having a lowest common denominator, and subtract the numerator of the subtrahend from that of the minuend for that of the difference. If several fractions are connected by + and – signs, reduce them to the lowest common denominator, and add the numerators, those whose fractions are preceded by a minus sign being taken negatively, i.e., with their signs changed.

See principle explained in addition.

177. Exercise in Subtracting Fractions.

1. $\frac{19}{20} - \frac{6}{17}$

2. $\frac{x-1}{3} - \frac{x+2}{5}$

3. $1 - \frac{x-a}{x+a}$

4. $\frac{a+b}{a-b} - \frac{c-d}{c+d}$

5. $\frac{1}{x-a} + \frac{1}{x+a}$

6. $\frac{1+x^2}{1-x^2} - \frac{4x^2}{1-x^4}$

7. $5x + \frac{4x-6}{7} - \left(2x + \frac{7x-12}{13}\right)$

8. $2a - 3x + \frac{a-x}{a} - \left(a - 5x + \frac{x-a}{x}\right)$

9. $x - \frac{x^2}{x-1} - \frac{x}{x+1}$

10. $\frac{2y^2 - 2y + 1}{y^2 - y} - \frac{y}{y-1}$

11. $\frac{x^2 + ax + a^2}{x^3 - a^3} - \frac{x^2 - ax + a^2}{x^3 + a^3}$

12. $\frac{1}{a-2x} - \frac{(a+2x)^2}{a^3 - 8x^3}$

13. $\frac{4}{4 - 7a - 2a^2} - \frac{5}{4 - 5a - 6a^2}$

14. $\frac{2}{x^2 - 3x + 2} + \frac{2}{x^2 - x - 2} - \frac{1}{x^2 - 1}$

15. $1 + \frac{a^2 - 2ab + b^2}{4ab}$

16. $\frac{a^2}{(a-b)(b-c)} + \frac{b^2}{(b-c)(b-a)}$

SUGGESTION. — We saw in **142, 14**, that $a-b$ and $b-a$ are the same quantity, but with opposite signs. If we change the $b-a$ in the second denominator (to make it the same as $a-b$ in the first denominator) the sign of the whole denominator is changed. For, changing the sign of one factor of a product changes the sign of the product (107, 1). Now, to leave the fraction as it was before, we change the sign of the fraction (157). Thus, we get,

$$\frac{a^2}{(a-b)(b-c)}-\frac{b^2}{(b-c)(a-b)}=\frac{a^2-b^2}{(a-b)(b-c)}=\frac{a+b}{b-c} \quad (160)$$

17. $\frac{b-a}{x-b}-\frac{a-2b}{b+x}-\frac{3x(a-b)}{b^2-x^2}$

18. $\frac{3}{1-2x}+\frac{7}{1+2x}+\frac{4-20x}{4x^2-1}$

19. $\frac{a^2}{(a-b)(a-c)}+\frac{b^2}{(b-c)(b-a)}+\frac{c^2}{(c-a)(c-b)}$

SUGGESTION.—For the sake of system, it is convenient to have regard to the *cyclic order of the letters.* To do this we think of them as placed in a circle so that they follow one another in regular order, the last being succeeded by the first. Thus, if there are four letters a, b, c, d, we go from a to b, from b to c, from c to d, then from d to a again. Of the six expressions in the example only three are written in cyclic order, viz., $a-b$, $b-c$, and $c-a$. Changing the others to this order as in the suggestion, Ex. 16, we get,

$$-\frac{a^2}{(a-b)(c-a)}-\frac{b^2}{(b-c)(a-b)}-\frac{c^2}{(c-a)(b-c)}$$

the l. c. d. of which is now easily seen.

20. $\frac{r-p-q}{(p-q)(r-p)}+\frac{q+r-p}{(q-r)(q-p)}+\frac{r+p-q}{(r-p)(r-q)}$

21. $\frac{y+z}{(x-y)(z-x)}-\frac{z+x}{(y-z)(y-x)}+\frac{x+y}{(x-z)(z-y)}$

SUGGESTION.—In the third expression both factors of the denominator will have to be changed, which will not change its sign (107, 1).

22. $\frac{1}{(b-c)(c-a)}+\frac{2}{(b-a)(a-c)}+\frac{3}{(a-b)(b-c)}$

23. $\frac{1}{x(x-y)(x-z)}+\frac{1}{y(y-x)(y-z)}-\frac{1}{xyz}.$

III. – MULTIPLICATION.

178. Multiplication of Fractions. Rule. (Ex. 21, Art. **255**.)

Factor the terms of the fractions, cancel common factors, and then multiply the numerators together for the numerator of the product and the denominators for the denominator of the product.

a. **Integral factors are to be regarded as numerators over the denominator 1 understood.**

b. **Quantities to be multiplied (or divided) must first be reduced to either integral quantities or simple fractions.**

179. Exercise in the Multiplication of Fractions.

1. $\frac{2b}{3a} \times 6a^2$ Operation. $\frac{2b}{\cancel{3a}} \times \frac{\overset{2a}{\cancel{6a^2}}}{1} = 4ab.$

2. $a \times \frac{x}{y} \times \frac{y}{a}$

3. $\frac{3a}{cd} \times ab$

4. $4xy \times -\frac{5ab}{4xy}$

5. $-\frac{15a^2x^3}{24c^2y} \times -\frac{6cy^2}{5bx^3}$

6. $\frac{a+c}{1-x}(a-c)$

7. $\frac{45a^2b^3c^4}{27x^4y^8z} \times \frac{243xy^2z^3}{180a^2bc^8}$

8. $\frac{m-n}{m+n} \times (m^2+2mn+n^2)$

9. $\frac{17x^2-4}{2abx+b} \times ab$

10. $-\frac{x^4-a}{3b+3c} \times 3y$

11. $\frac{4x+2}{3} \times \frac{5x}{2x+1}$

12. $\frac{a+b+c}{9(x+y)(x-y)} \times 3(x+y)$

13. $\left(b+\frac{bx}{a}\right) \times \frac{a}{x}$

14. $\frac{c-d}{5a} \times \frac{c+d}{2-a}$

15. $(y^2 - 7y + 10) \times \frac{2y+5}{y-2}$

16. $\frac{a+x}{a} \times \frac{a-x}{x} \times \frac{a^2+x^2}{3(a^2-x^2)}$

17. $\frac{a^2-121}{a^2-4} \times \frac{a+2}{a+11}$

18. $-\frac{x^2+5x+6}{x^2-1} \times -\frac{x^2-2x-3}{x^2-9}$

19. $\frac{a^2x-x^3}{a} \times \frac{3a}{2ax-2x^2}$

20. $\frac{x^2-13x+42}{x^2-5x} \times \frac{x^2-9x+20}{x^2-6x}$

21. $-3a \times -\frac{x+1}{2a} \times -\frac{x-1}{a+b}$

22. $\frac{x+1}{x-1} \times \frac{x+2}{x^2-1} \times \frac{x-1}{(x+2)^2}$

23. $\frac{(a+b)^4}{(m+n)^5} \times \frac{(m+n)^6}{(a+b)^8}$

24. $\left[\frac{1}{(a-b)^2} - \frac{1}{c^2}\right] \times \left[\frac{a^2-b^2}{a-b-c}\right]$

25. $(1-a+a^2)\left(1+\frac{1}{a}+\frac{1}{a^2}\right)$

26. Multiply together

$$\frac{x^2-x-20}{x^2 \times 25},\ \frac{x^2-x-2}{x^2+2x-8},\ \frac{x^2+5x}{x+1}$$

27. Multiply $\frac{1}{a}+\frac{1}{b}+\frac{1}{c}$ by $\frac{1}{a}-\frac{1}{b}+\frac{1}{c}$

IV. — DIVISION.

180. Division of Fractions. Rule. — Invert the divisor and proceed as in multiplication. See Ex. 22, Art. **255**.

181. Exercise in Division of Fractions.

1. Divide $\frac{4a^2}{9m}$ by $\frac{2a}{3}$. Inverting the divisor

$$\frac{\overset{2a}{\cancel{4a^2}}}{\underset{3}{\cancel{9}m}} \times \frac{\cancel{3}}{\cancel{2a}} = \frac{2a}{3m}$$

2. $-ab \div \frac{3d}{2ab}$

3. $\frac{3m^2x}{4a^2b} \div -3x$ (178, *a*)

4. $6 \div \frac{2a}{3b}$

5. $\left[-\frac{15b^2}{40c} \times -\frac{27c^2}{81d}\right] \div \frac{abc}{14d^3}$

6. $\left[\frac{m^2}{8n} \div \frac{15mpx^5}{27n^2x^3y}\right] \times \frac{36p^3q^2}{81mn}$

7. $\left(\frac{a^3}{b^3} \times \frac{xy^2}{ab} \times \frac{pb^2}{ax}\right) \div \frac{ap^2}{b^2}$

8. $\frac{a+1}{a} \div \frac{a^2-1}{a}$

9. $\frac{a^2-b^2}{x+y} \div (a-b)$

10. $\frac{1}{a^2-b^2} \div \frac{1}{a-b}$

11. Divide $\frac{b-c}{b+c}$ by $a+c$

12. Divide $1+x$ by $\frac{1}{x}(1+x)$

13. $\left(ab + \frac{b}{a}\right) \div \frac{1}{a^2 - 1}$

14. $\frac{6(a + x) + 2}{3} \div \frac{4}{3(a - x)}$

15. $\left[\frac{x^2}{a} - 8a + \frac{12a^3}{x^2}\right] \div \left[x - \frac{2a^2}{x}\right]$

16. $\left(x - \frac{3x}{1 + x}\right) \div \frac{x(x - 2)}{1 - x}$

17. $\left(\frac{x^2}{y^3} - \frac{1}{x}\right) \div \left(\frac{x}{y^2} + \frac{1}{y} + \frac{1}{x}\right)$

18. $\left(x^3 - \frac{1}{x^3}\right) \div \left(x - \frac{1}{x}\right)$

19. $\left[\frac{x^2 - 64}{x^2 + 24x + 128} \times \frac{x^2 + 12x - 64}{x^3 - 64}\right] \div \frac{x^2 - 16x + 64}{x^2 + 4x + 16}$

20. $\left[\frac{a^3 - 3a^2b + 3ab^2 - b^3}{a^2 - b^2} \div \frac{2ab - 2b^2}{3}\right] \div \frac{a^2 + ab}{a - b}$

182. Complex Fractions. — When the division of one fraction by another is indicated in the fractional form by writing the dividend over the divisor, there results what is called a *complex* fraction. **(155, 1.)**

To simplify such fractions write the numerator as the dividend, and the denominator as the divisor in a problem in division of fractions, and proceed according to the rule. **(180.)** Or, what comes to the same thing, multiply the extremes for the numerator and the means for the denominator, and then simplify the resulting fraction.

183. Exercise in the Reduction of Complex Fractions to Simple Fractions.

1. $\dfrac{\frac{5 - c}{x}}{\frac{7 - y}{a}}$

2. $\dfrac{\dfrac{a+b}{x}}{\dfrac{y-a}{b}}$

3. $\dfrac{\dfrac{ax+b}{a}}{\dfrac{bx-a}{b}}$

4. $\dfrac{\dfrac{3cy}{5a^2x}}{\dfrac{2b}{7axy}}$

5. $\dfrac{\dfrac{3a^2}{4b}}{\dfrac{3a}{7b^3}}$

6. $\dfrac{x+\dfrac{1}{y}}{x+\dfrac{1}{y+\dfrac{1}{z}}}$

SOLUTION. — We must reduce both the numerator and the denominator to the form of simple fractions by **167**.

$$\frac{\dfrac{xy+1}{y}}{x+\dfrac{1}{\dfrac{yz+1}{z}}}=\frac{\dfrac{xy+1}{y}}{x+\dfrac{z}{yz+1}}\quad\left(\text{Since } \frac{1}{\dfrac{yz+1}{z}}=\frac{z}{yz+1}\right)$$

$$=\frac{\dfrac{xy+1}{y}}{\dfrac{xyz+x+z}{yz+1}}\ \text{(By reducing the denominator, 167.)}$$

$$\text{Now, }\frac{\dfrac{xy+1}{y}}{\dfrac{xyz+x+z}{yz+1}}=\frac{xy+1}{y}\div\frac{xyz+x+z}{yz+1}=$$

$$\frac{(xy+1)(yz+1)}{y(xyz+x+z)}\quad \textit{Ans.}$$

REMARK. — Reduce continued fractions such as the denominator above, step by step, commencing with the last denominator (**167**), until simple fractions are attained in both terms.

7. $\dfrac{x}{y+\dfrac{a}{b}}$

8. $\dfrac{a}{b+\dfrac{m}{n}}$

9. $\dfrac{1+\dfrac{1}{a}}{m+\dfrac{1}{m}}$

10. $1-\dfrac{1}{1+\dfrac{1}{m}}$

11. $\dfrac{2\frac{1}{4}-\frac{3}{5}(x+2)}{1\frac{1}{3}+\frac{1}{2}(x-3)}$

12. $\dfrac{x+\dfrac{2x}{x-3}}{x-\dfrac{2x}{x-3}}$

13. $\dfrac{x-1+\dfrac{6}{x-6}}{x-2+\dfrac{3}{x-6}}$

14. $\dfrac{1}{2x+\dfrac{1}{3x+\dfrac{x}{4}}}$

15. $\dfrac{3}{x+1}-\dfrac{2x-1}{x^2+\dfrac{x}{2}-\dfrac{1}{2}}$

16. $\dfrac{\dfrac{1}{1+a}-\dfrac{1}{1-a}}{\dfrac{1}{1-a}+\dfrac{1}{1+a}}$

17. $\dfrac{1+\dfrac{a-x}{a+x}}{1-\dfrac{a-x}{a+x}}\div\dfrac{1+\dfrac{a^2-x^2}{a^2+x^2}}{1-\dfrac{a^2-x^2}{a^2+x^2}}$

18. $\dfrac{3}{1+\dfrac{3}{1+\dfrac{3}{1-x}}}$

184. General Exercise in Fractions. — Simplify the quantities of this article.

1. $\dfrac{ax}{a^2x^2-ax}$

2. $\dfrac{5a^3b+10a^2b^2}{3a^2b^2+6ab^3}$

3. $\dfrac{27a+a^4}{18a-6a^2+2a^3}$

4. $\dfrac{x^2-5x+6}{x^2-7x+12}$

5. $\dfrac{2x-3y}{3}+\dfrac{x+2y}{4}-\dfrac{3x-2y}{6}$

6. $\frac{1}{x-2}+\frac{2}{(x-2)^2}$

7. $\frac{a}{a-1}-1-\frac{1}{a(a-1)}$

8. $a+x-\frac{4ax-5x^2}{a-x}$

9. $\frac{a^3-3a^2b+3ab^2-b^3}{a^2-2ab+b^2}$

10. $\frac{1}{(a-b)(b-c)}+\frac{1}{(a-b)(a-c)}$

11. $\frac{3x-1}{x+2}+\frac{2x+5}{2x+4}+\frac{4x-1}{6x+12}$

12. $\frac{1}{x(x-1)}+\frac{2}{1-x^2}+\frac{1}{x(x+1)}$

13. $\frac{1}{x+3y}+\frac{6y}{x^2-9y^2}-\frac{1}{3y-x}$

14. $\frac{x^2-2x+1}{3x^3+7x^2-10}$

15. $\frac{a^2b^2+3ab}{4a^2-1}\div\frac{ab+3}{2a+1}$

16. $\frac{25a^2-b^2}{9a^2x^2-4x^2}\times\frac{x(3a+2)}{5a+b}$

17. $\left(\frac{a}{b}+\frac{b}{2a}\right)\times\left(\frac{b}{a}+\frac{a}{2b}\right)$

18. $\left(\frac{1}{1+x}\right)\div\left(1-\frac{1}{1+x}\right)$

19. $\left(1-\frac{2a}{1+a}\right)\times\left(1+\frac{2a}{1-a}\right)$

20. $\dfrac{3x-2+\frac{1}{x}}{\frac{3x-1}{x}}$

21. $\frac{p-q}{pq}+\frac{r-p}{pr}+\frac{q-r}{qr}$

22. $\frac{a^3-x^3}{a^3+x^3}\times\frac{(a+x)^2}{(a-x)^2}$

23. $\frac{3xy^4}{xy^2} - \frac{5y^2+7}{xy^3} - \frac{6x^2-11}{x^3y}$

24. $\frac{x^n}{y^m} + \frac{x^m}{y^n}$

25. $\frac{ab^mc^n}{qb^nc^m}$

26. $\frac{4a^2-16b^2}{a-2b} \times \frac{5a}{20a^2+8ab+8b^2}$

27. $\frac{a^{n+1}}{b^{n-1}} \times \frac{b^{n-2}}{a^{n-4}}$

28. $\frac{mx-nx}{(a+b)(m-n)}$

29. $\frac{x^3-4x^2+5}{x^3+1}$

30. $\frac{30(a^3-b^3)}{7(a+b)} \div 42(a-b)^2$

31. $-\frac{1-x^2}{1+y} \times \frac{1-y^2}{x^2+x} \times -\left(1+\frac{x}{1-x}\right)$

32. $\frac{a+b+\frac{b^2}{a^2}}{a+b+\frac{a^2}{b}}$

33. $\frac{4m-3n}{3(1-n)} + \frac{m+3n}{3(n-1)} + \frac{2n}{1-n}$

34. $\frac{y}{x(x^2-y^2)} + \frac{x}{y(x^2+y^2)}$

35. $\frac{4ax}{3by} \times -\frac{a^2-x^2}{c^2-x^2} \times -\frac{bc+bx}{a^2-ax}$

36. $\frac{3}{x+a} - \frac{1}{x+3a} + \frac{3}{a-x} + \frac{1}{x-3a}$

37. $\frac{2-5x}{x+3} - \frac{3+x}{3-x} + \frac{2x(2x-11)}{x^2-9}$

38. $\dfrac{1+a}{(a-b)(a-c)}+\dfrac{1+b}{(b-c)(b-a)}+\dfrac{1+c}{(c-a)(c-b)}$

39. $\dfrac{1}{x+1}-\dfrac{1}{(x+1)(x+2)}+\dfrac{1}{(x+1)(x+2)(x+3)}$

40. $\dfrac{x}{x^2-y^2}-\dfrac{y}{x^2+y^2}+\dfrac{x^3+y^3}{y^4-x^4}+\dfrac{xy}{(x+y)(x^2+y^2)}$

41. $\dfrac{\dfrac{m^2+n^2}{n}-m}{\dfrac{1}{n}-\dfrac{1}{m}}\times\dfrac{m^2-n^2}{m^3+n^3}$

42. $3a-[b+\{2a-(b-c)\}]+\dfrac{1}{2}+\dfrac{2c^2-\dfrac{1}{2}}{2c+1}$

43. $\dfrac{1}{x+\dfrac{1}{1+\dfrac{x+1}{3-x}}}$

44. $\dfrac{x-4+\dfrac{6}{x+1}}{x-\dfrac{6}{x-1}}\times\dfrac{1-\dfrac{x+5}{x^2-1}}{(x-1)(x-2)}$

45. $\dfrac{a^{m+n+p}}{y^p}\div\dfrac{a^{n+p+r}}{x^p}$

46. $\left(1-\dfrac{x-y}{x+y}\right)\left(2+\dfrac{2y}{x-y}\right)$

47. $\left(\dfrac{a}{x}+\dfrac{b}{y}+\dfrac{c}{z}\right)\div\left(\dfrac{m}{x}+\dfrac{n}{y}+\dfrac{p}{z}\right)$

48. $\left[\dfrac{a+x}{a-x}+\dfrac{a-x}{a+x}\right]\div\left[\dfrac{a+x}{a-x}-\dfrac{a-x}{a+x}\right]$

49. $\dfrac{1}{x-1}-\left[\dfrac{1}{2(x+1)}+\dfrac{x+3}{2(x^2+1)}\right]$

50. $\left(\dfrac{a}{b}+\dfrac{b}{a}\right)\left(\dfrac{c}{d}+\dfrac{d}{c}\right)\div\left(\dfrac{a}{b}-\dfrac{b}{a}\right)\left(\dfrac{c}{d}-\dfrac{d}{c}\right)$

SECOND GENERAL SUBJECT.[1] — SIMPLE EQUATIONS.

CHAPTER XIV.

SIMPLE EQUATIONS CONTAINING ONE UNKNOWN QUANTITY.

185. Definition. — An Equation is an Expression of the Equality of two Quantities.

By this is meant that the numerical values of the two quantities will be found upon reduction to be the same number, integral or fractional. (See **55**, where the meaning and use of the equality sign, =, was explained.)

Equations considered in a general way may be regarded as expressing the relations of numbers.

SECTION I.

General Definitions.

186. Equations are of two kinds, **Identical** and **Conditional.**

187. An equation in which one side is but the development of the operations indicated in the other, or is exactly the same on both sides, is called an **Identical Equation.**

Thus, $(x+3)(x-9) = x^2 - 6x - 27$,

and $ax + b = ax + b$,

illustrate the two kinds of identical equations.

[1] A discussion of powers and roots as forming the remaining subjects in the "Algebraic Notation" ought theoretically to be given before taking up equations. Other reasons exist, however, for changing this order.

The examples given in addition, subtraction, multiplication, division, factoring, and fractions *set equal to their answers* would be equations of this sort.

a. A prime characteristic of identical equations is the property that any letter may have *any value*, provided it has the same value on both sides of the equation. This is true because the two sides of the equation are either actually or virtually the same quantity, and consequently if for a letter the same number be substituted on both sides the two results are the same (84).

b. To indicate the relation of identity, or to show that one expression stands for another (113), the sign "$\equiv$" (read *identically equal to*) is used. Whenever the distinction remarked in this article needs to be brought to mind the identity sign will be used. All identities are, of course, equations, but not all equations are identities.

188. An equation in which a *definite number only* of numerical values of a letter will answer to maintain the equality is called a **Conditional Equation.**

This is the species of equation about to be investigated. For examples in which the unknown letter has but *one* numerical value, see **47** and **85.**

189. An Equation of Condition is regarded as containing along with known quantities an unknown quantity, (represented by a letter) whose numerical value is to be found such that when it is substituted, the equation may be verified. (**55**, *a.*)

Thus, if x represents some unknown quantity, and $6 + 9x = 51$, then $9x = 51 - 6 = 45$, and $x = 5$.

Therefore, substituting the value of x, $6 + 9 \times 5 = 51$, which is a true equation. If any value other than 5, as 7 or 12, etc., is substituted for x, the result is not a true equation.

a. Substituting the value of the unknown quantity for it, and then performing the operations indicated, showing the two sides of the equation to be equal, is called *verifying* the *equation.* When an equation is thus tested, and the sides found equal, the equation is said to be *satisfied.*

b. The simplest form of the equation occurs when the unknown number stands alone on one side of the equation. Such is the usual form of the equation in arithmetic, where, after the equality sign, an interrogation point is often substituted for the answer.

Thus, $16 + 9 \times 3 - 12 = ?$

In this case all that needs to be done is to find in the simplest form the numerical value of the given side of the equation. In algebra the unknown quantity may appear in any part of either side of an equation, or on both sides at once, and to find its value in such cases is far more difficult to accomplish.

c. The *Equation* takes such an important place in algebra that many writers define *algebra itself as the science of the equation.* The Germans call our arithmetic, *Rechnung,* reckoning, and literal arithmetic, *Buchstabenrechnung,* i.e., reckoning with letters, reserving the title, algebra, for the study of equations. The term *Arithmetik* is also used by them for literal arithmetic.

190. The Right Side of an equation is also called its **Right Member,** and the Left Side its **Left Member.**

191. A Numerical Equation is one in which the known quantities are represented by figures.

192. A Literal Equation is one in which the quantities regarded as known are represented by letters, or by letters with figures.

193. To Solve an Equation (i.e. to loose the unknown quantity from the others) is to find the value of the unknown quantity. This value is called the **Root** of the equation.

194. There are, as has been stated, two kinds of quantities present in a conditional equation, the **known** and the **unknown**. The latter are commonly represented by the last letters of the alphabet, (x, y, or z), while the former are represented by figures, and by the first letters of the alphabet. This, of course, is purely conventional.

195. The Degree of an Equation is the same as the highest power of the unknown quantity; or, in equations containing more than one unknown, it is the same as the degree of the term (see **77**) containing the greatest exponent, or the greatest sum of exponents of the unknown quantities.

Thus, $5x + 6 = 91$ is of the first degree.

$ax^2 + bx = c$ is of the second degree.

$11x + 9x^3 + 7x^2 = 42$ is of the third degree.

$4z^m + 3z^{m-1} - 2z^{m-2} = 7$ is of the mth degree.

$6x^3 + 9x^2y^3 + 14y^4 = 21xy$ is of the fifth degree.

196. Equations of the First Degree are also called **Simple Equations.** See heading of this chapter.

a. Of all equations those of the first degree are the easiest to solve. The treatment of simple equations only, and not of equations in general, will be taken up at this time.

SECTION II.

Algebraical Method of Treatment.

197. Nature of the Treatment. The method by which the problems of **47** and **85** were solved was of a special nature, similar to what is called analysis in arithmetic. Scarcely any two were reasoned out in the same way. The treatment to which the equation is now to be subjected is of a very different character, more comprehensive and more scientific.

I.—LOGICAL TERMS.

198. A Proposition is a statement presented for consideration. A proposition may be true or false, and may therefore be proved or refuted.

199. Propositions are of two kinds, *theorems* and *problems*.

200. A Problem in algebra is a question proposed for solution.

201. A Theorem is a proposition to be proved by a demonstration.

202. A Demonstration is a chain of reasoning by which the truth or falsity of a proposition is made evident.

203. A Corollary (cor.) is a truth inferred from a proposition or from something in its demonstration.

204. A Scholium (sch.) is a *remark* concerning a proposition or its proof.

205. An Hypothesis (hyp.) is a *supposition* made at the beginning or in the course of a proof or solution.

206. An Axiom (ax.) is a self-evident truth.

II.—AXIOMS.

207. The following are the Axioms used in the Solution of Equations.

a. Addition and Subtraction.

1. If the same quantity or equal quantities be added to equal quantities, the sums will be equal.

2. If the same quantity or equal quantities be subtracted from equal quantities, the remainders will be equal.

b. Multiplication and Division.

3. If equal quantities be multiplied by the same quantity or equal quantities, the products will be equal.

4. If equal quantities be divided by the same quantity or equal quantities, the quotients will be equal.

c. Powers and Roots.

5. If equal quantities, or the same quantity be raised to the same power the products are equal.

6. If equal quantities, or the same quantity have the same root extracted, the results (there being more than one) are respectively equal.

7. Quantities which are equal to the same quantity are equal to each other.

8. General Axiom.—If the same operation be performed on two equal quantities, the results will be equal.

NOTE.—It will often be convenient to refer to the addition and subtraction axioms *together*, as axiom *a*, and to the multiplication and division axioms as axiom *b*. When the addition axiom alone is referred to, it will be by the number, 1, and so for the others.

III.—SOLUTION OF EQUATIONS.

208. Solution of Equations of the Form $ax = b$. To solve this equation x must be made to stand alone on one side of the equation. (**193**.)

We have

$$ax = b \qquad \text{(Hypothesis. 205)}$$

Dividing both sides of the equation, that is the two equals by a,

$$x = \frac{b}{a} \qquad \text{(Axiom 4)}$$

Notice that this solution is genera. in its nature. For, a may stand for any co-efficient of x, and b for any quantity whatever on the right side of the equation (**113**). This element is characteristic of all algebraic solutions, and must be remembered to understand them. Notice further the use of the *axiom* to accomplish the end desired.

209. Exercise.

1. $9x = 36$.

Dividing the two equal quantities by 9, by axiom 4, the quotients are equal. $\therefore x = 4$, answer.

2. $16x = 64$.
3. $9x = 144$.
4. $11x = 29$.
5. $ax = 16$.
6. $3ax = 7c$.
7. $35 = 5x$.
8. $m = nx$.
9. $45y = 120$.
10. $17az = 3a$, $z = ?$
11. $6x + 5x = 33$.
12. $7x - 3x = 33$.
13. $2x + 3x + 9x = 28$.
14. $2x + 14x - 13x = 29 - 14 + 12 - 3$. **(87, 4.)**
15. $9y + 7y - 2y + 8y - y - 2y = 36 + 10 - 8$.
16. $25 + 8 - 3 + 20 - 9 + 1 = z + 2z + 11z - 7z$.
17. $(a + b + c)x = d + e + f$.
18. $mx + nx + px + qx = r$. **(97.)**
19. $ax + bx = a^2 - b^2$.
20. $8x + 3x - 15x = 28$.

NOTE. — It often happens that the coefficient of the unknown becomes minus on the left side of the equation. When this is the case the equation should be divided through (Ax. 4) by the coefficient of x taken with its negative sign. Thus, in the example before us, upon reduction

$$-4x = 28$$
$$x = -7 \quad \text{(by dividing through by } -4.)$$

21. $19y - (37y - 12y) = 14 - 34 - 4$.
22. $11y - (5y - 3y) = 25 - 88$.
23. $-y = a - b$.
24. $(a - b)x = a^3 - b^3$.

210. Solution of Integral Equations in which the known and unknown quantities appear promiscuously on both sides of the equation. Transposition.

To show how such equations can be solved, we may take the equation $ax - b = cx - d$ in which both a term containing x and a term independent of x appear on each side.

$$ax - b = cx - d. \qquad \text{(Hyp.)}$$

Adding b to each member of the equation,

$$ax - b + b = cx + b - d. \qquad \text{(Ax. 1, 207)}$$

Or, $$ax = cx + b - d. \qquad \text{(87, 4)}$$

Again, subtracting cx from each side of the last equation,

$$ax - cx = cx - cx + b - d. \qquad \text{(Ax. 2)}$$

Or, $$ax - cx = b - d \qquad \text{(87, 4)}$$

$$(a - c)\,x = b - d \qquad \text{(97)}$$

$$x = \frac{b - d}{a - c}. \qquad \text{(208)}$$

The solution shows that $-b$ was transferred or "transposed" from the left member to the right and its sign changed by adding $+b$ to both sides. Similarly cx was transposed from the right member to the left and its sign changed by subtracting cx from both sides. Evidently this would apply equally well to any quantities on either of the sides of the equation, and may be stated as a theorem as follows:—

THEOREM OF TRANSPOSITION. — *Any quantity may be transposed from one side of an equation to the other by changing its sign.*

SCHOLIUM. — By this theorem all the terms containing the unknown quantity can be transposed to the left member, and all the independent terms to the right member of an equation. When this is done, by uniting the terms on each side, the equation may be solved by **208**.

211. Exercise.

1. $6x = 3x + 24$.

By subtracting $3x$ from each side we get

$3x = 24$, whence $x = 8$. (**208**)

2. $9x - 2 = 7x + 16$.

MODEL SOLUTION.

$$9x - 2 = 7x + 16$$
$$9x - 7x = 2 + 16 \qquad (\text{Ax. } a)$$
$$2x = 18 \qquad (\mathbf{86})$$
$$x = 9. \quad \textit{Ans.} \qquad (\text{Ax. } 4)$$

The student should learn to follow this model, placing the appropriate reasons for the different operations in parentheses at the right margin. Positive terms are transposed by *subtracting* (Ax. 2), and negative terms by *adding* (Ax. 1).

3. $8x - 7 + 3x = 15 + 2x - 13$.

4. $9x - 14 = 17$.

5. $ax = mx - n$.

6. $6x + 4x - 13 - 2x - 3 = 0$.

7. $ax + bx + cx - d = 0$.

8. $mx + a + bx = b - px$.

9. $mx - nx = 0$.

SOLUTION. $(m - n)x = 0$; $\therefore x = 0$. (Ax. 4)

10. $15(x - 1) + 4(x + 3) = 2(7 + x)$.

SUGGESTION.—First perform the multiplications indicated.

11. $118 - 65x - 123 = 15x + 35 - 120x$.

12. $7x - 5[x - \{7 - 6(x - 3)\}] = 3x + 1$. (**106** and **102**)

13. $(x + 3)(2x + 3) - 14 = (x + 1)(2x + 1)$.

REMARK.—This equation has the *form* of one of the second degree (**195**), but upon multiplication and transposition, the x^2's cancel leaving a simple equation.

14. $3(x-1)^2 - 3(x^2-1) = x - 15.$

15. $2(x+2)(x-4) = x(2x+1) - 21.$

16. $2b - (b+c)x = (b-c)x$

$$-(b+c)x - (b-c)x = -2b \qquad \text{(Ax. 2)}$$
$$-2bx = -2b \qquad \textbf{(106, 86)}$$
$$x = 1. \quad \textit{Ans.} \qquad \text{(Ax. 4)}$$

17. $a + 2d + 3x - (2b + 4c) = x - (7c - 6d).$

18. $a^2x + bx - c = b^2x + cx - d.$ **(97)**

212. Solution of Fractional Equations in which the unknown may appear in any part of either side of the equation. Such equations can always be reduced to the integral form. For, an equation may be multiplied through by any factor whatever, and the new equation will hold true (Ax. 3). Now, if the equation be multiplied through by a common multiple of all the denominators of its terms, such denominators will all cancel out of the new equation. (See definition of common multiple **149**, *a*.)

THEOREM.—*Any equation can be cleared of fractions by multiplying both its members by a common multiple of its denominators.*

COROLLARY.—The signs of all the terms of both members of an equation may be changed from + to − and − to +, for this is equivalent to multiplying both members by −1, which is permissible by Ax. 3.

SCH. I.—*Any* common multiple of the denominators will answer to clear an equation of fractions. It is greatly preferable, however, to multiply through by the *lowest common denominator*, thereby obtaining simpler expressions, and that more readily.

SCH. II.—This proposition takes any equation and explains how to clear it of fractions. It is then ready for transposition, and ultimately for solution.

213. Exercise.

1. $\frac{5-3x}{2}=\frac{8x-9}{3}.$

SOLUTION. — The l. c. d. is 6, and we multiply through by it.

$$\frac{5-3x}{2}=\frac{8x-9}{3} \qquad \text{(Hyp.)}$$

$$15-9x=16x-18 \qquad \text{(Ax. 3.)}$$

$$-9x-16x=-15-18 \qquad \text{(Ax. 2)}$$

$$-25x=-33 \qquad \text{(86)}$$

$$x=\tfrac{33}{25}. \qquad \text{(Ax. 4)}$$

2. $\frac{x}{2}+\frac{x}{3}-\frac{x}{4}=\frac{x+2}{2}.$

3. $\frac{x-1}{2}-\frac{x-3}{4}=\frac{x-3}{2}.$

4. $5=8-\frac{21}{x}.$

5. $\frac{x}{5}-12\tfrac{3}{4}=42\tfrac{7}{18}.$

6. $\frac{x}{2}+\frac{x}{3}-\frac{x}{4}+\frac{x}{5}=\frac{x}{6}-\frac{x}{7}+319.$

7. $\frac{x}{2}-\frac{x-2}{3}=\frac{x+3}{4}-\frac{2}{3}.$

8. $3+\frac{x}{4}=\frac{1}{2}\left(4-\frac{x}{3}\right)-\frac{5}{6}+\frac{1}{3}\left(11-\frac{x}{2}\right).$

9. $3+\frac{x}{.5}=7-\frac{x}{.2}.$

10. $1.5=\frac{.36}{.2}-\frac{.9x-.18}{.9}.$

11. $ax+b=\frac{x}{a}+\frac{1}{b}.$

12. $\frac{x-a}{b-x}=\frac{x-b}{a-x}.$

13. $\frac{x}{x+2}+\frac{4}{x+6}=1.$

214. Examples of the Complete Solution of Equations.

1. $\frac{7x-8}{11}+\frac{15x+8}{13}=3x-\frac{31-x}{2}.$

MODEL SOLUTION. — l. c. d. $= 2\times 11\times 13.$

$$182x-208+330x+176=858x-4433+143x \quad \text{(Ax. 3)}$$
$$182x+330x-858x-143x=208-176-4433 \quad \text{(Ax. } a\text{)}$$
$$-489x=-4401 \quad \textbf{(86)}$$
$$x=9. \quad \text{(Ax. 4)}$$

VERIFICATION.

$$\frac{7\times 9-8}{11}+\frac{15\times 9+8}{13}=3\times 9-\frac{31-9}{2};$$

i.e., $5+11=27-11,$

or, $16=16.$ **(55, *a*)**

2. $\frac{x+4}{3x-8}=\frac{x+5}{3x-7}.$

3. $\frac{36}{a-5}=\frac{45}{a}.$

4. $\frac{3}{16}(x-1)-\frac{5}{12}(x-4)=\frac{2}{5}(x-6)+\frac{5}{48}.$

CAUTIONS. — The student should be careful not to make mistakes in addition, multiplication, fractions, signs, etc., and then attribute the failure to get correct results to something unknown about equations. We will refer to common sources of error by articles which the student would do well to look up and fix in mind. Reduce complex fractions to simple ones (182), and perform all indicated operations either before or after clearing as may be found convenient. See 100, 4; 168, *a*; and 176. Study carefully 171 *a*, and its application to equations. Reduce answers to their lowest terms (159). An answer may have to be changed to make it identical with that given (see 157). For factors in parentheses, see 111, 2. In clearing of fractions remember to multiply the integral quantities by the l. c. d. Other references might be given, but these will suggest such mistakes as are liable to be made.

215. General Rules for Solving Simple Equations with One Unknown.

1. Normal process.

(1. First clear the equation of fractions. Theorem, **212**.

(2. Next transpose all the terms containing x to the left member, and all the known terms to the right member of the equation. Theorem, **210**.

(3. Then by **87**, 4, collect the terms on both sides, and when necessary form the coefficient of x as in **97**.

(4. Finally, use axiom 4 to find the value of the unknown.

2. General process.

(1. Proceed *in any way* that seems advantageous by use of the axioms (**207**) to reduce the equation to the form $ax = b$.

(2. Finish the solution as in **208**.

REMARK.—The first rule is straightforward in its method, and so best for the beginner. The second is superior to the other for the expert student. It may be emphasized here *that any process in accordance with the axioms will always lead to correct results; and any process or operation not in accordance with them will lead inevitably to incorrect results.*

216. Find the Value of the Unknown Quantity in the following, and verify the answers by substituting them in the original equation (**189**, *a*).

1. $1\frac{2}{3}x = 8\frac{1}{18}$.

3. $\frac{2}{3}x + 12 = \frac{4}{5}x + 6$.

2. $\frac{x + 23}{x - 1} = 9$.

4. $\frac{1 - a}{x} - 4 = 5$.

5. $5 - 6x + \frac{7x + 14}{3} = \frac{17 - 3x}{5} - \frac{4x + 2}{3}$.

6. $3x + \frac{2x+6}{5} = 5 + \frac{11x-37}{2}.$

7. $\frac{1}{2}(27 - 2x) = \frac{9}{2} - \frac{1}{10}(7x - 54).$

8. $\frac{x}{7} - \frac{x-5}{11} + 5 = x - \left(\frac{2x}{77} + 1\right).$

9. $5x - \{8x - 3[16 - 6x + (4 - 5x)]\} = 6.$

10. $a(x-2) + 2x = 6 + a.$

11. $\frac{1}{2}\left(x - \frac{a}{3}\right) - \frac{1}{3}\left(x - \frac{a}{4}\right) + \frac{1}{4}\left(x - \frac{a}{5}\right) = 0.$

SOLUTION. — The l. c. d. of coefficients equals 12.

$$6\left(x - \frac{a}{3}\right) - 4\left(x - \frac{a}{4}\right) + 3\left(x - \frac{a}{5}\right) = 0 \qquad \text{(Ax. 3)}$$

$$6x - 2a - 4x + a + 3x - \frac{3a}{5} = 0 \qquad \text{(111, 2)}$$

$$5x = 2a - a + \frac{3}{5}a = \frac{8}{5}a \qquad \text{(Ax. } a\text{, 174)}$$

$$x = \frac{8a}{25}. \qquad \text{(Ax. 4)}$$

VERIFICATION.

$$\frac{1}{2}\left(\frac{8a}{25} - \frac{a}{3}\right) - \frac{1}{3}\left(\frac{8a}{25} - \frac{a}{4}\right) + \frac{1}{4}\left(\frac{8a}{25} - \frac{a}{5}\right) = 0$$

$$\frac{1}{2}\left(-\frac{a}{75}\right) - \frac{1}{3}\left(\frac{7a}{100}\right) + \frac{1}{4}\left(\frac{3a}{25}\right) = \frac{-2a}{300} - \frac{7a}{300} + \frac{9a}{300} = 0.$$

NOTE. — In some literal equations verification is easy, in others it is more or less difficult. In the following examples the verification is rather complex, though of course it can always be performed. In numerical problems verification is always comparatively easy.

12. $(a+x)(b+x) = (m+x)(n+x).$

13. $\frac{qx}{p} - \frac{s}{r} = \frac{p}{q} - \frac{rx}{s}.$

14. $\frac{9x+14}{12} - \frac{1}{7x} = \frac{5}{6x} - \frac{16-21x}{28}.$

SOLUTION. $\frac{3x}{4}+\frac{7}{6}-\frac{1}{7x}=\frac{5}{6x}-\frac{4}{7}+\frac{3x}{4}.$ (163, 159)

$\therefore\ \frac{7}{6}-\frac{1}{7x}=\frac{5}{6x}-\frac{4}{7}$ (Ax. 2)

$49x-6=35-24x$ (Ax. 3)

$49x+24x=35+6.$ (Ax. 1)

$x=\frac{41}{73}.$ *Ans.* (Ax. 4)

15. $8(x-1)+17(x-3)=4(4x-9)+4.$

16. $\frac{a}{c-x}=\frac{c}{a-x}.$

17. $\frac{x-1}{3}+\frac{4x-\frac{3}{4}}{5}-\frac{7x-6}{8}=2+\frac{x-2}{2}+\frac{3x-9}{10}.$

18. $6bx+4a^2+2ax=6a^2bx-3ax.$

19. $\frac{3x+1}{3(x-2)}=\frac{x-2}{x-1}.$

20. $\frac{x-7}{x+7}=\frac{2x-15}{2x-6}-\frac{1}{2(x+7)}.$

21. $\frac{4}{x+2}+\frac{7}{x+3}=\frac{37}{x^2+5x+6}.$

22. $\frac{7}{x-1}=\frac{6x+1}{x+1}-\frac{3(1+2x^2)}{x^2-1}.$

23. $\frac{x-3}{4(x-1)}=\frac{x-5}{6(x-1)}+\frac{1}{9}.$

24. $\frac{1}{a-b}+\frac{a-b}{x}=\frac{1}{a+b}+\frac{a+b}{x}.$

SOLUTION. $\frac{a-b}{x}-\frac{a+b}{x}=\frac{1}{a+b}-\frac{1}{a-b}.$ (Ax. 2)

$\frac{a-b-a-b}{x}=\frac{a-b-a-b}{a^2-b^2}$ (176)

$\frac{-2b}{x}=\frac{-2b}{a^2-b^2}$

$\frac{1}{x}=\frac{1}{a^2-b^2}$ (Ax. 4)

$x=a^2-b^2.$ *Ans.* (Ax. 3)

This solution is shorter and easier than if the equation had first been cleared of fractions (215, 2).

25. $\frac{ax+bx}{b} - x = \frac{1}{a-b} - \frac{1}{a}$.

SUGGESTION. — Add the members of the equation as they stand before clearing.

26. $(x-5)(x-2) - (x-5)(2x-5) + (x+7)(x-2) = 0$.

27. $\frac{1}{5}(2x-10) - \frac{1}{11}(3x-40) = 15 - \frac{1}{5}(57-x)$.

28. $3\frac{1}{3}\left\{28 - \left(\frac{x}{8} + 24\right)\right\} = 3\frac{1}{2}\left\{2\frac{1}{3} + \frac{x}{4}\right\}$.

29. $(a+x)(b+x) - a(b+c) = \frac{a^2c}{b} + x^2$.

30. $(x+1)^2 = \{6 - (1-x)\}x - 2$.

31. $.5x - .3x = .25x - 1$.

32. $4.8x - \frac{.72x - .05}{.5} = 1.6x + 8.9$.

33. $\frac{1}{5} - \frac{3}{x-1} = \frac{2 + \frac{x+4}{1-x}}{3}$.

34. $\frac{x-2}{5} + \frac{.301}{.5} = .001x + .6 - \frac{x-2}{.05}$.

35. $\frac{1}{ab-ay} + \frac{1}{bc-by} = \frac{1}{ac-ay}$.

36. $\frac{1}{x-2} - \frac{1}{x-4} = \frac{1}{x-6} - \frac{1}{x-8}$.

SUGGESTION. — Add the members (176).

37. $\frac{ax^2+bx+c}{px^2+qx+r} = \frac{ax+b}{px+q}$.

38. $ax^2 - bx = bx^2 - cx$.

SUGGESTION. — Divide through by x.

39. $\frac{9x+3}{27} + \frac{3x-6}{2x-5} = \frac{2}{3} + \frac{3x+22}{9}$.

SOLUTION. — Clearing of monomial denominators *only* (215, 2),

$$9x + 3 + \frac{81x - 162}{2x - 5} = 18 + 9x + 66 \qquad \text{(Ax. 3)}$$

$$\frac{81x - 162}{2x - 5} = 81 \qquad \text{(Ax. 2, 86)}$$

$$81x - 162 = 162x - 405 \qquad \text{(Ax. 3)}$$

$$-81x = -243 \qquad \text{(Ax. } a\text{)}$$

$$x = 3. \qquad \text{(Ax. 4)}$$

40. $\dfrac{9(2x - 3)}{14} + \dfrac{11x - 1}{3x + 1} = \dfrac{9x + 11}{7}.$

41. $\dfrac{7x - 6}{35} - \dfrac{x - 5}{6x - 101} = \dfrac{x}{5}.$

42. $\dfrac{6 - 5x}{15} - \dfrac{7 - 2x^2}{14(x - 1)} = \dfrac{1 + 3x}{21} - \dfrac{10x - 11}{30} + \dfrac{1}{105}.$

43. $\dfrac{x - 1}{x - 2} - \dfrac{x - 2}{x - 3} = \dfrac{x - 5}{x - 6} - \dfrac{x - 6}{x - 7}.$

SOLUTION.

$$1 + \frac{1}{x - 2} - 1 - \frac{1}{x - 3} = 1 + \frac{1}{x - 6} - 1 - \frac{1}{x - 7} \qquad (161)$$

$$\frac{1}{x - 2} - \frac{1}{x - 3} = \frac{1}{x - 6} - \frac{1}{x - 7}$$

$$\frac{-1}{(x - 2)(x - 3)} = \frac{-1}{(x - 6)(x - 7)} \qquad (176)$$

$$-x^2 + 13x - 42 = -x^2 + 5x - 6 \qquad \text{(Ax. 3)}$$

$$8x = 36 \qquad \text{(Ax. } a\text{, 86)}$$

$$x = 4\tfrac{1}{2}. \qquad \text{(Ax. 4)}$$

44. Solve $br = p$ regarding, first b, then r, and lastly p as the unknown quantity.

First,	Second,	Third,
$br = p$ (Hyp.)	$br = p$ (Hyp.)	$br = p$ (Hyp.)
$b = \dfrac{p}{r}$ (Ax. 4)	$r = \dfrac{p}{b}$ (Ax. 4)	or $p = br$.

This equation $br = p$ is the fundamental formula in Percentage in arithmetic. b standing for base, r for rate expressed decimally, and p for the percentage. The two equations obtained, $b = \frac{p}{r}$, and $r = \frac{b}{p}$, are the formulae for the cases, given the percentage and rate to find the base, and given the base and percentage to find the rate.

45. Solve $a = b(1 + r)$ for b, and r. (a = "amount").

46. Solve $d = b(1 - r)$ for b, and r. (d = "difference").

47. Solve $i = prn$ for p, r, and n in succession.

By the definition of interest, (i = interest, p = principal, r = rate, and n = time in years), $i = prn$; we are to find the principal when the rate, time, and interest are given; next, to find the rate, when the principal, time, and interest are given; lastly, to find the time when the principal, rate, and interest are given.

48. (1) Solve $t = \frac{s}{c}$ for s and c. (2) Solve $t = \frac{1}{c}$ for c.

49. Solve $\frac{a}{b - c} + \frac{d}{b - d} = 0$ for a, b, c and d in succession.

50. Solve $\frac{ab}{cd} + 1 = 0$ in the same manner.

217. General Remarks on the Solution of Equations.—By Rule 2 the student is at liberty to pursue *any course* in consonance with axiomatic principles. Certain of the examples of the preceding article have exhibited an advantage gained by deviating from the normal process. The following precepts have already been exemplified in one or another of the exercises.

1. Mark at the outset and after every reduction whether the equation can be *divided* through by any factor, monomial or other, and remove it. (Axiom 4.)

2. Study the original equation and each derived equation to see if by uniting certain terms as they stand, or after transposition, quantities can be made to disappear, and the equation be thereby simplified.

3. Examine whether any simple reduction, as that of an improper fraction to a mixed number, will pave the way for the simplification suggested in 2.

4. Consider, before applying axiom 3, whether it would not be better to clear the equation of only a part of its denominators, and then simplify, clearing of the remaining denominators afterwards.

SECTION III.

Problems Involving Simple Equations containing One Unknown.

218. The Problems of Algebra (see **47** and **85**) are quite like those of arithmetic, except that the former are usually much more difficult; but the algebraic treatment of them is different. In arithmetic we were accustomed to start with what was given and work towards the result; while in algebra we conceive the problem to be solved, and proceed as if we were verifying our answer, using meanwhile a letter to stand for it, since we do not know its value. Going through the form of a verification, the letter used everywhere taking the place of the unknown number, *an equation results*, the solution of which gives the value of the letter, and solves the problem. This is called the *analytic method.*

219. The Solution of a Problem involves two operations.

1. *The Statement of the Problem.*—The statement of a problem is its expression in algebraic symbols in the form of an equation.

2. *The Solution of the Equation.* — Section II. of this chapter dealt with the systematic solution of equations, and the student is supposed to be familiar with this part of the process.

220. In the Enunciation of every problem we notice two things. (See those in **47** and **85**.)

1. The description of an unknown number (or quantity) together with one or more others dependent upon the first for their values.

2. The assertion that two numbers or quantities, obtained in different ways, are equal.

The rule to be given explains the translation of the problem into algebraic language. It will be convenient before giving the rule to define the term "*function* of a quantity."

221. A Function of a Quantity is one which depends upon it for its value.

Thus, $3x + 5$ is a function of x. If $x = 6$, then the function equals $3 \times 6 + 5 = 23$; if $x = 9$, the function equals $3 \times 9 + 5 = 32$, and so on. From this we can see that any quantity which contains x is a function of x, and any quantity which contains a is a function of a, and so on.

222. General Rule for Solving Problems. — Compare with the solutions of Arts. **47** and **85**.

1. Let x, or some convenient function of x, represent the unknown quantity.

2. Express the different functions of x to be used in forming the equation in terms of x.

3. Construct the two expressions described as equal, and make them the two members of an equation.

4. Solve the equation.

5. Obtain the values of such functions of the root as the problem asks for.

a. Illustrations. — In the examples as set down in **47** and **85** we first wrote x for the unknown quantity; then underneath its functions. Afterwards the equation was formed, which was solved according to the best light we then had. To particularize, let us take Ex. 5, Art. 47. The functions of x are $2x$ and $2x + 7$. The equation is then formed, after which it is solved, giving $x = 8$. But the problem requires two other answers, viz., $2x$ and $2x + 7$, which are found equal to 16 and 23 respectively.

b. The *two sides* of an equation must be expressed in terms of the *same unit.* Of course, one side cannot be expressed as a certain number of dollars, and the other as an equal number of cents, neither can one side denote metres and the other centimetres, and so on. Careless thinking sometimes leads a student even into such absurdities as these.

223. Proportions are Transformed into Equations by means of the fundamental property that *the product of the means equals the product of the extremes.*

224. Exercise in the Solution of Problems in One Unknown Quantity.

1. A gentleman divides two dollars among twelve children, giving to some 18 cts. each, and to the others 14 cts. each. How many were there of each class?

Let $x =$ the number receiving 18 cts. each, **(222, 1)**
Then $12 - x =$ the number receiving 14 cts. each, **(222, 2)**
$18x =$ the number of cents the first class received, **(222, 2)**

$(12 - x)\,14 =$ the number of cents the second class received, **(222, 2)**

$$18x + (12 - x)\,14 = 200, \quad \textbf{(222, 3, and } b\textbf{)}$$
$$18x + 168 - 14x = 200 \quad \textbf{(110)}$$
$$4x = 200 - 168 = 32 \quad \text{(Ax. 2)}$$
$$x = 8 \quad \text{(Ax. 4)}$$
$$12 - x = 4 \quad \textbf{(222, 5)}$$

Verification. $8 \times 18 + 4 \times 14 = 144 + 56 = 200.$

2. There are three consecutive numbers such that if they be divided by 10, 17, and 26, respectively, the sum of the quotients will be 10; find them.

Here the student may have trouble in expressing the three numbers (220, 1). The idea is that the three numbers follow one another, the second being one greater than the first, and the third one greater than the second.

Let x = the least number, (222, 1)
then $x + 1$ = the next greater, (222, 2)
$x + 2$ = the greatest number, (222, 2)

$$\frac{x}{10} + \frac{x+1}{17} + \frac{x+2}{26} = 10 \qquad (222, 3)$$

$$442\,x + 260\,x + 260 + 170\,x + 340 = 44200, \qquad (\text{Ax. } 3)$$

$$872\,x = 44200 - 260 - 340 = 43600, \qquad (\text{Ax. } 2)$$

$$\left.\begin{aligned} x &= 50 \\ x + 1 &= 51 \\ x + 2 &= 52 \end{aligned}\right\} \qquad \begin{aligned} &(\text{Ax. } 4) \\ &(222, 5) \\ &(222, 5) \end{aligned}$$

VERIFICATION. $\frac{50}{10} + \frac{51}{17} + \frac{52}{26} \equiv 5 + 3 + 2 = 10$

3. An estate is divided among four children in such a manner that

the first has \$200, more than $\frac{1}{4}$ of the whole,
" second 340, " " $\frac{1}{5}$ " "
" third 300, " " $\frac{1}{6}$ " "
" fourth 400, " " $\frac{1}{8}$ " "

what is the value of the estate?

Here the beginner may not see how the statement of 220, 2 is true of this problem. It must be taken rather by implication than by the direct statement of the problem.

Let x = the value of the estate (222, 1)

$\frac{x}{4} + 200$ = the number of dollars of the first child's portion, (222, 2)

$\frac{x}{5} + 340$ = " " " " " second " " (222, 2)

$\frac{x}{6} + 300$ = " " " " " third " " (222, 2)

$\frac{x}{8} + 400$ = " " " " " fourth " " (222, 2)

$$\frac{x}{4} + 200 + \frac{x}{5} + 340 + \frac{x}{6} + 300 + \frac{x}{8} + 400 = x \qquad (222, 3)$$

$$\frac{x}{4} + \frac{x}{5} + \frac{x}{6} + \frac{x}{8} - x = -1240$$

$$30x + 24x + 20x + 15x - 120x = -148800$$

$$-31x = -148800$$

$$x = 4800.$$

VERIFICATION.

$$\frac{4800}{4} + 200 + \frac{4800}{5} + 340 + \frac{4800}{6} + 300 + \frac{4800}{8} + 400 = 4800.$$

4. A boy bought an equal number of apples, lemons, and oranges for 56 cts.; for the apples he gave 1 ct. apiece, for the lemons, 2 cts. apiece, and for the oranges 5 cts. apiece. How many of each did he purchase?

5. Divide the number 181 into two such parts that 4 times the greater may exceed 5 times the less by 49.

6. A boy ate $\frac{1}{4}$ of his plums and gave away $\frac{1}{8}$ of them. The difference between the number he ate and the number he gave away was 3. How many had he?

7. What number is that whose half, third, and fourth parts together equal 63? Verify the answer.

8. A father aged 54 years has a son aged 9 years: in how many years will the age of the father be just 4 times that of the son?

9. A father is now 40 years of age and his daughter 13. How many years ago was the father's age 10 times that of the daughter's?

10. A man's age and his wife's age now bear to each other the ratio of 6 : 5. But 15 years hence they will be to each other as 9 : 8. How old are they now?

In this problem we might let x = the man's age, and $\frac{5x}{6}$ his wife's age; however, we can avoid fractions in the following preferable way:

Let $6x$ = the man's age, (222, 1)

then $5x$ = the wife's age, (222, 1)

$6x + 15$ = the man's age 15 years later, (222, 2)

$5x + 15$ = the wife's age 15 years later, (222, 2)

$$6x + 15 : 5x + 15 :: 9 : 8$$

$$48x + 120 = 45x + 135 \qquad (223)$$

$$x = 5$$

$$\left.\begin{array}{l} 6x = 30 \\ 5x = 25 \end{array}\right\} \textit{Ans.}$$

VERIFICATION. $30 + 15 : 25 + 15 :: 9 : 8.$

11. A gentleman is now twenty-five years old, and his youngest brother is 15. How many years must elapse before their ages will be as 5 : 4 ?

12. The difference between two numbers is 12, and the greater is to the less as 11 : 5. What are the numbers ?

13. What two numbers are as 3 : 4, to each of which if 4 be added the sums will be to each other as 5 : 6 ?

14. A farm contains 26 acres. Three times A's part is 6 acres less than 4 times B's part. How many acres had each ?

15. Find two numbers differing by 10 whose sum is equal to twice their difference.

16. A man has four children the sum of whose ages is 48 years, and the common difference of their ages is equal to twice that of the youngest. Find their ages.

17. Three boys are talking of their money. A says to B, "I have three times as many cents as you have." Says C to A and B, "I have as much as the difference between your sums." Now, A's money added to twice B's and twice C's makes 63 cents. How much had each ?

18. A grocer has two sorts of sugar, one worth 9 cents, and the other 13 cents a pound. How many pounds of each sort must be taken to make a mixture of a hundred pounds worth 12 cents per pound ?

19. A paymaster wishing to draw for use on pay day the sum of $11360 requested the teller to make up the sum with bank-bills of different denominations as follows: a certain number of 100's, twice as many 50's, twice as many 20's as 50's, twice as many 10's as 20's, twice as many 5's as 10's, twice as many 2's as 5's, and twice as many 1's as 2's. How many of each denomination were needed ?

20. A grocer bought two casks full of oil, one of which held twice as much as the other. From one cask he drew 10 gal. and from the other 1 gal., and then drew from the larger $\frac{1}{3}$ of the remaining oil, and found that the two casks now contained equal quantities of oil. How much did each hold when full?

21. The garrison of a certain town consisted of 125 men, partly cavalry and partly infantry. The monthly pay of a cavalryman is $20, and that of an infantryman is $15, and the whole garrison receives $2050 a month. What is the number of cavalry and what of infantry?

22. Divide 88 dollars between A, B, and C, giving to B $\frac{2}{3}$, and to C $\frac{3}{7}$ as much as to A.

23. A man having completed $\frac{2}{5}$ of his journey, finds that after traveling 30 miles farther only $\frac{2}{7}$ of the journey remains. Required the length of the journey.

24. Two-fifths of a pole is in the water, one-tenth in the mud, and the remainder out of the water. There are 6 ft. less of it in the water than in the mud and above water. How long is the pole?

25. There are two silver cups and one cover for both. The first weighs 12 oz., and with the cover twice as much as the other without it; but the second with the cover weighs $\frac{1}{3}$ more than the first without it. Find the weight of the cover.

26. I have a certain number in mind. I multiply it by 7, add 3 to the product, and divide the sum by 2; I then find that if I subtract 4 from the quotient I get 15. What number am I thinking of?

27. A certain number multiplied by 5, 24 subtracted from the product, the remainder divided by 6, and this quotient increased by 13, results in the number itself. What is the number?

28. A person spent $\frac{1}{4}$ of his money, after which he earned \$3. Then he lost $\frac{1}{3}$ of what he then had, receiving afterwards \$2 back. Lastly he gave away $\frac{1}{7}$ of what he now had, when he found he had only \$12 remaining. What had he at first?

Let x = the number of dollars he had at first, (222, 1)

then $\frac{3}{4}x$ = remainder after spending one fourth, (222, 2)

$\frac{2}{3}\left(\frac{3}{4}x + 3\right)$ = the second remainder, (222, 2)

$\frac{6}{7}\left[\frac{2}{3}\left(\frac{3}{4}x + 3\right) + 2\right]$ = the third remainder = 12. (222, 3)

$$\frac{12}{21}\left(\frac{3}{4}x + 3\right) + \frac{12}{7} = 12 \qquad (109, 5)$$

$$\frac{3x}{7} + \frac{12}{7} + \frac{12}{7} = 12$$

$$3x + 12 + 12 = 84$$

$$x = 20 \quad \textit{Ans.}$$

VERIFICATION. $\frac{3}{4}$ of $20 = 15$; $15 + 3 = 18$; $\frac{2}{3}$ of $18 = 12$; $12 + 2 = 14$; $\frac{6}{7}$ of $14 = 12 = 12$.

29. A basket of oranges is emptied by one person taking half of them and one more, a second person taking half of the remainder and one more, and a third person taking half of this remainder and 6 more. How many did the basket contain at first.

30. In making a journey a traveler went on the first day $\frac{1}{5}$ of the distance and 8 miles more; on the second day he went $\frac{1}{6}$ of the distance that remained and 15 miles more; and the third day he went $\frac{1}{3}$ of the distance that remained and 12 miles more; on the fourth day he went 35 miles and finished his journey. What was the distance traveled?

31. A man who has \$2000 invested in a mill from which he receives a certain per cent, and \$1000 in real estate from which he derives only $\frac{3}{4}$ of the previous rate, has an income from both of \$330. What rate per cent does he receive?

Let x = rate % on the \$2000 investment,

$\frac{3}{4}x$ = rate on the \$1000 investment;

$\therefore\ 2000 \times \frac{x}{100} = 20x$ = the interest on the first, and

$1000 \times \frac{\frac{3}{4}x}{100} = \frac{30x}{4}$ = interest on the second, by definition of rate per cent.

$$20x + \frac{30x}{4} = 330$$

$$80x + 30x = 1320$$

$$110x = 1320$$

$$\left.\begin{aligned} x &= 12\% \\ \tfrac{3}{4}x &= 9\% \end{aligned}\right\}$$

32. What is the property of a person whose income is \$430, when he has $\frac{2}{3}$ of it invested at 4%, $\frac{1}{4}$ at 3%, and the remainder at 2%.

33. A person who is worth \$12000 invests a certain amount in railroad stock from which he receives 6%. One-third of the difference pays 4% interest, and the other two-thirds 5% interest, and from all his funds he clears \$896. What amount was invested in railroad stock? On what amount had he realized but \$600? (See **252**, 7.)

34. A man has \$5000 invested at a certain rate, \$2000 at double the former rate, and \$1000 at triple the first rate. He received from a property an income of \$1900, but pays out \$2000 for improvements, personal expenses, insurance, etc. He finds he has \$500 remaining at the end of the year. What does he receive on the \$2000?

35. A workman engaged for 48 days at the rate of \$2 per day and his board. But for every day he might be idle he was to pay \$1 for his board. At the end of the time he received only \$42. How many days did he work?

36. A man hired a laborer for one year at the wages of \$90 and a suit of clothes. At the end of 7 months the laborer quit his service and received \$33.75. At what price were the clothes estimated?

37. A boy engaged to carry 30 glass vessels to a certain place on condition of receiving 5 cents for every one he delivered safe, and forfeiting 12 cents for every one he should break. On settlement he received 99 cents. How many did he break?

38. A tree standing vertically on level ground is 60 feet high. Upon being broken over in a storm the upper part reached from the top of the trunk to the ground just 30 feet from the foot of the trunk. What was the length of the part broken off?

NOTE.—This problem depends upon the theorem proved in geometry and stated in most arithmetics, that the square on the longest side of a right-angled triangle is equal to the sum of the squares on the other two sides.

39. A can perform a piece of work in 6 days, B can perform the same work in 8 days, and C in 24 days. In how many days can they finish it if all work together?

Let x = the number of days when all work together. Now, A does $\frac{1}{6}$ of the work in one day, and in x days he will do x times $\frac{1}{6}$ or $\frac{x}{6}$ of the work. For a like reason B will do $\frac{x}{8}$, and C, $\frac{x}{24}$ of the work in one day. But they do one times the work in x days, therefore

$$\frac{x}{6}+\frac{x}{8}+\frac{x}{24}=1 \qquad (219,\ 1)$$

$$\therefore \quad x=3.^{1} \quad Ans. \qquad (219,\ 1)$$

NOTE.—The whole work, as is customary in intellectual arithmetic, is called one; then the fractions of the work which each one does added together make the whole work, or one.

The equation $\frac{1}{6}+\frac{1}{8}+\frac{1}{24}=\frac{1}{x}$ (virtually the same as the one just given, being obtained from it by dividing through by x, Ax. 4) is found by saying that if it takes all of them x days, in one day they will do $\frac{1}{x}$ of the work. Solving this equation $x=3$ as before.

40. A and B together can do a piece of work in 12 days, A and C in 15 days, and B and C in 20 days. In what time can they do it all working together?

[1] Hereafter, when it is convenient to leave the solution of an equation to the student, the answer will be given with a reference to **219**, 2.

Let $x =$ the number of days in which all can do the work. Then $\frac{1}{x} - \frac{1}{12} =$ the part C can do in one day, etc.

41. A cistern can be filled by one pipe in 15 minutes, by another in 12 minutes, and by a third in 10 minutes. In what time can it be filled if all are left open?

42. A tank can be filled by two pipes in 24 and 30 minutes respectively, and emptied by a third in 20 minutes. What time will it be in filling if all are left open?

43. A man who can perform a piece of work in 14 days, works 4 days when he calls in a boy, and together they finish the job in 6 days. In how many days could the boy do the work alone?

44. A cistern can be filled by two pipes in 9 minutes and 12 minutes, and emptied by two others in 15 and 18 minutes respectively. The first pipe is left open one minute, then the third is opened. At the end of the second minute the second pipe is opened, and at the end of the third minute the fourth is opened. How soon will the cistern be in filling counting from the time the first pipe was opened?

45. A boy buys apples at the rate of 5 for 2 cts. He sells half of them at the rate of 2 for 1 ct., and the rest at the rate of 3 for 1 ct., and clears 1 ct. by the transaction. How many did he buy?

46. A market woman bought some eggs at the rate of 2 for 1 ct., and as many more at the rate of 3 for 1 ct. She sold them all at the rate of 5 for 2 cts., and found she had lost 4 cts. How many of each sort did she buy?

47. A train on the Northwestern line passes from London to Birmingham in 3 hours; a train on the Great Western line, which is 15 miles longer, traveling at a speed which is less by 1 mile per hour, passes from one place to the other in $3\frac{1}{2}$ hours. Find the length of each line.

REMARK. — Problems of distance, time, and rate occur so frequently, that it may be helpful to the student to be familiar with the answers to the following general questions: —

Given the time and rate, how is the distance found?

Given the distance and rate, how is the time found?

Given the distance and time, how is the rate found?

Similarly in problems involving price, number of articles, and total cost,

Given the price and the number of things, how is the total cost found?

Given the price and the total cost, how is the number of things found?

Given the number of things and the total cost, how is the price found?

So also we might discuss problems involving length, breadth, and thickness, area and solid contents.

The student must decide in every problem what kind of units the two sides of the equation are to contain, and prepare the functions accordingly. This will tend to make the process of solution clear.

48. A person walks to the top of a mountain at the rate of $2\frac{1}{3}$ miles an hour, and down the same way at the rate of $3\frac{1}{2}$ miles an hour, and is out 5 hours. How far is it to the top of the mountain?

49. A boy who runs at the rate of 12 feet per second, starts 20 yards behind another whose rate is $10\frac{1}{2}$ feet. How soon will the first boy be 10 yards ahead of the second?

50. The distance from M. to L. is $31\frac{1}{2}$ miles. The express down train leaves M. at 11.30 A.M., and arrives in L. at 12.30 P.M. The up train leaves L. at 11.45 A.M., and arrives at M. at 12.35 P.M. Supposing the speed of each train to be uniform, when will they meet?

51. A bought eggs at 18 cts. a doz. Had he bought 5 more for the same money, they would have cost him $2\frac{1}{2}$ cts. a dozen less. How many eggs did he buy?

52. A rows 4 miles and B 3 miles an hour. A is 14 miles farther up stream than B, and they row towards each other till they meet 4 miles above B's position. How rapid is the current?

Suggestion. — Let $x =$ the rate of the current, then $x + 4$ equals A's rate down, $3 - x$ B's rate up.

53. A boatman who can row 6 miles an hour in still water rows a certain distance down stream and back again in 3 hours. How far did he row, supposing the stream to flow 3 miles in two hours?

54. There are two numbers in the ratio of $\frac{1}{2}$ to $\frac{2}{3}$, which being increased respectively by 6 and 5, are in the ratio of $\frac{2}{3}$ to $\frac{1}{2}$. What are the numbers?

55. A farmer makes a mixture of rye, oats, and barley, using 3 bushels of rye as often as 4 of oats, and 5 of barley. The whole amount of grain used was 66 bushels. How many bushels were there of each?

56. The estate of a bankrupt, valued at $21,000, is to be divided among 4 creditors proportionately to what is due them. The debts due A and B are as 2 and 3; B's and C's claims are in the ratio of 6 : 7; and D's claim is equal to A's. What sum must each receive?

Suggestion. — Let $4x =$ A's and $6x =$ B's money.

57. A general arranging his men in the form of a solid square finds he has 21 men over, but attempting to add one man to each side of the square finds he wants 200 men to fill up the square. How many men had he?

58. The length of a room exceeds its breadth by 8 feet. If each had been increased by 2 feet, the area would have been increased by 60 square feet. Find the original dimensions of the room.

CHAPTER XV.

SIMPLE EQUATIONS CONTAINING TWO OR MORE UNKNOWN QUANTITIES. ELIMINATION.

225. Simultaneous Equations are sets of equations which are to be satisfied by the same values of the unknown quantities contained in them.

a. All proper conditional problems (188) give rise either to a single equation containing one unknown quantity, or to two or more equations containing as many unknown quantities. In the latter case some of the *functions* (222, 2) of the unknown are regarded as *other unknown quantities.* Consequently we may solve problems in one of two ways: either regard the problem as containing but one unknown (x) and express everything in terms of it, as we did in the last chapter; or regard the problem as containing two or more unknowns (x, y, z, etc.), and form *two or more equations.* To make this plain let us take Ex. 18, 224.

1st. With one unknown.

Let $x =$ the number of lbs. of 9 ct. sugar, (222, 1)

then $100 - x$ the number of lbs. of 13 ct. sugar, (222, 2)

$$9x + 13(100 - x) = 1200, \qquad (219, 1)$$

$$\left.\begin{aligned} x &= 25 \\ 100 - x &= 75 \end{aligned}\right\} \qquad (219, 2)$$

2d. With two unknowns.

Let $x =$ the number of lbs. of 9 ct. sugar,

and $y =$ the number of lbs. of 13 ct. sugar,

$$\left.\begin{aligned} &\text{Equation (1)}\ x + y = 100 \\ &\text{Equation (2)}\ 9x + 13y = 1200 \end{aligned}\right\} \qquad (219, 1)$$

$y = 100 - x$ (From Eq. (1) by Ax. 2)

$9x + 13(100 - x) = 1200$ (Substituting $100 - x$ for y in the second equation)

$x = 25$ (219, 2)

$y = 75$ (By substituting 25 for x in (2))

In the second solution, a new unknown quantity, y, is first introduced and then removed from the second equation by substituting its value obtained from the first. *The resulting equation is the same as the equation of the first solution.* In this problem it makes little difference whether we solve it with two or with one unknown quantity; but if the second quantity is a very complex function of the first, the second method has an advantage over the first.

b. A single equation containing two unknown quantities is "*indeterminate*," since the value of either of the unknowns cannot be found until that of the other is known.

To illustrate this, suppose

$$5x - 3y = 18$$
$$5x = 18 + 3y \qquad \text{(Ax. 1)}$$
$$x = \frac{18 + 3y}{5} \qquad \text{(Ax. 4)}$$

If $y = 1$, then $x = \dfrac{18 + 3}{5} = 4\tfrac{1}{5}$

If $y = 2$, then $x = \dfrac{18 + 3 \times 2}{5} = 4\tfrac{4}{5}$

If $y = 3$, then $x = \dfrac{18 + 9}{5} = 5\tfrac{2}{5}$ and so on.

If, however, *two* equations are given, such as

$$(1)\quad 3x + 4y = 9$$
$$(2)\quad 2x + 9y = 17$$

the value of y can be found from equation (1)

$$(1)\quad 3x + 4y = 9,\ 4y = 9 - 3x,\ y = \frac{9 - 3x}{4},$$

and then placed in Eq. (2), obtaining

$$2x + 9\ \frac{9 - 3x}{4} = 17,$$

and we have resulting *one* equation containing but *one* unknown, which solved, gives $x = \frac{13}{19}$. Similarly, the value of x might have been found from (1) and substituted in (2), whence the value of y might have been found.

c Any proper axiomatic operation performed on an equation will not change the value of any letter, and so will not change the "*identity*" of the equation. It will still be the same equation, only under a new guise. (Consult Art. 360.)

It will be convenient to use a system of marking equations by which it can be indicated that a certain equation appears in a new form. The figures written at the left in parentheses will be used to designate the equations, while the same figures with subscripts will distinguish the several equations *in their new forms*.

Thus,

(1) $3x + 4y = 27$

(2) $2x - 3\frac{1}{4}y = 21$

(1_1) $9x + 12y = 81$ [Eq. (1) multiplied by 3, Ax. 3]

(2_1) $8x - 13y = 84$ [Eq. (2) multiplied by 4, Ax. 3]

(1_2) $x + \frac{4}{3}y = 9$ [Eq. (1) divided by 3, Ax. 4]

and so on.

226. Elimination is the process of deducing from two or more equations containing two or more unknown quantities a single equation with but one unknown quantity.

a. There are three kinds of elimination usually given: elimination by *substitution*; by *comparison*; and by *addition* and *subtraction*.

We shall take up in section I. the case in which there are but two unknowns. Then in the next section that in which there are three or more unknowns.

SECTION I.

Solution of Simple Equations containing Two Unknown Quantities.

227. Equations Containing Two Unknown Quantities are solved by the methods of elimination. We give first the methods of elimination with exercises, and then problems involving such simultaneous equations.

I.—ELIMINATION BY SUBSTITUTION.

228. Elimination by Substitution is performed by finding the value of one of the unknowns from either of the given equations and substituting it in the other.

229. After the value of one of the unknowns has been found, it is substituted in one of the foregoing equations, thus giving an equation containing only the other unknown whose value can then be found.

230. Exercise in Elimination by Substitution.

1. Given (1) $8x + 5y = 68$, and (2) $12x + 7y = 100$

$$(1_1)\quad 8x = 68 - 5y \qquad \text{(Ax. 2)}$$

$$(1_2)\quad x = \frac{68 - 5y}{8} \qquad \text{(Ax. 4)}$$

$$(2)\quad 12\left(\frac{68 - 5y}{8}\right) + 7y = 100 \qquad (\mathbf{228})$$

$$\frac{204 - 15y}{2} + 7y = 100 \qquad (\mathbf{178})$$

$$204 - 15y + 14y = 200 \qquad \text{(Ax. 3)}$$

$$-y = -4$$

$$y = 4 \quad \textit{Ans.}$$

$$(1)\quad 8x + 5 \times 4 = 68 \qquad (\mathbf{229})$$

$$8x = 48$$

$$x = 6. \quad \textit{Ans.}$$

2. Given $\begin{cases} (1)\ \dfrac{2y}{15} - \dfrac{3x}{10} = \dfrac{3y - 4x + 7}{5} - \dfrac{5}{6} \\ (2)\ \dfrac{y-1}{3} + \dfrac{x}{2} - \dfrac{3y}{20} - 1 = \dfrac{y - x}{15} + \dfrac{x}{6} + \dfrac{1}{10} \end{cases}$

$$(1_1)\quad 4y - 9x = 18y - 24x + 42 - 25 \qquad \text{(Ax. 3)}$$

$$(1_2)\quad 15x - 14y = 17 \qquad (\text{Ax. } a, \text{ and } \mathbf{86})$$

$$(1_3)\quad y = \frac{15x - 17}{14} \qquad (\text{Ax's } a, \text{ and } \mathbf{4})$$

$$(2_1)\quad 20y - 20 + 30x - 9y - 60 = 4y - 4x + 10x + 6 \qquad \text{(Ax. 3)}$$

$$(2_2)\quad 24x + 7y = 86$$

$$(2_2)\quad 24x + 7\left(\frac{15x - 17}{14}\right) = 86 \qquad (\mathbf{228})$$

$$48x + 15x - 17 = 172$$

$$63x = 189$$

$$x = 3 \quad \textit{Ans.}$$

$$(1_3)\quad y = \frac{15 \times 3 - 17}{14} = 2. \quad \textit{Ans.} \qquad (\mathbf{229})$$

NOTE. — As either unknown may be eliminated the student is left to discover which one it will be advisable to eliminate first. As a rule that unknown should be eliminated which has *the least coefficient.* In substituting in one of the preceding equations (229), choose the *simplest form.* Complex equations, like Ex. 2, must be simplified before proceeding with the elimination.

3. Given (1) $3x + 4y = 10$, and $4x + y = 9$.

SUGGESTION. — Find the value of y in (2) and substitute it in (1).

4. $\begin{cases} (1) \quad x + 2y = 13 \\ (2) \quad 2x + 3y = 21. \end{cases}$

5. $\begin{cases} (1) \quad 4x - 3y = 26 \\ (2) \quad 3x - 4y = 16. \end{cases}$

6. $\begin{cases} (1) \quad x - y = 10 \\ (2) \quad \frac{x}{5} + \frac{y}{3} = 10. \end{cases}$

7. $\begin{cases} (1) \quad \frac{x}{5} - \frac{y}{4} = 1 \\ (2) \quad 5x - 3y = 10. \end{cases}$

8. $\begin{cases} (1) \quad 35x + 17y = 86 \\ (2) \quad 56x - 13y = 17. \end{cases}$

9. $\begin{cases} (1) \quad \frac{1}{2}x + \frac{1}{3}y = 6 \\ (2) \quad \frac{1}{4}x + \frac{1}{2}y = 5. \end{cases}$

10. $\begin{cases} (1) \quad \frac{2}{3}x + y = 16 \\ (2) \quad x + \frac{y}{4} = 14. \end{cases}$

11. $\begin{cases} (1) \quad 8 - \frac{y}{3} = \frac{x}{2} \\ (2) \quad \frac{x}{3} = \frac{y}{5} - 1. \end{cases}$

12. $\begin{cases} (1) \quad 3x - 7y = 0 \\ (2) \quad \frac{2}{7}x + \frac{5}{3}y = 7. \end{cases}$

13. Given (1) $ax + by = c$ and (2) $mx + ny = p$.

$(1_1)\ ax = c - by$

$(1_2)\ x = \dfrac{c - by}{a}$

$$(2)\ m\left(\frac{c - by}{a}\right) + ny = p \qquad (228)$$

$$mc - mby + any = ap \qquad (\text{Ax. } 3)$$

$$(an - bm)y = ap - cm \qquad (97,\ \text{Ax. } 2)$$

$$y = \frac{ap - cm}{an - bm}. \quad \textit{Ans.}$$

The value of x may be found by substituting this value of y, but in *literal* exercises it is often simpler to go back and eliminate the other unknown.

$(1_3)\ y = \dfrac{c - ax}{b}$

$$(2)\ mx + n\left(\frac{c - ax}{b}\right) = p \qquad (228)$$

$$\text{whence } x = \frac{bp - cn}{bm - an}. \quad \textit{Ans.} \qquad (219,\ 2)$$

Or, substituting the value of y in (1) we have

$$ax + b\left(\frac{ap - mc}{an - bm}\right) = c \qquad (229)$$

$$\text{whence } x = \frac{bp - cn}{bm - an} \text{ as before.} \qquad (219,\ 2)$$

a. When equations are *symmetrical* (i.e., such that when corresponding letters are *interchanged*, the equation is not altered) one answer may be obtained from the other by inspection. Thus, to get the value of x from y, change y to x, a to b, b to a, m to n, and n to m, which can be quickly done.

14. (1) $cx + dy = 1$ and (2) $ex + fy = 1$.

15. (1) $ax = by$ and (2) $bx + ay = c$.

II. — ELIMINATION BY COMPARISON.

231. Elimination by Comparison is performed by solving for the same unknown in both equations, and setting the two values thus obtained equal to each other (Ax. 7).

232. Exercise in Elimination by Comparison.

1. (1) $\dfrac{1-3x}{7}+\dfrac{3y-1}{5}=2$ (2) $\dfrac{3x+y}{11}+y=9$

(1_1) $5-15x+21y-7=70$ (2_1) $3x+y+11y=99$ (Ax. 3)

(1_2) $-15x=72-21y$ (2_2) $3x=99-12y$ (Ax. *a*)

(1_3) $x=\dfrac{21y-72}{15}=\dfrac{7y-24}{5}$ (2_3) $x=33-4y$ (Ax. 4)

$$\frac{7y-24}{5}=33-4y \qquad \text{(Ax. 7)}$$

$$7y-24=165-20y$$

$$27y=189$$

$$y=7 \quad \textit{Ans.}$$

(2_3) $x=33-4\times7=5.$ *Ans.*

2. $\begin{cases} (1) \quad 2x+7y=65 \\ (2) \quad 6x-2y=34. \end{cases}$

3. $\begin{cases} (1) \quad 3x+5y=29 \\ (2) \quad 3x-5y=19. \end{cases}$

4. $\begin{cases} (1) \quad \dfrac{x}{2}-\dfrac{y}{3}=2 \\ (2) \quad \dfrac{x}{3}-y=-1. \end{cases}$

5. $\begin{cases} (1) \quad \dfrac{2x}{7}+\dfrac{3y}{5}=27 \\ (2) \quad \dfrac{3x}{9}-\dfrac{y}{7}=2. \end{cases}$

6. $\begin{cases} (1) \quad 8-\dfrac{x+3}{4}=7-\dfrac{3x-2y}{5} \\ (2) \quad 4x-\dfrac{8-y}{3}=24\tfrac{1}{2}-\dfrac{2x+1}{2}. \end{cases}$

7. $\begin{cases} (1) \quad 21y+20x=165 \\ (2) \quad 77y-30x=295. \end{cases}$

8. $\begin{cases} (1) \quad \dfrac{2}{x+3}=\dfrac{3}{y-2} \\ (2) \quad 5(x+3)=3(y-2)+2. \end{cases}$

9. $\begin{cases} (1) & 1\frac{1}{2}x = 1\frac{1}{3}y + 4\frac{5}{12}. \\ (2) & 4\frac{1}{2}x = \frac{1}{3}y - 21\frac{7}{12}. \end{cases}$

10. $\begin{cases} (1) & x - 2y = a. \\ (2) & 2x + 8y = b. \end{cases}$

11. (1) $\frac{x}{a} - \frac{y}{b} = 1$ (2) $\frac{x}{3a} - \frac{y}{6b} = \frac{2}{3}$

(1_1) $\frac{x}{a} = 1 + \frac{y}{b}$ (2_1) $\frac{x}{3a} = \frac{2}{3} + \frac{y}{6b}$

(1_2) $x = a + a\frac{y}{b}$ (2_2) $x = 2a + \frac{ay}{2b}$

$$\therefore \quad a + \frac{ay}{b} = 2a + \frac{ay}{2b} \qquad \text{(Ax. 7)}$$

$$\frac{ay}{2b} = a \qquad \text{(Ax. 2)}$$

$$y = 2b. \quad \textit{Ans.} \qquad \text{(Ax. 5)}$$

$$(1_2)\ x = a + \frac{2ab}{b} = 3a. \quad \textit{Ans.} \qquad \textbf{(229)}$$

12. (1) $\frac{x}{a} + \frac{y}{b} = m$ (2) $\frac{x}{c} + \frac{y}{d} = n.$

III.—ELIMINATION BY ADDITION AND SUBTRACTION.

233. Elimination by Addition or Subtraction is performed by making the coefficients of one of the unknowns the same in both equations, and then adding or subtracting the equations member by member according as these coefficients have unlike or like signs.

1. In order to make coefficients the same in both equations, proceed as follows: —

(1. Reduce each equation to the form $ax + by = c$, i.e. collect into the first term all the terms containing x, into the second term all the terms containing y, and into the right member all the known terms.

(2. Find the l. c. m. of the two coefficients of the unknown to be eliminated.

(3. Multiply both members of each equation by the quotient obtained from dividing the l. c. m. by the coefficient in that equation of the unknown to be eliminated.

2. Upon adding or subtracting the new equations member by member, the terms with equal coefficients *cancel* and one unknown is thus eliminated.

234. Exercise in Elimination by Addition and Subtraction.

1. Given (1) $\frac{3x-5y}{2}+3=\frac{2x+y}{5}$

(2) $8-\frac{x-2y}{4}=\frac{x}{2}+\frac{y}{3}$

(1_1) $15x-25y+30=4x+2y$

(1_2) $11x-27y=-30$ **(233**, 1, (1)

(2_1) $96-3x+6y=6x+4y$

(2_2) $-9x+2y=-96$ **(233**, 1, (1)

(1_3) $99x-243y=-270$ **(233**, 1, (3)

(2_3) $-99x+22y=-1056$ (Ax. 1)

$-221y=-1326$

$y=6.$ *Ans.*

(2_2) $-9x+2\times 6=-96 \therefore x=12.$ *Ans.* **(229)**

2. (1) $8x-21y=33$ (2) $6x+35y=177$

(1_1) $24x-63y=99$

(2_1) $24x+140y=708$

$203y=609$

$y=3$ *Ans.*

(1) $8x-21\times 3=33 \therefore x=12.$ *Ans.* **(229)**

3. (1) $16x+17y=500$ (2) $17x-3y=110.$

(1_1) $48x+51y=1500$

(2_1) $289x-51y=1870$

$337x=3370$ (Ax. 1)

$x=10.$ *Ans.*

$y=20.$ *Ans.* **(229)**

4. $\begin{cases} (1)\ 3x + 2y = 19 \\ (2)\ 2x - 3y = 4. \end{cases}$

5. $\begin{cases} (1)\ \dfrac{x}{4} + \dfrac{y}{5} = 8 \\ (2)\ \dfrac{x}{5} + \dfrac{y}{3} = 9. \end{cases}$

6. $\begin{cases} (1)\ \dfrac{x+y}{5} - \dfrac{x-y}{2} = 3 \\ (2)\ \dfrac{x-y}{2} + \dfrac{x+y}{10} = 0. \end{cases}$

7. $\begin{cases} (1)\ \dfrac{2x}{3} + \dfrac{y}{4} = 16 \\ (2)\ x + \dfrac{y}{4} = 14. \end{cases}$

8. $\begin{cases} (1)\ \dfrac{2}{5}x - \dfrac{1}{12}y = 3 \\ (2)\ 4x - y = 20. \end{cases}$

9. $\begin{cases} (1)\ \dfrac{x}{7} + \dfrac{y}{5} = 1\frac{3}{7} \\ (2)\ x + \dfrac{y}{3} = 4\frac{2}{3}. \end{cases}$

10. $\begin{cases} (1)\ \dfrac{4x}{5} - \dfrac{2y}{5} = 4 \\ (2)\ 6x = 9y. \end{cases}$

11. $\begin{cases} (1)\ \dfrac{x+3y}{2} = 7\frac{1}{2} \\ (2)\ \dfrac{4x+5y}{4} = 8. \end{cases}$

12. $\begin{cases} (1)\ \dfrac{3}{8}x - \dfrac{2}{5}y = \dfrac{1}{10}y \\ (2)\ \dfrac{5}{8}x + \dfrac{1}{6}y = 7. \end{cases}$

13. $\begin{cases} (1)\ 4x + 8y = 2.4 \\ (2)\ 10.2x - 6y = 3.48. \end{cases}$

14. $\begin{cases} (1) & ax + by = p \\ (2) & ax + dy = q. \end{cases}$

15. $\begin{cases} (1) & mx + ny = 0 \\ (2) & y = ax + b. \end{cases}$

16. $\begin{cases} (1) & \dfrac{a}{b+y} = \dfrac{b}{3a+x} \\ (2) & ax + 2by = d. \end{cases}$

17. $\begin{cases} (1) & x + \dfrac{1}{2}(3x - y - 1) = \dfrac{1}{4} + \dfrac{3}{4}(y-1) \\ (2) & \dfrac{1}{5}(4x + 3y) = \dfrac{7}{10}y + 2. \end{cases}$

18. $\begin{cases} (1) & \dfrac{1}{x} + \dfrac{2}{y} = \dfrac{11}{15} \\ (2) & \dfrac{3}{x} + \dfrac{4}{y} = \dfrac{9}{5}. \end{cases}$

SOLUTION. — In equations containing reciprocals it is usually best *not to clear of fractions*, but to solve directly for $\frac{1}{x}$ and $\frac{1}{y}$.

$$(1_1)\quad \frac{3}{x} + \frac{6}{y} = \frac{11}{5} \qquad (233,\ 1)$$

$$(2)\quad \frac{3}{x} + \frac{4}{y} = \frac{9}{5}$$

$$\frac{2}{y} = \frac{2}{5} \qquad (\text{Ax. } 2)$$

$$10 = 2y \qquad (\text{Ax. } 3)$$

$$y = 5 \quad \textit{Ans.}$$

$$(1_1)\quad \frac{1}{x} = \frac{11}{15} - \frac{2}{5} = \frac{1}{3} \therefore x = 3 \qquad (229)$$

19. $\begin{cases} (1) & \dfrac{5}{x} + \dfrac{6}{y} = 3 \\ (2) & \dfrac{15}{x} + \dfrac{3}{y} = 4. \end{cases}$

20. $\begin{cases} (1) & 21 \cdot \dfrac{1}{x} + 12 \cdot \dfrac{1}{y} = 5 \\ (2) & \dfrac{1}{y} - \dfrac{1}{x} = \dfrac{1}{42}. \end{cases}$

21. $\begin{cases} (1)\ 4 \cdot \dfrac{1}{x} + 27 \cdot \dfrac{1}{y} = 42 \\ (2)\ 14 \cdot \dfrac{1}{x} - 15 \cdot \dfrac{1}{y} = 1. \end{cases}$

22. $\begin{cases} (1)\ 2x^{-1} + 3y^{-1} = \frac{3}{2} \\ (2)\ 4x^{-1} + 8y^{-1} = \frac{10}{3}. \end{cases}$

23. $\begin{cases} (1)\ 2y - x = 4xy \\ (2)\ \dfrac{4}{y} - \dfrac{3}{x} = 9. \end{cases}$

24. $\begin{cases} (1)\ \dfrac{3}{x} + \dfrac{5}{y} = \dfrac{8}{15} \\ (2)\ 9y - 22x = \dfrac{3xy}{25}. \end{cases}$

IV.—SPECIAL METHODS OF ELIMINATION.

235. Special Methods of Elimination. In these the elimination is attained by one or other of the three processes already explained. The difference lies in the preparation of the equations for the elimination.

A. Elimination by Undetermined Multipliers.

236. The General Process of Elimination by Undetermined Multipliers for equations of any degrees is called Bezout's Method.

1. Given (1) $5x - 7y = 20$ (2) $9x - 11y = 44$.

(1_1) $5mx - 7my = 20m$ (Ax. 3)

(2) $9x - 11y = 44$

(3) $(5m - 9)x - (7m - 11)y = 20m - 44$ (Ax. 2)

Now, if $5m - 9 = 0$, then x will be eliminated from the equation.

If $5m - 9 = 0$, $m = \frac{9}{5}$. Substituting this value in (3), we have

$$0 - \left(7 \times \frac{9}{5} - 11\right)y = 20 \times \frac{9}{5} - 44;$$

Or,
$$-\frac{8}{5}y = -\frac{40}{5}$$
$$y = 5$$
$$\therefore x = 11.$$
(229)

If $7m - 11$ be put equal to nought, y will be eliminated.

Placing $7m - 11 = 0$, $m = \frac{11}{7}$. Substituting this value in (3)

$$\left(5 \times \frac{11}{7} - 9\right) x - 0 = 20 \times \frac{11}{7} - 44$$

$$-\frac{8}{7}x = -\frac{88}{7}$$

$$x = 11$$

$$\therefore y = 5. \qquad (229)$$

NOTE. — We readily see that this method amounts to the multiplication of one of the equations by some number integral or fractional which makes the coefficients of one of the unknowns the same in both, the multiplers being the values of m. (233, 1.)

2. Given $3x - 5y = 45$ and $18x - 24y = 324$.
3. Given $5x - 7y = 20$ and $9x - 11y = 70$.
4. Given $\frac{x}{9} + \frac{y}{7} = 6.3$ and $\frac{x}{3} + \frac{53y}{56} = 39.2$.
5. Given $x + ay = b$ and $ax + by = 1$.

B Elimination by Means of one or more Derived Equations.

237. When One of the Regular Processes of Elimination gives rise to large products, the work can often be materially shortened by means of **derived equations.**

If either or both of the given equations be multiplied by some number and the resulting equations added or subtracted member by member, we get a new equation which will be satisfied by the same values of x and y as the given equations (Ax. a.) This new equation, it is true, has both x and y still in it, but with smaller coefficients. Now, this equation used with either one of the given equations, or with another similarly derived, since two equations suffice, will give the values of x and y by ordinary methods of elimination.

1. $(1)\ \ 9x + 13y = 184 \qquad (2)\ 13x + 19y = 268$

$$\begin{array}{lrr}
(1_1) & 27x + 39y = 552 & \\
(2_1) & 26x + 38y = 536 & \\
\hline
(3\ ) & x + \ \ y = \ \ 16 & \text{(Ax. 2)} \\
(1\ ) & 9x + 13y = 184 & \\
(3_1) & 9x + \ \ 9y = 144 & \\
\hline
 & 4y = 40 & \text{(Ax. 2)} \\
 & y = 10 & \\
 & \therefore x = \ \ 6. & \textbf{(229)}
\end{array}$$

2. $(1)\ 14x - 17y = 159 \quad (2)\ 29x - 37y = 324.$

SUGGESTION. — Multiply (1) by 2 and subtract from (2).

3. $(1)\ 139x + 152y = 1377 \quad (2)\ 35x + 37y = 348.$
4. $(1)\ 755x - 564y = 2074 \quad (2)\ 1133x - 847y = 3113.$
5. $(1)\ x + 50y = 557 \quad (2)\ 50x + y = 361.$

SUGGESTION. — Add the two equations and divide through by 51; then subtract the one from the other and divide through by 49.

6. $(1)\ 59x + 73y = 390\frac{5}{6} \quad (2)\ 73x + 59y = 379\frac{1}{6}.$
7. $(1)\ 23x - 41y = -2\frac{7}{12} \quad (2)\ 39x - 25y = 6\frac{3}{4}.$
8. $(1)\ 28x - 35y = 56 \quad (2)\ 29x - 13y = 151.$
9. $(1)\ (a + 2b)x - (a - 2b)y = 6ac$
   $(2)\ (a + 3c)y - (a - 3c)x = 4ab.$

C. Elimination by Method of finding H. C. F.

238. This Method depends upon the Same Principles as the last and is really only a special way of applying them.

1. Given $(1)\ 5x - 7y = -8 \qquad (2)\ 5y + 32 = 7x$

$$(1_1)\ 5x - 7y + 8 = 0 \qquad (2_1)\ 7x - 5y - 32 = 0$$

$$\begin{array}{l}
7x - 5y - 32)\ 5x - 7y + 8\ (1 \\
\qquad\quad 5x - 7y + 8 \\
\hline
2)\ 2x + 2y - 40 \\
\hline
\quad x + \ \ y - 20)\ 5x - 7y + 8\ \ (5 \\
\qquad\qquad\qquad 5x + 5y - 100 \\
\hline
\qquad\qquad\qquad - 12y + 108 = 0 \\
\qquad\qquad\qquad\qquad - 12y = - 108 \\
\qquad\qquad\qquad\qquad\qquad y = 9 \\
\qquad\qquad\qquad\qquad\ \therefore x = 11. \qquad\qquad \text{(229)}
\end{array}$$

EXPLANATION. — Each equation is so transposed that one member is zero. Then the divisor is multiplied through by some factor, and the result subtracted from the other given equation, giving a new equation, which is used with the divisor (or dividend), and so on. The last remainder does not contain x; i.e., *it has been eliminated.* This remainder, equal to zero, gives the value of y.

2. (1) $8x - 21y = 33$ (2) $6x + 35y = 177$.

3. (1) $16x + 17y = 500$ (2) $17x - 3y = 110$.

239. Exercise in Elimination. The choice of the method is left to the student. Addition and Subtraction usually gives a more elegant form to the solution than the other two methods.

1. $4x + 9y = 51$ and $8x - 13y = 9$.
2. $7x - 9y = 7$ and $3x + 10y = 100$.
3. $4x + 8y = 2.4$ and $10.2x - 6y = 3.48$.
4. $x + y = 24$ and $x = 5y$.
5. $2y + 79 = 5x$ and $3x - 7 = 4 + x + y$.
6. $\frac{x}{3} + \frac{2y}{5} = 6$ and $\frac{2x}{3} + \frac{y}{5} = 6$.
7. $\frac{x}{6} + \frac{y}{5} = \frac{x}{2} + 2$ and $\frac{x}{4} + \frac{y}{3} = \frac{3y}{10} + 4$.
8. $\frac{4}{5 + y} = \frac{5}{12 + x}$ and $2x + 5y = 35$.
9. $10x = 2 + 2y$ and $4y = 20 - 4x$.
10. $\frac{x + 2}{3} + 8y = 31$ and $\frac{y + 5}{4} + 10x = 192$.
11. $ax + by = c - d$ and $mx = ny$.
12. $\frac{2x + 6}{3y + 2} = \frac{8}{7}$ and $8x - 4 = 9y$.
13. $\frac{x + 2}{x - y} = 2$ and $\frac{2x + 3y}{x + a} = b$.
14. $5x + 7y : 3x + 11 :: 13 : 7$ and $11x + 27 : 7x + 5y :: 19 : 11$.

15. $y+1 : x :: 5 : 3$ and $\frac{2y}{3} - \frac{5-x}{2} = \frac{41}{12} - \frac{2y-1}{4}$.

16. $2(2x+3y) = 3(2x-3y)+10$ and $4x-3y = 4(6y-2x)+3$.

17. $\frac{1-3x}{7} + \frac{3y-1}{5} = 2$ and $\frac{3x+y}{11} + y = 9$.

18. $\frac{1}{7}(x+2) + \frac{1}{4}(y-x) = 2x - 8$ and $\frac{1}{3}(2y-3x) + \frac{1}{5}(8x+6y-4) = 3x+4$.

19. $2x : y :: 29 : 4$ and $y+4x+6 = \frac{4y^2+13xy-12x^2}{4y-3x-1}$.

20. $\frac{2}{x} + \frac{3}{y} + \frac{1}{4} = \frac{18}{y}$ and $\frac{2}{x} - \frac{1}{y} = \frac{1}{5}\left(\frac{1}{x} + \frac{1}{y}\right) + \frac{1}{12}$.

(See Ex. 18 in **234**.)

21. $\frac{x}{a+b} + \frac{y}{a-b} = 2a$ and $\frac{x-y}{4ab} = 1$.

22. $\frac{3x+4y+3}{10} - \frac{2x+7-y}{15} = 5 + \frac{y-8}{5}$ and $\frac{9y+5x-8}{12} - \frac{x+y}{4} = \frac{7x+6}{11}$.

23. $3ax - 2by = c$ and $a^2x + b^2y = 5bc$.

24. $\frac{10}{2x+3y-29} + \frac{9}{7x-8y+24} = 8$ and $\frac{2x+3y-29}{2} = \frac{7x-8y}{3} + 8$.

SUGGESTION. — Put $2x+3y-29 \equiv u$ and $7x-8y+24 \equiv v$; then the two equations become

$$\frac{10}{u} - \frac{9}{v} = 8 \text{ and } \frac{u}{2} = \frac{v}{3},$$

from which (**228**) $u = 2$, and $v = 3$.
Substituting these values, we now have

$$2x+3y-29 = 2, \text{ and } 7x-8y+24 = 3,$$

which, being solved, give the values of x and y.

25. $\frac{8}{2x-3y+17}+5x-8y+44=5$ and $\frac{5}{2x-3y+17}$ $+16y=10x+88\frac{1}{2}$.

SUGGESTION. — Put $\frac{1}{2x-3y+17}=u$ and $5x-8y+44=v$, transposing $16y$.

240. Problems containing Two Unknown Quantities.

1. When the greater of two numbers is divided by the less the quotient is 4 and the remainder 3; and when the sum of the two numbers is increased by 38 and the result divided by the greater of the two numbers the quotient is 2 and the remainder 2.

Let x equal the greater number
and y equal the less. (**225.** *a.*)

Then (1) $\frac{x-3}{y}=4$ and (2) $\frac{x+y+38-2}{x}=2$ (**222.** 3)

$$(1_1)\quad x-4y=3$$
$$(2_1)\quad x-y=36$$
$$3y=33 \qquad \text{(Ax. 2)}$$
$$y=11$$
$$\therefore\quad x=47 \qquad \textbf{(229)}$$

2. A farmer paid four men and six boys 72 dimes for laboring one day, and afterwards at the same rate he paid three men and nine boys 81 dimes for one day. What were the wages of each?

3. Find two numbers whose sum is 857142 and whose difference is 571428.

4. If A's money were increased by \$360 he would have three times as much as B, but if B's money were diminished by \$50 he would have half as much as A. Find the sum possessed by each.

5. In an assembly of 325 people, a measure was adopted by a majority of 35. How many voted aye and how many nay?

6. The planet Venus and the Earth complete their revolutions round the sun in different times. At one time they are together on one side of the sun, that is in *conjunction*; at another time they are in *opposition*, with the sun between them. When in opposition they are 159900000 miles apart, when in conjunction only 25700000 miles apart. Supposing both to move in circular orbits, what are their distances from the sun?

7. A said to B, if $\frac{1}{4}$ of your money were added to $\frac{1}{6}$ of mine the sum would be \$6. B replied, if $\frac{1}{4}$ of yours were added to $\frac{1}{6}$ of mine the sum would be \$$5\frac{2}{3}$. What sum had each?

8. A and B are in trade together with different sums. If \$50 be added to A's money and \$20 be taken from B's they will have the same sum; but if A's money were 3 times and B's 5 times as great as each really is, they would together have \$2350. How much has each?

9. If the smaller of two numbers be divided by the greater the quotient is .21 and the remainder .0057; but if the greater be divided by the smaller the quotient is 4 and the remainder 1.742. What are the numbers?

10. When Mr. Smith was married his age was $\frac{4}{3}$ of his wife's; 12 years afterwards his age was $\frac{6}{5}$ of his wife's. How old were they when married?

11. Fifty laborers were engaged to remove an obstruction on a railroad: some of them by agreement were to receive 90 cts. per day, and others \$1.50. There was paid them just \$48. No memorandum having been made, it is required to find how many worked at each rate?

12. A son asked his father how old he was, and was answered thus: if you take away 5 from my years and divide the remainder by 8 the quotient will be $\frac{1}{3}$ of your age. But if you add 2 to your age, multiply the sum by 3, and then subtract 6 from the product, you will have the number of years of my age. What were the ages of the father and son?

13. What fraction is that whose numerator being doubled and denominator increased by 7 the value becomes $\frac{2}{3}$; but the denominator being doubled and the numerator increased by 2 the value becomes $\frac{3}{5}$?

14. An apple-woman bought a lot of apples at 1 ct. each and a lot of pears at 2 cts. each, paying $1.70 for the whole; 11 of the apples and 7 of the pears were bad, but she sold the good apples at 2 cts. each and the good pears at 3 cts. each, realizing $2.60. How many of each fruit did she buy?

15. If a certain number be divided by the sum of its two digits, the quotient is 6 and the remainder 3; if the digits be interchanged and the resulting number be divided by the sum of the digits, the quotient is 4 and the remainder 9. What is the number?

SOLUTION. — In order to represent numbers in the Arabic scale of 10 *by letters*, we must say,

Let $x =$ the number denoted by the figure in unit's place,
and $y =$ the number denoted by the figure in ten's place,
and so on, using other letters for higher orders.

Then, $10y + x$ represents a number of two figures,
$100z + 10y + x$ represents a number of three figures,
and so on.

When the order of the digits is reversed,
$10x + y =$ the number of two places.

The equations of the above problem are

$$(1)\quad \frac{10y + x - 3}{x + y} = 6 \text{ and } (2)\quad \frac{10x + y - 9}{x + y} = 4.$$

16. Find a number which is greater by 2 than 5 times the sum of its digits, and if 9 be added to it the digits will be reversed.

17. Find that number of two digits, to which if the number found by changing its digits be added, the sum is 121; and if the less of the same two be taken from the greater the remainder is 9.

18. In 11 hours C walks $12\frac{1}{2}$ miles less than D walks in 12 hours; and in 5 hours D walks $3\frac{1}{4}$ miles less than C does in 7 hours. How many miles does each walk per hour?

19. A man has two measures. Nine of the first or fifteen of the second will fill a certain vessel. Using both measures 13 times in all, how many times is each used?

20. A banker has two kinds of money. It takes a pieces of the one or b pieces of the other to make a dollar. If c pieces be given for a dollar, how many of each will be used?

21. A pound of tea and 3 pounds of sugar cost $1.20. But if tea were to rise 50% and sugar 10% they would cost $1.56. Find the price per pound of each.

22. A grocer knows neither the weight nor the first cost of a box of tea he had purchased. He only recollects that if he had sold the whole at 30 cts. per pound he would have gained $1. But if he had sold it at 22 cts. per pound he would have lost $3. Required the number of pounds in the box and the first cost per pound.

23. Two persons 27 miles apart setting out at the same time are together in 9 hours if they walk in the same direction; but if they walk in opposite directions towards each other, in 3 hours. Find their rates.

24. Find two numbers in the ratio of $5:7$ to which two other required numbers in the ratio $3:5$ being respectively added, the sums shall be in the ratio of $9:13$, and the difference of whose sums equals 16.

25. Two trains set out at the same moment, the one to go from Boston to Springfield, the other from Springfield to Boston. The distance between the two cities is 98 miles.

They meet each other at the end of one hour and 24 minutes, and the train from Boston travels as far in 4 hours as the other does in 3. What was the speed of each train?

26. If the sides of a rectangular field were each increased by 2 yards, the area would be increased by 220 square yards. If the length were increased and the breadth diminished each by 5 yards, the area would be diminished by 185 square yards. What is the area?

SECTION II.

SIMULTANEOUS EQUATIONS CONTAINING THREE OR MORE UNKNOWN QUANTITIES.

241. When Three or More Unknown Quantities appear in as Many Equations the process of eliminating, properly conducted, will cause the unknowns to successively disappear, and ultimately lead to the solution of the problem.

a. When there are three equations containing three unknown quantities, we first select that one of the unknowns which can be most easily made to disappear, and proceed to eliminate it by using, say, the first and second equations, obtaining a new (fourth) equation; then the same unknown is eliminated by using the first and third (or, if preferable, the second and third) equations, giving another new (fifth) equation. These two equations, the fourth and fifth, contain but two unknowns, and can be solved as in Section I. of this chapter. The whole process becomes more intelligible by studying an example.

242. Examples of Equations containing Three or More Unknown Quantities.

1. $$\left.\begin{aligned} (1)\quad x - 2y + 3z &= 6 \\ (2)\quad 2x + 3y - 4z &= 20 \\ (3)\quad 3x - 2y + 5z &= 26 \end{aligned}\right\}$$

On inspection we perceive that of the three unknowns x can most readily be eliminated, having the smallest coefficients.

$$
\begin{array}{lrr}
(1_1) & 2x - 4y + \ 6z = 12 & \text{(Ax. 3)} \\
(2) & \underline{2x + 3y - \ 4z = 20} & \\
(4) & 7y - 10z = \ 8 & \text{(Ax. 2)} \\
(1_2) & 3x - 6y + \ 9z = 18 & \\
(3) & \underline{3x - 2y + \ 5z = 26} & \\
(5) & 4y - \ 4z = \ 8 & \text{(Ax. 2)} \\
(5_1) & y - \ z = \ 2 & \text{(Ax. 4)} \\
(5_2) & 7y - \ 7z = 14 & \text{(Ax. 3)} \\
(4) & \underline{7y - 10z = \ 8} & \\
 & 3z = \ 6 & \text{(Ax. 2)} \\
 & z = \ 2. \ \textit{Ans.} & \\
(5_1) & y - 2 = 2 \therefore y = 4. \ \textit{Ans.} & \textbf{(229)} \\
(1) & x - 2 \times 4 + 3 \times 2 = 6 \therefore x = 8. \ \textit{Ans.} & \textbf{(229)}
\end{array}
$$

NOTE.—In deriving equations (4) and (5) *any two* pairs of the three equations may be used.

2.
$$
\left.
\begin{array}{ll}
(1) & x + 2y - 3u + \ z = 4 \\
(2) & 2x - \ y + 2u - 3z = 1 \\
(3) & 5x - 3y - \ u - 2z = 11 \\
(4) & 3x + 4y - 5u + 6z = -9
\end{array}
\right\}
$$

$$
\begin{array}{lrr}
(1_1) & 3x + 6y - 9u + 3z = 12 & \\
(2) & \underline{2x - \ y + 2u - 3z = \ 1} & \\
(5) & 5x + 5y - 7u \qquad = 13; & \text{(Ax. 1)} \\
(1_2) & 2x + 4y - 6u + 2z = \ 8 & \\
(3) & \underline{5x - 3y - \ u - 2z = 11} & \\
(6) & 7x + \ y - 7u \qquad = 19; & \text{(Ax. 1)} \\
(2_1) & 4x - 2y + 4u - 6z = \ 2 & \\
(4) & \underline{3x + 4y - 5u + 6z = -9} & \\
(7) & 7x + 2y - u \qquad = -7; & \text{(Ax. 1)}
\end{array}
$$

Equations (5), (6), and (7) constitute a new set.

$$
\begin{array}{lrr}
(5) & 5x + 5y - \ 7u = 13 & \\
(6_1) & \underline{35x + 5y - 35u = 95} & \\
(8) & 30x \qquad - 28u = 82 & \text{(Ax. 2)} \\
(8_1) & 15x \qquad - 14u = 41 & \text{(Ax. 4)}
\end{array}
$$

$$
\begin{array}{lll}
(6_2) & 14x + 2y - 14u = 38 & \\
(7) & 7x + 2y - u = -7 & \\
\hline
(9) & 7x - 13u = 45 & \text{(Ax. 2)}
\end{array}
$$

Equations (8) and (9) constitute the third set.

$$
\begin{array}{lll}
(9) & 7x - 13u = 45 & \\
(8_1) & 15x - 14u = 41 & \\
\hline
(10) & 8x - u = -4 & \text{(Ax. 2, 237)} \\
(10_1) & 104x - 13u = -52. & \\
(9) & 7x - 13u = 45 & \\
\hline
 & 97x = -97 & \text{(Ax. 2)} \\
 & x = -1. \quad \textit{Ans.} & \\
(10) & 8 \times -1 - u = -4 \therefore u = -4. \quad \textit{Ans.} & (229) \\
(6) & 7 \times -1 + y - 7 \times -4 = 19 & \\
 & \therefore y = -2 \quad \textit{Ans.} & (229) \\
(1) & -1 + 2 \times -2 - 3 \times -4 + z = 4 & \\
 & \therefore z = -3. \quad \textit{Ans.} & (229)
\end{array}
$$

243. Rule.[1] — **The Normal Process of Elimination,** where there are three or more unknowns, may be described as follows: —

1. Select the unknown to be first eliminated. Then combine the equations in pairs in the most desirable manner to each time eliminate this unknown. However, the derived equations must be *independent*, i.e., such that no one of them can be derived from one or more of the others. (Cf. **246**, *a*.) The number of new equations is one less than the given equations.

2. Proceed in the same manner with the derived equations to eliminate one of the remaining unknowns. Then, in like manner, with the set thus obtained, and so on, until the value of the last remaining unknown is found.

3. Substitute the value of the last unknown in one of the preceding equations containing but two unknown quantities

[1] A far more abridged and elegant method of solving these problems is by means of *Determinants*.

and the value of the other is immediately known. Substitute these values in an equation which contains a third unknown, and so on.

a. If in any case one or more of the equations do not contain one of the unknowns, it will usually be better to regard them as belonging to the second set.

b. Special methods of elimination in particular problems frequently far surpass the normal process in elegance and brevity.

244. Exercise in Equations containing Three or More Unknown Quantities.

1. $\begin{cases} (1)\ x + y + z = 6 \\ (2)\ 5x + 4y + 3z = 22 \\ (3)\ 15x + 10y + 6z = 53. \end{cases}$

2. $\begin{cases} (1)\ x + 2y = 23 \\ (2)\ 3x + 4z = 57 \\ (3)\ 5y + 6z = 94. \end{cases}$

SUGGESTION. — Eliminate z between (2) and (3), and combine the new equation with (1).

3. $\begin{cases} (1)\ x + y - z = 132 \\ (2)\ x - y + z = 65.4 \\ (3)\ -x + y + z = -1.2. \end{cases}$

4. $\begin{cases} (1)\ \dfrac{x+y}{3} + 2z = 21 \\ (2)\ \dfrac{y+z}{2} - 3x = -65 \\ (3)\ \dfrac{3x+y-z}{2} = 38. \end{cases}$

5. $\begin{cases} (1)\ \frac{1}{2}(x + z - 5) = y - z \\ (2)\ \qquad\qquad = 2x - 11 \\ (3)\ \qquad\qquad = 9 - (x + 2z) \end{cases}$

6. $\begin{cases} (1) \text{ and } (2)\ \dfrac{y+z}{4} = \dfrac{z+x}{3} = \dfrac{x+y}{2} \\ (3)\ x + y + z = 27. \end{cases}$

7. $\begin{cases} (1), (2)\ x : y : z = 5 : 12 : 13 \\ \text{(i.e., } x : y :: 5 : 12 \text{ and } x : z :: 5 : 13) \\ (3)\ x + y + z = 27. \end{cases}$

8. $\begin{cases} (1)\ x + \frac{1}{2} y = 10 - \frac{1}{3} z \\ (2)\ \frac{1}{2} (x + z) = 9 - y \\ (3)\ \frac{1}{4} (x - z) = 2 y - 7. \end{cases}$

9. $\begin{cases} (1)\ x + a = y + z \\ (2)\ y + a = 2 x + 2 z \\ (3)\ z + a = 3 x + 3 y. \end{cases}$

10. $\begin{cases} (1)\ ax + by = c \\ (2)\ cx + az = b \\ (3)\ bz + cy = a. \end{cases}$

11. $\begin{cases} (1)\ \frac{2}{x} - \frac{3}{y} + \frac{4}{z} = 2.9 \\ (2)\ \frac{5}{x} - \frac{6}{y} - \frac{7}{z} = -10.4 \\ (3)\ \frac{9}{y} + \frac{10}{z} - \frac{8}{x} = 14.9. \end{cases}$

12. $\begin{cases} (1)\ \frac{3}{x} - \frac{4}{5 y} + \frac{1}{z} = 7\frac{3}{8} \\ (2)\ \frac{1}{3 x} + \frac{1}{2 y} + \frac{2}{z} = 10\frac{1}{4} \\ (3)\ \frac{4}{5 x} - \frac{1}{2 y} + \frac{4}{z} = 16\frac{1}{10}. \end{cases}$

Suggestion. — Examples 11 and 12 are to be solved as reciprocals. Clear 12 of numerical denominators. (See Ex. 18, **234**.)

13. (1) $xy + yz + zx = 9 xyz$; (2) $yz + 2 zx - 3 xy = - 4 xyz$; (3) $3 yz - 2 zx + xy = 4 xyz$.

Suggestion. — Divide each of these equations through by xyz.

14. (1) $x + y = 12$; (2) $y - z = 3$; (3) $z + u = 7$; (4) $u + x = 8$.

Suggestion. — First add the equations, and divide by 2.

15. (1) $x + y = 16$; (2) $z + x = 22$; (3) $y + z = 28$.

245. Problems Involving Three or More Unknown Quantities.

1. Determine three numbers such that their sum is 9; the sum of the first, twice the second, and three times the third, 22; and the sum of the first, four times the second, and nine times the third, 58.

2. A and B together possess only $\frac{2}{3}$ as much money as C; B and C together have 6 times as much as A; and B has $680 less than A and C together; how much has each?

3. A boy bought at one time 2 apples and 5 pears for 12 cts.; at another 3 pears and 4 peaches for 18 cts.; at another 4 pears and 5 oranges for 28 cts.; and at another 5 peaches and 6 oranges for 39 cts.; required the cost of each kind of fruit.

4. A gentleman divided a sum of money among his four sons, so that the share of the oldest was $\frac{1}{2}$ of the sum of the shares of the other three, the share of the second $\frac{1}{3}$ of the sum of the other three, the share of the third $\frac{1}{4}$ of the sum of the other three; and it was found that the share of the eldest exceeded that of the youngest by $14. What was the whole sum, and what was the share of each son?

5. A person has two horses and two saddles, all of which are worth $265. The poorer horse and better saddle are worth $5 less than the better horse and poorer saddle, while the better horse and better saddle are worth $45 more than the poorer horse and poorer saddle; and the horses are worth 5 times as much as the saddles. What is the value of each horse and saddle?

6. The average age of three persons is 41 years. The average age of the first and second is 37 years, and of the second and third is 29 years. Find their ages.

7. The average age of A, B, and C is a. The average age of A and B is b, and of B and C is c. What are their ages?

8. Three numbers are in the ratio 3 : 4 : 5. If to fivefold the first number we add fourfold the second number and

threefold the third number, the sum will be 345. Name the three numbers.

9. If A and B can perform a certain work in 12 days, and A and C in 15 days, and B and C in 20 days, in what time could each do it alone?

10. A number is expressed by three figures whose sum is 11. The figure in the place of units is double that in the place of hundreds, and when 297 is added to this number the sum obtained is expressed by the figures of this number reversed. What is the number?

11. Roads join four cities, A, B, C, D, thus forming a quadrangle. If I go from A to D through B and C, I must pay $6.10 hack fare. If I go from A to B through D and C, I must pay $5.50. Going from A to C through B, I pay the same as from A to C through D. On the other hand, from B to D through A costs 40 cts. less than from B to D through C. What are the distances A B, B C, C D, and D A, if the fare is 10 cts. per mile?

12. A, B, and C in a hunting excursion killed 96 birds, which they wish to share equally; in order to do this A, who has most, gives to B and C as many as they already have; next, B gives to A and C as many as they had after the first division; and lastly, C gives to A and B as many as they each had after the second division. It was then found that each had the same number. How many had each at first?

13. A piece of work can be completed by A, B, and C in 10 days; by A and B together in 11 days, supposing C to have worked 5 days and left off; by B and C if B works 15 days and C works 30 days. How long will it take each alone to do the work?

CHAPTER XVI.

INDETERMINATE AND REDUNDANT EQUATIONS.— PROBLEMS IN ALGEBRA.

SECTION I.

INDETERMINATE EQUATIONS.

246. Indeterminate Equations occur when the number of unknown quantities exceeds the number of Independent Equations. (See **225**, *b*.)

a. Equations seemingly different sometimes reduce to the same equation. Hence two given simultaneous equations might be *indeterminate*. Thus, (1) $2(3x - 17\frac{1}{2}) = -4\frac{1}{3}y$.

(2) $10x + 7y = 58\frac{1}{3}$.

Upon clearing of fractions and transposing, both become

$$60x + 42y = 350.$$

247. Examples of Indeterminate Equations. These may be classified into two kinds: those on which are imposed no restrictions; and those in which only positive integral values are permissible. The latter are often called *Diophantine Equations*.

a. An example of the former kind was given in **225**, *b*. It is there made plain that any number of results can be obtained from such an equation, every one of which is a solution. Also two or more equations may be reduced by elimination to a single equation containing two or more unknowns, whose solution would be like that in **225**, *b*. All that can be done with these equations, then, is to assign arbitrary values to one or more of the unknowns, and derive the corresponding values of the remaining one. These equations find an appropriate use in analytic geometry.

b. **If, however, the answers be confined to integral positive numbers, the problem becomes definite, and may be solved, although the method of solution is very unlike that of other equations.**

1. Given $5x + 11\ y = 47$; required to find the pairs of positive integral values of x and y which satisfy the equation.

$$5x = 47 - 11y \qquad \text{(Ax. 2)}$$

$$x = \frac{47 - 11y}{5} = 9 - 2y + \frac{2 - y}{5} \qquad \textbf{(162)}$$

$$x - 9 + 2y = \frac{2 - y}{5} \qquad \text{(Ax. 2)}$$

Since x and y are to be whole numbers, the integral quantity, $x - 9 + 2y$, must be a whole number; and if the *left* side of the equation is a whole number, the *right* side must be a whole number also. Consequently values of y must be selected which will make $\frac{2 - y}{5}$ an integer (positive, negative, or zero).

Such values are $y = 2$, making

$$\frac{2 - y}{5} = 0 \text{ (for } 0 \div 5 = 0)$$

$$y = 7, \text{ making } \frac{2 - y}{5} = -1$$

$$y = 12, \text{ making } \frac{2 - y}{5} = -2$$

Finding the corresponding values of x by substitution (**229**), we have $y = 2, x = 5$; $y = 7, x = -6$. But negative values are excluded. Furthermore, larger values of y would give still larger negative values of x. Hence $x = 5$, $y = 2$, are the only pair of integral roots.

2. Given $41x + 11y = 790$

$$y = \frac{790 - 41x}{11} = 71 - 3x + \frac{9 - 8x}{11}. \qquad \textbf{(162)}$$

As before $\frac{9-8x}{11}$ must be an integer, and the value of x must be found from it by trial. We may, however, derive a *simpler* expression from which to find the value of x. For, if $\frac{9-8x}{11}$ is an integer, any whole number of times it must also be an integer. Let us multiply this fraction by 7 (since 7 times 8 is one greater than a multiple of 11), and reduce the new improper fraction to a mixed quantity.

$$\frac{63-56x}{11}=5-5x+\frac{8-x}{11} \tag{162}$$

Now, if $5-5x+\frac{8-x}{11}$ is integral, since the part $5-5x$ is integral, the fraction $\frac{8-x}{11}$ is likewise integral. From this fraction, then, values of x may be found which must make it integral.

Putting

$$x=8, \left(\frac{8-x}{11}=0\right), \; y=\frac{790-41\times 8}{11}=42. \quad \textit{Ans.}$$

$$x=19, \left(\frac{8-x}{11}=-1\right), y=\frac{790-41\times 19}{11}=1. \quad \textit{Ans.}$$

No other values of x will give y positive.

248. Rule for the Solution of Indeterminate Equations.

1. When there are more than one of the given equations, reduce them by elimination to a single equation containing two or more unknown quantities.

2. If more than two unknowns remain in the single equation thus left, arbitrary values must be assigned to all but two of them.

3. Solve for that one of the unknowns most readily found in terms of the other, and reduce its value (when

divisible) to the form of a mixed quantity. The fraction thus obtained can frequently be transformed into a simpler fraction by multiplying it by some number (always less than its denominator) which will make the coefficient of the unknown one greater than some multiple of the denominator, and then reducing this new fraction to a mixed quantity. The *fractional part* of this mixed quantity may be used instead of the original fraction.

4. Supply successively such values to the unknown in the numerator of the *fraction* as will make it zero, or some multiple of the denominator, and each time find the corresponding value of the other. Such pairs of values will be answers so long as they are both positive.

REMARK. — The method here given is very elementary. Other letters used as auxiliaries in the solution are often employed, but their use is not essential, and they have not been introduced in the explanations given above.

249. **Exercise in the Solution of Indeterminate Equations.**

1. $2x + 3y = 25$. 2. $5x + 7y + 4 = 56$.

3. $x + 3y + 2z = 10$.

SUGGESTION. — Here it will be necessary to assign arbitrary values to one of the letters, say z. Zero values being excluded, z may equal 1, or 2, or 3. These values give three different indeterminate equations to be solved for x and y.

4. $2373 = 13x + 24y$.

5. $17x + 53y - 123 = 441 - 19x + 15y$.

6. $\begin{cases} (1)\ x + y = 17 \\ (2)\ 3x + 3y = 51. \end{cases}$ (See **246**, *a*)

7. $\begin{cases} (1)\ x + 2y + 3z = 50 \\ (2)\ 4x - 5y - 6z = -66. \end{cases}$

8. $\begin{cases} (1)\ 29y = 8x - 4 \\ (2)\ 452 = 17x - z. \end{cases}$

9. $\begin{cases} (1)\ 2x + 3y - 7z = 19 \\ (2)\ 5x + 8y + 11z = 24 \\ (3)\ 7x + 11y + 4z = 43. \end{cases}$ (See **246**, *a*)

250. Problems involving Diophantine Equations.

1. Separate 71 into parts of which the one is divisible by 5 and the other by 8 without remainders.

2. "Had I two times as many eggs as I now have" said one peasant girl to another, "and you seven times as many as you now have, and I were then to give you one egg, we should each have the same number." How many eggs had each?

3. In how many ways may 100 be divided into two parts one of which shall be a multiple of 7 and the other of 9?

4. What is the *least* number which when divided by 5 and 3 leaves remainders 3 and 2 respectively. How many such numbers are less than 100?

SUGGESTION. — Let z be the number and x and y the quotients.

5. A person bought 40 animals consisting of pigs, geese, and chickens for $40. The pigs cost $5 a piece, the geese $1, and the chickens 25 cts. each. Find the number he bought of each.

6. Divide 17 into three such parts that if the first be multiplied by 5, the second by 4, and the third by 7, the sum of these products is 80.

7. A number consisting of three digits, of which the middle one is 2, has its digits inverted by adding 198. What is the number?

8. A farmer buys oxen, sheep, and hens. The whole number bought is 100 and the total price $100. If the oxen cost $35, the sheep $3, and the hens 25 cts. each, how many of each did he buy?

9. A boy sees that he can buy oranges at 2 cts., 3 cts., 4 cts., 5 cts., or 6 cts. apiece and spend all his money; but if he buys at 7 cts. apiece he will have 5 cts. remaining. How much money has he if he has the least sum possible?

SUGGESTION. — The l. c. m. of 2, 3, 4, 5, and 6 is 60. Let $60x =$ his money.

10. Find four integral numbers such that the sum of the first three shall be 18; the sum of the first, second, and fourth 16; and the sum of the first, third and fourth 14.

SECTION II.

REDUNDANT EQUATIONS.

251. Redundant Equations occur when the number of equations exceeds the number of Unknown Quantities. These equations may be classed into two kinds:

1. *Compatible Redundant Equations* are such as are satisfied by values of the unknowns derived from the other equations of the set to which they belong.

a. Redundant equations were used in **237**. The new equations formed out of the old were satisfied by the same values of the unknowns, i.e., were compatible. When new equations were constructed an equal number of old ones were dropped.

We can also have *independent* redundant equations. Thus, (1) $7x + 2y = 53$, (2) $12x + 5y = 94$, (3) $14x - 11y = 76$, are all satisfied by $x = 7$, $y = 2$. All that can be done with such equations is to test for compatibility, or incompatibility.

2. **Incompatible Redundant Equations** are such as are not satisfied by the values of the unknowns obtained from the other equations of the set to which they belong. Thus, in (1) $2x + 3y = 23$ (2) $4x - 5y = -9$ (3) $11x + 17y = 6$, (1) and (2) give $x = 4$, $y = 5$; (2) and (3) give $x = -1$, $y = 1$; (1) and (3) give $x = 373$, $y = -241$.

SECTION III.

NEGATIVE AND INCONSISTENT SOLUTIONS IN PROBLEMS INVOLVING SIMPLE EQUATIONS.

252. Problems involving Arithmetical Inconsistencies.

1. If from $\frac{5}{8}$ of a certain number 1 be subtracted, the remainder equals the sum of twice the number divided by 7,

five times the number divided by 14 and 3. What is the number?

$$\frac{5}{8}x - 1 = \frac{2x}{7} + \frac{5x}{14} + 3 \qquad (219, 1)$$

$$x = -224 \qquad (219, 2)$$

Here the negative answer points to some absurdity in the problem viewed as an arithmetical one. On examination we find that $\frac{2x}{7} + \frac{5x}{14}$ is greater than $\frac{5x}{8}$, and should be diminished rather than increased to produce $\frac{5x}{8}$.

2. A father's age is 40 years; his son's age is 13; in how many years will the age of the father be four times that of the son?

$$40 + x = 4(13 + x) \qquad (219, 1)$$

$$x = -4 \qquad (219, 2)$$

In this question the negative answer shows that the tacit assumption that the epoch named would be in the future was wrong. The question should have read "how long ago," and the equation have been

$$40 - x = 4(13 - x), \text{ whence } x = +4.$$

3. A and B went into business agreeing to divide the profits in a certain way. Twice B's money diminished by \$4250 indicated A's financial standing in the company when they began. A made \$4000 and B made \$750 when it was found that A had \$500 more than B. How much had each when they began?

(1) $x + 4250 = 2y$

(2) $x + 4000 = y + 750 + 500$ (219, 1)

$x = -1250 =$ A's, and $y = 1500 =$ B's (219, 2)

The result indicates that A was *in debt* when he first began business.

4. A man worked 7 days and had his son with him 3 days, and received as pay $2.20. He afterwards worked 5 days and had his son with him one day, and received for wages $1.80. What was the father's daily wages, and what was the effect of the son's presence?

$$(1)\quad 7x + 3y = 220$$
$$(2)\quad 5x + y = 180$$

$x = 40$ cts. father's wages, $y = 20$ cts. son's expense.
i.e., the father paid out 20 cts. a day for the son.

5. A and B travel in the same direction at the rate of 6 mi. and 4 mi. respectively per hour. A arrives at a certain place P at a certain time, and at the end of 8 hours from that time B arrives at a certain place Q. Find when A and B meet.

1. Suppose the distance P Q 50 miles.

Let x = the number of hours from the time A is at P, till they meet at R. Then since A travels at the rate of 6 miles per hour, the distance P R is $6x$ miles. Also B goes over the distance Q R in $x - 8$ hours, so that Q R equals $4(x - 8)$ miles. Now P R = P Q + Q R. Hence

$$6x = 50 + 4(x - 8)$$
$$x = 9 \qquad (219, 2)$$

2. But if the distance P Q is 20 miles the equation would be

$$6x = 20 + 4(x - 8)$$

Whence $x = -6$.

The negative value of x indicates that they met to the *left* of P instead of at the right. It is plain that in 8 hours B would walk 32 miles, a number greater than 20. Consequently they met *before* A reached P.

3. Next suppose A travels 4 miles, and B 6 miles per hour, and suppose P Q = 50 again. Then

$$4x = 50 + 6(x - 8)$$
$$\therefore x = -1$$

In 8 hours B travels 48 miles. He would therefore be just 2 miles beyond P at the time A arrived there. They had met one hour *before* A arrived at P.

4. Lastly suppose A and B travel in as 3, but that P Q is 20 miles.

$$4x = 20 + 6(x - 8)$$
$$\therefore x = 14$$

Cases (1 and (4 are in accord with the idea of meeting in the future as implied in the statement, while (2 and (3 *contradict* this supposition.

6. A grocer has two barrels of molasses, one of which contains twice as much as the other. From the larger cask he draws 16 gals. and from the smaller 10 gals. Then after a fourth of what remained in the larger cask had been withdrawn, the two casks were found to contain an equal number of gallons. How much did each cask hold?

Let $2x$ = the number of gals. in the larger
and x = " " " " " " smaller.

Then $\frac{3}{4}(2x - 16) = x - 10$
$$x = 4$$
$$2x = 8$$

Here the answers are positive, but that does not save us from the absurdity of drawing 16 gals. from an 8 gal. cask, and 10 gals. from a 4 gal. cask!

Verifying the equation, $\frac{3}{4} \times -8 = -6$. Thus we learn that the existence of the minus quantities *in the process of solution* vitiates the result for the arithmetical problem.

7. Ex. 33, Art. **224**, illustrates the point brought out in Ex. 6 still further. By the statement of that problem we

are led to suppose that the amount invested at 6 % is only a part of the $12000. The solution shows that the man borrowed $13200, on part of which he paid 4 % and the other part 5 %. He got 6 % on the whole $25200. Six per cent on $12000 amounts to only $720.

253. The Problems of the preceding Article illustrate,

1. That a *negative* result indicates either some arithmetical incongruity in the statement of the problem, or at least that the number obtained as the answer should be taken in a sense contrary to that implied in the statement of the problem.

2. That positive results do not necessarily prove that problems are satisfactorily solved. Hence all problems which do not admit of algebraic interpretation (i.e., those which do not admit of both positive and negative values for their quantities), when solved by algebraic methods should have their answers tested for arithmetical consistency.

SECTION IV.

Literal Problems. — Generalization.

254. The Generalization of a Problem is attained by replacing the known numbers by letters and then deriving its solution.

a. Problems are commonly first studied by using particular numbers; afterwards, on account of the frequency with which they occur, they are generalized. After such a solution has been worked out, all that is necessary afterwards is to substitute the numbers of any particular case of the problem in the answer of the generalized solution, and reduce as in 81.

255. Exercise in the Solution of Literal Problems and Generalization.

a. In some of the first problems, and also those difficult to state, reference to a particular form of the given problem, or one similar to it, will be made. The student will find these references very helpful if aid is needed in the statement.

b. The following set of problems includes such as contain sometimes one, at other times *more* unknown quantities. The student is left to decide to which class any given problem ought to be assigned.

1. (1. Particular form. — Divide \$183 between two men so that $\frac{4}{7}$ of what the first receives shall equal $\frac{3}{10}$ of what the second receives.

$$\tfrac{4}{7}x = \tfrac{3}{10}(183 - x) \qquad (219, 1)$$
$$x = 63;\ 183 - x = 120 \qquad (219, 2)$$

(2. Generalized form. — Divide \$$a$ between two men so that $\frac{m}{n}$ of what the first receives shall be equal to $\frac{p}{q}$ of what the second receives.

$$\frac{m}{n}x = \frac{p}{q}(a - x) \qquad (219, 1)$$

$$x = \frac{anp}{np + mq} \qquad (219, 2)$$

(3. Special case. Solution by substitution. — Divide 168 into two parts so that $\frac{5}{9}$ of one shall equal $\frac{5}{12}$ of the other.

Here $a = 168$, $m = 5$, $n = 9$, $p = 5$, $q = 12$. Hence

$$x = \frac{168 \times 9 \times 5}{9 \times 5 + 5 \times 12} = 72. \quad \textit{Ans.} \quad 168 - 72 = 96. \quad \textit{Ans.}$$

VERIFICATION. $\frac{5}{9}$ of $72 = \frac{5}{12}$ of 96.

2. A boy bought an equal number of apples, lemons, and oranges for c cts.; for the apples he gave l cts., for the lemons m cts., and for the oranges n cts. apiece. How many of each did he purchase? See Ex. 4, **224**.

3. A father is now a years of age, and his daughter b years of age. How many years ago was the father's age n times that of the daughter? See ex. 9, **224**.

Obtain the answer in the following cases by substitution.

$a = 51$, $b = 24$, $n = 2\frac{1}{2}$; $a = 75$, $b = 19$, $n = 5$; $a = 37$, $b = 12$, $n = 3$.

4. The sum of two numbers is s, and their difference d; what are the numbers?

Solution, Let $x + y = s$ and $x - y = d$

$$\begin{aligned} x + y &= s \\ x - y &= d \\ \hline 2x &= s + d \\ x &= \frac{s+d}{2} \end{aligned} \qquad \begin{aligned} x + y &= s \\ x - y &= d \\ \hline 2y &= s - d \quad (\text{Ax. } a) \\ y &= \frac{s-d}{2} \quad (\text{Ax. } 4) \end{aligned}$$

These formulae may be stated in the form of *theorems.*

To find the greater of the two numbers take half the sum of the sum and difference.

To find the less of the two numbers take half the difference of the sum and difference.

Given $s = 64$, $d = 26$; $s = 195$, $d = 14$; $s = 300$, $d = 312$.

5. Divide a into two parts such that the difference between one part and b, shall equal n times the difference between the other part and c. Put $a = 4$, $b = 3$, $n = 2$, $c = 1$.

6. Two men a miles apart travel towards each other, one m miles, the other n miles an hour. In how many hours will they meet? Put $a = 49$, $m = 4\frac{1}{2}$, $n = 3\frac{2}{3}$.

7. A farmer sells a horses and b cows for c dollars. And at the same prices a' horses, and b' cows for c' dollars. What is the price of each? Given in a particular problem that $a = 10$, $b = 9$, $c = \$1600$, $a' = 12$, $b' = 7$, $c' = \$1720$. Find the prices.

8. A person has a hours at his disposal; how far may he ride in a coach which travels b miles per hour, and yet have time to return on foot walking c miles per hour? See Ex. 48, Art. **224**.

9. A can do a piece of work in p days, B in q days, and C in r days. In how many days will they finish it all working together? Given $p = 10$, $q = 15$, $r = 30$.

10. The sum of two numbers is equal to a and their sum is to their difference as m is to n. Required the numbers. Given $a = 17$, $m = 6$, $n = 5$.

11. Divide the number n into two such parts that the quotient of the greater divided by the less shall be q with a remainder r. Put $n = 485$, $q = 79$, $r = 5$.

12. Divide the number d into three such parts, that the second shall exceed the first by b, and the third exceed the second by c. Put $d = 588$, $b = 64$, $c = 91$.

13. A and B together can perform a piece of work in d days, A and C together in e days, and B and C together in f days. In what time can each person alone perform the work?

14. A merchant bought p gallons of two kinds of oil giving for one m cents a gallon and for the other n cents a gallon. By mixing them and selling the mixture at r cents a gallon he gained \$$a$. How many gallons of each did he buy?

15. In an old Chinese arithmetic called Kiu Tschang, which was completed about 2600 B.C. and elucidated and enlarged about 1250 A.D. by Tsin Kiu Tshaou, there is found the following problem: In the middle point of a square pond m feet each way, there grows a reed which rises p feet above the water. Now if the reed be pulled to the middle point of one of the edges of the pond it just reaches to the top of the water. What is the depth of the water? See Ex. 38, Art. **224**.

1.) Put $m = 10$, $p = 1$.

2.) One side of a right angled triangle is 25 inches, and the other increased by 12 inches equals the hypothenuse.

What is the hypothenuse equal to? Solve by substitution. What does $m =$, what p?

3.) One side of a rectangular tract of land is one mile, and the other side increased by .414 mile is equal to the diagonal. What is the length of the diagonal, and what is the length of the other side? What does p equal here?

16. If the selling price of a certain product is p a merchant gains $n\%$. How many per cent is gained or lost if the selling price is p'?

17. Two bodies move towards each other from points l meters apart. The one p meters a minute, the other q meters. In how many minutes will they be r meters from each other. Put $l = 500$, $p = 30$, $q = 25$, $r = 60$.

18. The fore wheel of a wagon is a feet and the hind wheel b feet in circumference. Through what distance must the wagon pass in order that the fore wheel shall have made n more revolutions than the hind wheel?

19. A person distributed a cents among n beggars, giving b cents to some and c cents to the others. How many were there of each?

1.) A father divides \$8500 among 7 children, giving to each son \$1750, and to each daughter \$500. How many of his children were sons and how many daughters?

20. A, B, and C hold a pasture in common for which they pay q dollars a year. A puts in a cows for m months, B, b cows for n months, C, c cows for p months. Required each one's share of the rent.

1.) $q = \$181.20$, $a = 6$, $b = 5$, $c = 8$, $m = 30$, $n = 40$, $p = 28$.

20'. A boy who had three studies, Latin, Greek, and mathematics, received at the end of a certain term 95 in mathematics, 90 in Latin, and 85 in Greek. The mathematics recited 5 times a week, the Latin 4 times, and the Greek 3 times. What was his average for the term in all his studies?

To make a convenient formula for calculating an average grade x, let a be an *approximate* value of the average grade (as 90, or 80, or 60), and let l, m, n be the differences (taken with proper signs) between the given grades and the assumed average grade. Also let b, c, and d be the number of hours per week the respective studies recite. Then

$$x = \frac{(a+l)\,b + (a+m)\,c + (a+n)\,d}{b+c+d} = a + \frac{lb+mc+nd}{b+c+d}$$

which expression is more convenient for calculating the value of x.

21. To derive the rule for multiplication of fractions.

Let $\frac{a}{b}$ and $\frac{c}{d}$ be the factors and x their product.

Then $x = \frac{a}{b} \times \frac{c}{d}$

$$bdx = \frac{a}{b} \cdot b \times \frac{c}{d} \cdot d \qquad \text{(Ax. 3 and } \mathbf{38}, 2)$$

$$= a \times c \qquad \text{(by definition of division, } \mathbf{43})$$

$$\therefore x = \frac{ac}{bd} \qquad \text{(Ax. 4)}$$

Or, $\frac{a}{b} \times \frac{c}{d} = \frac{ac}{bd}$

Hence, to multiply fractions multiply their numerators for a new numerator, and their denominators for a new denominator. The result is then to be reduced to its lowest terms, by striking out equal factors in the numerator and denominator. To save unnecessary writing this cancellation is done before and not after the terms are multiplied.

22. To derive the rule for division of fractions. Let $\frac{a}{b}$ and $\frac{c}{d}$ be the fractions and x their quotient. Then

$$x = \frac{a}{b} \div \frac{c}{d}$$

$$\frac{c}{d} \times \frac{d}{c} x = \left(\frac{a}{b} \div \frac{c}{d}\right) \times \frac{c}{d} \times \frac{d}{c} \qquad \text{(Ax. 3)}$$

But by the definition of division $\left(\frac{a}{b} \div \frac{c}{d}\right) \times \frac{c}{d} = \frac{a}{b}$, and by the preceding $\frac{c}{d} \times \frac{d}{c} = 1$. Then

$$x = \frac{a}{b} \times \frac{d}{c}$$

Or, $\frac{a}{b} \div \frac{c}{d} = \frac{a}{b} \times \frac{d}{c} = \frac{ad}{bc}$

Hence, to divide one fraction by another, invert the divisor and multiply.

REMARK.—It was thought that the student would be able to understand these proofs better after studying equations. They might have been given just as they stand in articles 178 and 180. They include as particular cases the principles of division often given:

Multiplying the $\begin{cases}\text{dividend}\\\text{numerator}\end{cases}$ or dividing the $\begin{cases}\text{divisor}\\\text{denominator}\end{cases}$

multiplies the $\begin{cases}\text{quotient}\\\text{value of the fraction}\end{cases}$.

Dividing the $\begin{cases}\text{dividend}\\\text{numerator}\end{cases}$ or multiplying the $\begin{cases}\text{divisor}\\\text{denominator}\end{cases}$

divides the fraction.

THIRD GENERAL SUBJECT.

(THE NOTATION CONCLUDED.)

POWERS, ROOTS, AND RADICALS.

CHAPTER XVII.

OF POWERS.

256. Involution as a term in algebra signifies raising quantities to powers.

a. Involution is merely one case in multiplication (**56, 57**) where the numbers multiplied are equal. And so some exercises in involution have been given under multiplication (**117**). For present purposes it will be convenient to treat this subject under three heads: monomial powers, binomial powers, and polynomial powers of three or more terms.

SECTION I. — Monomial Powers.

257. Exercise in Involving Monomials to Powers. — The student should familiarize himself anew with **117** and **128**.

a. To develop a fraction both terms must be raised to the proper power. (**178.**)

1. Square a^3c, $11\,b^2c^3$, $-4\,a^4b^5x^2$, $-\frac{2}{3}\,a^2x^4$, $-\frac{4}{3\,x^2y}$, $\frac{7\,a^m}{2}$, a^mb^m, $\left(\frac{7\,a^2}{3\,b}\right)^m$, a^xb^{-4}, $2^{-3}\cdot 2^{-5}$, $(3\frac{3}{4})^{-2}$, $(-\,a^2x)^6$, $(-\,a)^2\,(-\,b)^3$ $(-\,c)^4$, $\frac{2^3}{2^{-5}}$, $\frac{2}{5}\,a^3x^{m+2}$, y^{p-1}, $y^{n-1}\,y^{m+1}$

2. Cube, $-6a^6$, a^x, b^4y, $a^{11} \times a^5$, $-2a^7c^2$, $\left(\frac{3ab}{5cd}\right)^n$, $\frac{m^{4a+b} \cdot m^{6a-8b}}{m^{2a-6b} \cdot n^{4a-7b}} \cdot \frac{c^{13b-7a}}{c^{14b-18a}}$, $\left(\frac{x^2y^8z^4}{2}\right)^4$

3. Develop $(x^2y^3)^4$, $\left(\frac{3a^2y}{2a^8b^2}\right)^5$, $(-3xy^8)^6$, $(a^2bc^8)^n$, $\left(\frac{1}{3}\right)^{-2}$, $\left(\frac{1}{3^{-8}}\right)^2$, $\left[\left(\frac{3}{4}\right)^{-2}\right]^8$, $[-a^2b^5c^3]^m$.

NOTE. — In the last expression it is not known whether m is even or odd, and, consequently, whether the result is + or —. In such cases $(-1)^m$ may be written as the *sign coefficient*.

4. Raise $-(abc)^2$ to the fifth power; raise $-a^2b^5c^3$ to the mth power, when m is even.

5. Find the cube of $(-ab^4c^2)(-a^2b^8c^m)$; the square of $4a^2b^8c(-x^2y^8z^4)$; the fifth power of $-\frac{2}{3}x^n$; the cube of $-\frac{3}{5} \cdot \frac{a^n}{b^{-2}}$; the seventh power of abc^{-1}; the sixth power of $\left(\frac{m}{p}\right)^2 \cdot \left(\frac{n}{q}\right)^{-8}$.

6. Simplify $2a(-3b^2mn^3)^8$; $(4ab^n)^r$; $5(m^n)^n$; $a(5a)^3$; $7(7pq^2r^3)^3$; $3(-2ab)^{-2}$; $(-4)^2 \times 6(a)^2$; $a(a^ma^n)^2$; $(ab^{-1}c^{-2})^{-6}$; $(x^ry^{-r})^{-r}$; $\frac{7(a+b)^n}{(3(a+b))^{-2}}$; $(-b)^{2n+1} \times (-1)^{n-1}$; $2(2n^{-1}x^{-2})^6$; $a^n(3a^2x^3)^n$; $3x^6(6x^2)^{-2}$; $a^2b^2(a^{-3}b^6)^{-1}$; $\frac{27}{8}x^{-2}y^{-6}a^6\left(-\frac{2x^3y}{3abc}\right)^6$

SECTION II. — Binomial Powers.

258. Investigation of Binomial Powers. — A study of the development of binomials leads to Newton's theorem.

The development of the square of a binomial was studied in Theorems I. and II. in multiplication; then the other simple cases of the cube and the fourth power of a binomial in **116**, 4 and 6. But each of these led to a special theorem. We are now to seek for a general law applicable alike to all powers.

Let A and B represent any two quantities. Then A + B or A — B is a binomial. We proceed to form the powers by multiplication.

$$
\begin{array}{l}
A + B \\
A + B \\
\hline
A^2 + AB \\
\quad + AB + B^2 \\
\hline
A^2 + 2\,AB + B^2 \\
A + \;B \\
\hline
A^3 + 2\,A^2B + AB^2 \\
\quad + A^2B + 2\,AB^2 + B^3 \\
\hline
A^3 + 3\,A^2B + 3\,AB^2 + B^3 \\
A + \;B \\
\hline
A^4 + 3\,A^3B + 3\,A^2B^2 + AB^3 \\
\quad + \;A^3B + 3\,A^2B^2 + 3\,AB^3 + B^4 \\
\hline
A^4 + 4\,A^3B + 6\,A^2B^2 + 4\,AB^3 + B^4 \\
A + \;B \\
\hline
A^5 + 4\,A^4B + 6\,A^3B^2 + 4\,A^2B^3 + AB^4 \\
\quad + \;A^4B + 4\,A^3B^2 + 6\,A^2B^3 + 4\,AB^4 + B^5 \\
\hline
A^5 + 5\,A^4B + 10\,A^3B^2 + 10\,A^2B^3 + 5\,AB^4 + B^5 = (A + B)^5
\end{array}
$$

$$
\begin{array}{l}
A - B \\
A - B \\
\hline
A^2 - AB \\
\quad - AB + B^2 \\
\hline
A^2 - 2\,AB + B^2 \\
A - \;B \\
\hline
A^3 - 2\,A^2B + AB^2 \\
\quad - \;A^2B + 2\,AB^2 - B^3 \\
\hline
A^3 - 3\,A^2B + 3\,AB^2 - B^3 = (A - B)^3 \\
A - \;B \\
\hline
A^4 - 3\,A^3B + 3\,A^2B^2 - AB^3 \\
\quad - \;A^3B + 3\,A^2B^2 - 3\,AB^3 + B^4 \\
\hline
A^4 - 4\,A^3B + 6\,A^2B^2 - 4\,AB^3 + B^4
\end{array}
$$

A careful scrutiny of these multiplications leads us to the following conclusions, which if true in all the above cases we may guess to be true of all powers. The *demonstration* of Newton's theorem which is rather too difficult to be inserted here does justify the conclusions about to be given.

In the development of a binomial, as $(a+b)^n$, we learn

1. The number of terms is always one greater than the exponent of the power to which the binomial is raised.

Thus there are five terms in the fourth power.

2. The exponent of the leading letter in the first term of any power is the same as the exponent of the power of the binomial. In the second term it is one less, in the third term one less than in the second term, and so on, the leading letter not appearing in the last term. On the other hand the other letter begins in the second term with the exponent 1 which increases by unity each time, until in the last term it is the same as that of the power of the binomial. Thus the product is symmetrical with respect to the letters.

3. When the binomial is a sum the signs of all the terms are positive. But when the binomial is a residual, every term which contains an odd power of the second letter is minus.

4. The coefficient of the first term is 1 (understood), as also that of the last term. The coefficient of the second term is the same as the exponent of the power of the binomial, as also that of the next to the last term. Furthermore, the coefficients increase to the middle term, or terms, and then diminish to the last, those equally distant from the extreme terms being equal to each other.

5. The third coefficient can always be derived from the second by multiplying the second by 1 less than itself and

dividing the product by 2. The fourth coefficient may be derived from the third by multiplying it by 1 less than the previous multiplier, and dividing by 1 greater than the previous divisor (i.e. by 3), and so on, the multiplier becoming 1 less and the divisor 1 greater than the last for every new term. Thus the coefficient 6 in the third term of $(A + B)^4$ is derived from the 4 by multiplying 4 by $4 - 1 = 3$, and dividing by 2; the coefficient 4 following the 6, by multiplying 6 by 2 (1 less than 3, the previous multiplier) and dividing by 3 (1 greater than 2, the previous divisor). The last coefficient, 1, from the preceding coefficient, 4, by multiplying it by 1, and dividing the product by 4.

Unlike the preceding laws, the fifth is probably too complex for the student to have perceived it unaided.

a. There is a memoriter rule which can be more easily followed when A and B stand for two first powers: — Multiply each coefficient by the exponent of the leading letter in that term and divide the product by the number of that term from the left. The quotient will be the coefficient of the next term.

Thus in the fifth term of $(A + B)^5$, $5\ AB^4$, the 5 is found from $10\ A^2B^3$ by multiplying 10 by 2, the exponent of A, and dividing the product by 4, the number of the term $10\ A^2B^3$ from the beginning.

It may be observed that the divisor is always 1 greater than the exponent of the second letter.

b. The student will find it advantageous to *remember* the binomial coefficients up to the fifth or sixth powers.

259. Examples of the Use of the Binomial Theorem.

1. Develop $(A + B)^5$ by the theorem.

In this expansion there will be 6 terms, and hence when *three* coefficients are found the others can be written down directly, being those already obtained in *reverse order*. The reasons in parentheses refer to sections in the previous article.

(4, 2) (4, 2) (5, or a, 2) (4, 2)

$$(A + B)^5 = A^5 + 5\,A^4B + \frac{5 \times 4}{2}\,A^3B^2 + 10\,A^2B^3$$

(4, 2) (4, 2)

$$+ 5\,AB^4 + B^5$$

2. Develop $(x - y)^8$.

Here there will be 9 terms in all, and it will be necessary to determine 5 coefficients.

(3, 4) (5, or a) (3, 5)

$$(x - y)^8 = x^8 - 8\,x^7y + \frac{8 \times 7}{2}\,(= 28)\,x^6y^2 - \frac{28 \times 6}{3}$$

(5, or a) (4)

$$(= 56)\,x^5y^3 + \frac{56 \times 5}{4}\,(= 70)\,x^4y^4 - 56\,x^3y^5 + 28\,x^2y^6 -$$

$$8\,xy^7 + y^8$$

3. Develop $(2\,x^2 - 3\,ay)^4$.

Here $A \equiv 2\,x^2$, and $B \equiv 3\,ay$. To preserve their identity they are written *in parentheses*, treating $(2\,x^2)$ as A, and $(3\,ay)$ as B.

$$(2\,x^2 - 3\,ay)^4 = (2\,x^2)^4 - 4\,(2\,x^2)^3\,(3\,ay) + \frac{4 \times 3}{2}\,(2\,x^2)^2$$

$$(3\,ay)^2 - \frac{6 \times 2}{3}\,(2\,x^2)\,(3\,ay)^3 + (3\,ay)^4$$

The single terms must now be simplified.

$(2\,x^2)^4 = 16\,x^8$; $4\,(2\,x^2)^3\,(3\,ay) = 96\,ax^6y$; $\frac{4 \times 3}{2}\,(2\,x^2)^2$ $(3\,ay)^2 = 216\,a^2x^4y^2$

$\frac{6 \times 2}{3}\,(2\,x^2)\,(3\,ay)^3 = 216\,a^3x^2y^3$; $(3\,ay)^4 = 81\,a^4y^4$. Hence

$$(2\,x^2 - 3\,ay)^4 = 16\,x^8 - 96\,ax^6y + 216\,a^2x^4y^2 - 216\,a^3x^2y^3 + 81\,a^4y^4.$$

NOTE. — The student must remember to keep distinct the *binomial* coefficients, and any numbers that may come into the result from *numerical* coefficients in the quantities developed. We could never, for instance, obtain the third coefficient 216 in the expansion just given by multiplying 96 (!) by 9.5 and dividing by 2. For the *binomial* coefficient of that term is not 96 but 4.

260. Exercise in Raising Binomials to Powers.

1. $(pq - r)^2$.
2. $(2m - p)^3$.
3. $(x^2 + 4y^2)^3$.
4. $(5a - bc)^3$.
5. $(2a^2 + ax)^4$.
6. $(1 - 2b)^4$.
7. $(a + 1)^9$.
8. $\left(a - \frac{2b}{3}\right)^4$.
9. $(5 - 4x)^4$.
10. $(2r - 6m)^3$.
11. $(\frac{1}{2}a - 3b)^4$.
12. $(5x^5 - 4x^4)^3$.
13. $(a - 1)^7$.
14. $(\frac{3}{5}x - \frac{5}{3}y)^4$.
15. $(x^2 - 2xy^2)^6$.
16. $\left(7\frac{y^2}{x^2} - 8\frac{x^2}{y^2}\right)^4$.
17. $(x^p - 2)^5$.
18. $(x^m - y^n)^6$.
19. Find the first four terms of $(2am + 1)^8$.
20. Verify $99^3 = (100 - 1)^3$.
21. Cube by the theorem $999 = (1000 - 1)$.
22. Raise 9999 to the fourth power.
23. Raise $12\frac{1}{2} = 12 + \frac{1}{2}$ to the fourth power.
24. Raise $1892 = (1900 - 8)$ to the third power.

SECTION III.—Polynomial Powers.

261. Development of Polynomial Powers.—Polynomials are most readily developed by regarding them as Binomials, and using Newton's theorem.

a. Polynomial squares are easily written out by the rule given in 116, 3.

1. Required to cube $a + b + c$.

To make $a + b + c$ a binomial it is written $(a + b) + c$. Next the binomial thus formed is developed, and afterwards the different powers of $a + b$ are expanded. Last of all the resulting indicated operations are performed.

By section II. we have

$$[(a+b)+c]^3 = (a+b)^3 + 3(a+b)^2c + 3(a+b)c^2 + c^3$$
$$= a^3 + 3a^2b + 3ab^2 + b^3 + 3(a^2 + 2ab + b^2)c + 3(a+b)c^2 + c^3$$
$$= a^3 + 3a^2b + 3ab^2 + b^3 + 3a^2c + 6abc + 3b^2c + 3ac^2 + 3bc^2 + c^3$$

Ans.

$$= a^3 + b^3 + c^3 + 3(a^2b + a^2c + ab^2 + ac^2 + b^2c + bc^2) + 6abc.$$

State this result in the form of a theorem.

2. Required the fourth power of $3a + 2b - c + \frac{d}{2}$.

$$\left[(3a+2b) - \left(c - \frac{d}{2}\right)\right]^4 = (3a+2b)^4 - 4(3a+2b)^3\left(c - \frac{d}{2}\right)$$
$$+ 6(3a+2b)^2\left(c - \frac{d}{2}\right)^2 - 4(3a+2b)\left(c - \frac{d}{2}\right)^3$$
$$+ \left(c - \frac{d}{2}\right)^4$$

$$= (3a)^4 + 4(3a)^3(2b) + 6(3a)^2(2b)^2 + 4(3a)(2b)^3 + (2b)^4$$
$$- 4\left[(3a)^3 + 3(3a)^2(2b) + 3(3a)(2b)^2 + (2b)^3\right]\left[c - \frac{d}{2}\right]$$
$$+ 6\left[9a^2 + 12ab + 4b^2\right]\left[c^2 - cd + \frac{d^2}{4}\right]$$
$$- 4[3a + 2b]\left[c^3 - 3c^2\frac{d}{2} + 3c\frac{d^2}{4} - \frac{d^3}{8}\right]$$
$$+ c^4 - 4c^3\frac{d}{2} + 6c^2\frac{d^2}{4} - 4c\frac{d^3}{8} + \frac{d^4}{16}.$$

Upon developing the monomial expressions and performing the multiplications.

$$\left(3a + 2b - c + \frac{a}{2}\right)^4 = 81a^4 + 216a^3b + 216a^2b^2 + 96ab^3 + 16b^4$$
$$- 108a^3c - 216a^2bc$$
$$- 144ab^2c - 32b^3c + 54a^3d + 108a^2bd + 72ab^2d$$
$$+ 16b^3d + 54a^2c^2 + 72abc^2 + 24b^2c^2 - 54a^2cd$$
$$- 72abcd - 24b^2cd + \frac{27}{2}a^2d^2 + 18abd^2 + 6b^2d^2 - 12ac^3$$
$$+ 18ac^2d - 9acd^2 + \frac{3}{2}ad^3 - 8bc^3 + 12bc^2d - 6bcd^2$$
$$+ bd^3 + c^4 - 2c^3d + \frac{3}{2}c^2d^2 - \frac{1}{2}cd^3 + \frac{d^4}{16}.$$

3. Square $xy + yz + zx$.

4. Develop $(1 + x + x^2 + x^3)^2$.

5. Cube $a + b - c$.

6. Develop $(1 - a - a^2)^3$.

7. Develop $(1 - 2x + x^2)^3$.

8. $(2a^m - b^n + c^p)^3 = ?$

9. Cube $ax + by + 1$.

10. Develop $(a + 2b - c)^4$.

11. Develop $(1 + x + x^2)^4$.

12. Develop $(ax + by + 1)^4$.

13. $(x + y + 1)^5 = ?$

14. Cube $1799 = 1000 + 800 - 1$, and verify the answer.

CHAPTER XVIII.

OF EXACT ROOTS.

262. Evolution as a term in algebra signifies extracting the roots of quantities.

a. This chapter will be separated into the following sections: roots of monomials; the square root of polynomials; the cube root of polynomials; other roots of polynomials.

SECTION I.—Monomial Roots.

263. Roots of Monomials.—As in simple multiplication and division, there are three things to be considered: the coefficient, the sign, and the literal part. Consult articles **45, 46**, and **129**. See also **58, 59, 60**.

264. Signs of Roots. (45.) The reason for the existence of two square roots so different in character may at first puzzle the beginner. It is a consequence, however, of the rule of multiplication that minus by minus as well as plus by plus gives plus. Properly speaking, the sign $\pm$ should be read "plus *and* minus," instead of "plus *or* minus." If, now, it is known that a power has been produced by multiplying two positive numbers, then its root is positive; or, if by two negative factors, it is negative. In the solution of arithmetical problems, the nature of the problem may admit of only positive answers; then only the positive root is taken for the answer.

265. Imaginary Quantities.—It can be shown that a real meaning may be assigned to what have been called "imaginary," or "impossible" quantities (**45**). But it is only by going out of the realm of algebra proper that it can be done. Whenever an imaginary value is found for the unknown in a problem, it shows that the problem is algebraically impossible, just as a negative result shows its problem to be arithmetically impossible.

266. Exercise in Extracting the Real Roots of Monomials.

1. $\sqrt{a^6b^4}$, $\sqrt[4]{x^8}$, $\sqrt[5]{c^{20}}$, $\sqrt{-4a^6b^2}$, $\sqrt[3]{-64}$, $\sqrt[4]{81x^{12}}$.

2. $\sqrt[3]{\frac{8}{27}}$, $\sqrt{\frac{36}{a^{36}}}$, $\sqrt{\frac{81x^{10}}{25c^4}}$, $\sqrt[3]{-x^9}$, $\sqrt{-2\frac{47}{121}}$, $\sqrt{\frac{81a^{18}}{36b^{12}}}$.

3. Find the cube root of $\frac{125a^3b^6}{216x^6y^9}$; of $\frac{-27\times 64}{8}$.

4. $\sqrt[3]{-343a^{12}x^9}$, $\sqrt[3]{-\frac{27x^{27}}{64y^{63}}}$, $\sqrt{-a^4b^6c^2}$, $\sqrt[4]{-16a^8x^{12}}$.

5. $\sqrt[9]{-\frac{m^{18}}{x^{27}y^9}}$, $\sqrt[5]{-32a^{15}}$, $\sqrt[7]{\frac{128}{a^{63}b^{56}}}$, $\sqrt[3]{-1728c^6d^{12}x^3}$.

6. $\sqrt[6]{729x^6y^{12}}$, $\sqrt{a^{2n}b^2}$, $\sqrt[n]{a^nb^{2n}}$, $\sqrt[m]{a^{2m}b^{5m}c^{3m}}$, $\sqrt[5]{-a^{-5}b^5c^{5n}}$

7. $\sqrt[m]{a^mb^{-2m}}$, $\sqrt[p]{\frac{a^p}{b^{2p}}}$.

8. Express the nth root of x, of 7.

9. Extract the roots and simplify

$$\sqrt{25a^2b^4c^2}+\sqrt[3]{-8a^3b^6c^3}-\sqrt[4]{81a^4b^8c^4}-\sqrt[5]{-32a^5b^{10}c^5}.$$

10. $\sqrt[3]{-27x^3y^6}\times\sqrt[5]{243y^5z^5}\times\sqrt{16x^4z^2}=?$

SECTION II.—Square Root of Polynomials.

267. Extraction of the Square Root of Polynomials.—How effected.

a. If the square root of a polynomial is not a binomial it may readily be put in the form of one (as in **261**). Now, just as the binomial theorem would serve to develop the powers of any polynomial, so here the investigation of binomial roots will include the general case.

1. Required to extract the Square root of $A^2 \pm 2\,AB + B^2$.

Referring to **258** we see that when a power is arranged the root of its first term is the first term in the root. Hence, in extracting roots in order to make a beginning, the quantity is arranged with reference to the exponents of its leading letter, and the root of its first term is then extracted.

$$\begin{array}{r|l} & A^2 + 2\,AB + B^2 \quad (A + B \\ & A^2 \\ \hline 2\,A + B & 2\,AB + B^2 \\ & 2\,AB + B^2 \\ \hline \end{array}$$

Explanation. — The first term of the root A having been found, as just explained, it remains to find the other term of the root; and since A^2 is the square of A alone, A is squared and subtracted from the polynomial, leaving $2\,AB + B^2 = (2\,A + B)B$.

Of course, in this example, we know perfectly well that the square root of $A^2 \pm 2\,AB + B^2$ is $A \pm B$. (See the theorems in multiplication.) Hence B is the second term of the root sought. Manifestly, now, if $2\,AB$ is divided by $2\,A$ (double the root term already found), B is the quotient. Similarly $-2\,AB \div 2\,A = -B$, which is the second term of the root in this case.

Lastly we observe that if B be now annexed to $2\,A$ and the sum multiplied by B, the product, $2\,AB + B^2$, subtracted from the previous remainder leaves zero, and the process is complete.

To show how this method applies when there are more than two terms in the root, let us take the square of $a + b + c = a^2 + 2ab + b^2 + 2\,ac + 2\,bc + c^2$ which can be written $(a + b)^2 + 2\,(a + b)\,c + c^2$. Thus, when the first two terms, $a + b$, are given, the third can be found by dividing the first term of the remainder, $2\,(a + b)\,c$, by twice the root already found. Now the root terms, $a + b$, must be found from the first terms of the polynomial in a preliminary operation similar to that given above for $A + B$, and its square subtracted leaves the remainder $2\,(a + b)\,c + c^2$. If there are still other terms in the root, to find each succeeding term all of those already found are considered as one quantity. To better illustrate this, we give another example:

2. Required to extract the square root of $4\,x^2 + a^2y^2 + 9\,z^4 - 4\,axy + 12\,xz^2 - 6\,ayz^2$.

Arranging this with reference to the powers of x (and y and z), we have

$$\begin{array}{r|l}
 & 4x^2 - 4axy + 12xz^2 + a^2y^2 - 6ayz^2 + 9z^4 \;((2x - ay) + 3z^2 \\
 & 4x^2 \\
\hline
4x - ay & -4axy + 12xz^2 + a^2y^2 \\
 & \underline{-4axy \qquad\qquad + a^2y^2} \\
4x - 2ay + 3z^2 & 12xz^2 \qquad\qquad - 6ayz^2 + 9z^4 \\
 & \underline{12xz^2 \qquad\qquad - 6ayz^2 + 9z^4}
\end{array}$$

In this problem after two terms of the root are secured, they are considered as one quantity and doubled for the trial divisor. Dividing we get the third term of the root, which, when found, is annexed to the others with its proper sign. And so in all cases, the root already found is treated as one quantity.

268. Rule for Extracting the Square Root of Polynomials.

1. Arrange the polynomial with reference to its leading letter, extract the square root of the first term, and subtract its square from the polynomial.

2. Divide the first term of the remainder by double the root quantity, and the quotient will be the second term of the root, which is written at the right of the previous or *trial* divisor, as well as in the root. This *complete* divisor is then multiplied by the root term just found and the product taken from the remainder.

3. Double the two terms of the root for the next trial divisor and annex to it the new root term, when found, before multiplying. Continue the operation until there is no remainder, or as far as desired.

269. Exercise in Extracting the Square Root of Polynomials.

1. $B^2 - 2AB + A^2 \,(= A^2 - 2AB + B^2)$.

The root is $B - A$, which differs in sign only from $A - B$, the root previously obtained. Thus, $B - A = -(A - B)$. (See **45**, 4, (2.) Moreover, this is true of all polynomial square roots.

If a polynomial is a square root of a quantity, *the same polynomial with all of its signs changed is also a root.*

2. $f^6 + 6f^3x^4 + 9x^8$.

3. $\dfrac{9a^8}{4} + 2a^4n^3 + \dfrac{4n^6}{9}$.

4. $a^2 - 2 + a^{-2}$.

REMARK. — Trinomial squares can usually be resolved mentally.

5. $\dfrac{a^2x^2 + 2ab^2x^3 + b^4x^4}{a^{2m} + 2a^mx^n + x^{2n}}$.

6. $\sqrt{x^4 - 2x^3 + 3x^2 - 2x + 1}$.

7. $\sqrt{9x^4 - 12x^3 + 16x^2 - 8x + 4}$.

8. $\sqrt[4]{a^4 + 8a^3b + 24a^2b^2 + 32ab^3 + 16b^4}$.

SUGGESTION. — Extract the square root twice. (See **45**, 3.)

9. $\left(x + \dfrac{1}{x}\right)^2 - 4\left(x + \dfrac{1}{x} - 1\right)$.

10. $1 - 2z + 2z^2 - z^3 + \dfrac{z^4}{4}$.

11. $x^2 - 2x + 1 + 2xy - 2y + y^2$.

12. $x(x + 1)(x + 2)(x + 3) + 1$.

13. $(13x^2)^2 + (4x^3)^2 + (7x)^2 + 210x^3 - 120x^5$.

14. Obtain four terms of $\sqrt{a^2 + x^2}$.

15. $\dfrac{n^{2r+2}}{a^2} + \dfrac{bn^{2r}}{6a} + \dfrac{b^2n^{2r-2}}{144} + \dfrac{n^{r+2}}{a} + \dfrac{bn^r}{12} + \dfrac{n^2}{4}$.

270. Extraction of the Square Root of Arithmetical Numbers as Polynomials. — The rule for the extraction of numbers (written in the Arabic notation) is a specialized form of that for polynomials. We proceed to illustrate this by examples.

1. Required to extract the square root of 344569.

We write the number as a polynomial whose terms have respectively an even number of ciphers.

```
                 340000 + 4500 + 69 | 500 + 80 + 7
                 250000
      1000 + 80 | 90000 + 4500 = 94500
       (1080)   |                86400
                   1160 + 7      | 8100 + 69 = 8169
                         (1167)  |             8169
```

EXPLANATION. — Starting at the decimal point the number is separated into terms of two figures each, annexing an even number of ciphers to fill out the vacant orders. Evidently two ciphers in the power will give one in the root, four in the power will give two in the root, and so on. This enables us to get the first figure in the root. For the first term of this number polynomial will give the first term in the root. In the present example, the first figure of the root cannot be 6 for 600 squared is 360,000, which is greater than the given number 344569. Taking 500 as the greatest number of hundreds and subtracting its square from the first term gives 9000, which added to the next term makes 94500.

Doubling the root found, 500, the trial divisor is found to be 1000. This is contained in the remainder 90 times; but upon adding 90 to the trial divisor and multiplying by 90 the product is 98100, which is too great. Hence eighty is the second term of the root instead of 90. Writing 80 in the root and adding it to the trial divisor, we have 1080 for the complete divisor. Multiplying the complete divisor by the second term of the root and subtracting the product from 94500 the remainder is 8100, to which the last term of the polynomial, 69, is added.

Doubling the root already found we have for a new trial divisor 1160, from which 7 is found, and the process is completed by adding 7 to the trial divisor and multiplying. The 500, of course, corresponds to A, and 80 to B in the first operation; then 580 to A and 7 to B in the second.

The process here explained succeeds in separating the number into the terms of a polynomial, which is a perfect square. Thus, $344569 = 250000 + 86400 + 8169 = (500 + 87)^2$.

2. We next solve *the same problem, abridging the work as much as possible.*

```
   34'45'69( 587
   25
108 )945
     864
1167 )8169
      8169
```

EXPLANATION. — The number is separated into periods of two figures each for the reason given above. Having found the first figure of the root from the first period, it is squared and subtracted, and the next period of two figures is *annexed* to the remainder. Then the first figure of the root is doubled for a trial divisor, and divided into 945, or rather 94. It is contained, as we saw above, not 9 but 8 times. This figure 8 is written after 5 in the root, and also *annexed* to the trial divisor. After multiplying by 8 and subtracting, the next period is brought down, and so on.

A careful comparison of this solution with the preceding will make the whole process plain. The difference is that in the second solution all the ciphers are omitted.

2. Extract the square root of the decimal fraction .0003627.

```
    .00'03'62'70( .01904 +
    .00 01
29 ).000262
       261
    3804 )170'00
          152 16
          178400  etc.
```

EXPLANATION. — Here as before the number is divided off into periods of two figures each commencing at the decimal point. The first figure of the root having been found, it is squared and the result subtracted, after which the next period is brought down. The first figure 1 is then doubled for a trial divisor and the process is continued.

A cipher is annexed to 7 to make its period full. Thereafter two new ciphers would be used for each new period. The second trial divisor 38 not being contained in 17, a cipher is placed in the root and also at the right of the trial divisor, after which a new period of two ciphers is annexed to the remainder. The trial divisor is then contained 4 times.

271. Rule for Extracting the Square Root of Numbers written in the Arabic Notation.

1. Commencing at the decimal point divide the number off into periods of two figures each.

2. Find the greatest square in the left hand period, place the root at the right subtract and bring down the next period.

3. Double the root already found and find how many times it is contained in the remainder exclusive of the right hand figure, and place the quotient in the root and at the right of the trial divisor. Multiply by the figure found, and subtract, after which proceed as before.

a. Fill out decimal periods with ciphers as needed.

b. If at any time a trial divisor is not contained once in the remainder place a cipher in the root and one at the right of the divisor.

c. Point off into periods commencing at the decimal point (*both* ways in a mixed decimal), and place the decimal point in the root at the time a decimal period is taken.

272. Exercise in Extracting the Square Root of Numbers.

1. Find the square root of 2601; 47089; 1772.41; 433.4724.

2. $\sqrt{41.2164}$; $\sqrt{965.9664}$; $\sqrt{17.338896}$; $\sqrt{.00103041}$.

3. Extract the square root of $\frac{1}{4}$; $\frac{289}{324}$; $\frac{7225}{17689}$; $\frac{2704}{4225}$.

Remark. — Sometimes fractions need to be reduced to their lowest terms before attempting to extract the root indicated.

4. Find to within less than .001 the square root of 4.64; of 2; of 9.999; of .00111.

5. Reduce to decimals and extract the square root of the following to within .00001; i.e., to five places: —

$\frac{1}{2}$; $1\frac{2}{3}$; $\frac{3}{4}$; $\frac{7}{116}$; $5\frac{2}{7}$; $3\frac{5}{8}$.

SECTION III.

Cube Root of Polynomials.

273. Extraction of the Cube Root of Polynomials. How Effected.

1. Required to extract the cube root of $A^3 + 3A^2B + 3AB^2 + B^3$.

$$A^3 + 3A^2B + 3AB^2 + B^3 \,(\, A + B \quad \text{Root}$$
$$A^3$$
$$3A^2 + 3AB + B^2\,)\, 3A^2B + 3AB^2 + B^3$$
$$3A^2B + 3AB^2 + B^3$$

EXPLANATION. — The first term of the root having been obtained by extracting the cube root of the first term of the polynomial, it is cubed and the result subtracted. The remainder is $3A^2B + 3AB^2 + B^3$.

By reference to **258** we know that $A + B$ is the cube root of the polynomial, and that the second term of the root is B. Now we see that the second term of the root can be gotten by dividing the first term of the remainder by three times the square of the first term of the root already found, or, $3A^2B$ by $3A^2$.

If to the $3A^2$, which becomes a trial divisor for the next term of the root, we now annex $3AB + B^2$, i.e., three times the first by the second, plus the square of the second, and multiply the sum by the second term B, upon subtracting, nothing remains, and the operation is complete.

It can be shown, as in **267**, that the binomial solution is of general application, and that when the root is to consist of three or more terms, we first find two and then proceed in the next operation as if they were one quantity, and so on.

2. Extract the cube root of $8x^6 - 36x^5 + 114x^4 - 207x^3 + 285x^2 - 225x + 125$.

$$8x^6 - 36x^5 + 114x^4 - 207x^3 + 285x^2 - 225x + 125 \;\underline{|\, 2x^2 - 3x + 5}$$
$$8x^6$$
$$12x^4 - 18x^3 + 9x^2 \;\Big|\; -36x^5 + 114x^4 - 207x^3$$
$$-36x^5 + 54x^4 - 27x^3$$
$$12x^4 - 36x^3 + 27x^2 \qquad 60x^4 - 180x^3 + 285x^2 - 225x + 125$$
$$30x^2 - 45x$$
$$+ 25$$
$$12x^4 - 36x^3 + 57x^2 - 45x + 25 \qquad 60x^4 - 180x^3 + 285x^2 - 225x + 125$$

EXPLANATION. — The $12x^4 = 3(2x^2)^2$; the $-18x^3 = 3 \times -3x \times 2x^2$; $9x^2 = (3x)^2$

In the next operation, $12x^4 - 36x^3 + 27x^2 = 3(2x^2 - 3x)^2$; $30x^2 - 45x = 3 \times 5(2x^2 - 3x)$; $25 = 5^2$.

274. Rule for Extracting the Cube Root of Polynomials.

1. Arrange the polynomial with reference to the exponents of the leading letters, extract the cube root of the first term, and subtract its cube from the polynomial.

2. Divide the first term of the remainder by 3 times the square of the root quantity, and the quotient will be the second term of the root. Now to the trial divisor is added 3 times the product of the first term of the root by the second, plus the square of the second ($3\,AB + B^2$), thus forming the complete divisor, which is multiplied by the second root term, and the product subtracted from the first remainder.

3. If this does not finish the operation, square the root quantity (of two terms) already found, and multiply by 3 for a trial divisor, and continue as before.

275. Exercise in Extracting the Cube Root of Polynomials.

1. $8\,a^3 - 36\,a^2 b + 54\,ab^2 - 27\,b^3$.
2. $8\,a^3 - 84\,a^2 x + 294\,ax^2 - 343\,x^3$.
3. $x^6 + 3\,x^5 + 6\,x^4 + 7\,x^3 + 6\,x^2 + 3\,x + 1$.
4. $a^6 - 3\,a^5 b + 6\,a^4 b^2 - 7\,a^3 b^3 + 6\,a^2 b^4 - 3\,ab^5 + b^6$.
5. $a^3 - b^3 + c^3 - 3\,(a^2 b - a^2 c - ab^2 - ac^2 - b^2 c + bc^2) - 6\,abc$.
6. $\dfrac{x^3}{y^3} - \dfrac{3\,x^2}{y} + 3\,xy - y^3$.
7. $8\,x^9 - 60\,x^6 z^2 + 150\,x^3 z^4 - 125\,z^6$.
8. $1 - 6\,x^m + 12\,x^{2m} - 8\,x^{3m}$.
9. $1 - x$ to three terms in the answer.
10. $60\,c^2 x^4 + 48\,cx^5 - 27\,c^6 + 108\,c^5 x - 90\,c^4 x^2 + 8\,x^6 - 80\,c^3 x^3$.

276. Extraction of the Cube Root of Numbers in the Arabic Notation as Polynomials.

1. Required to extract the cube root of 158252632.929.

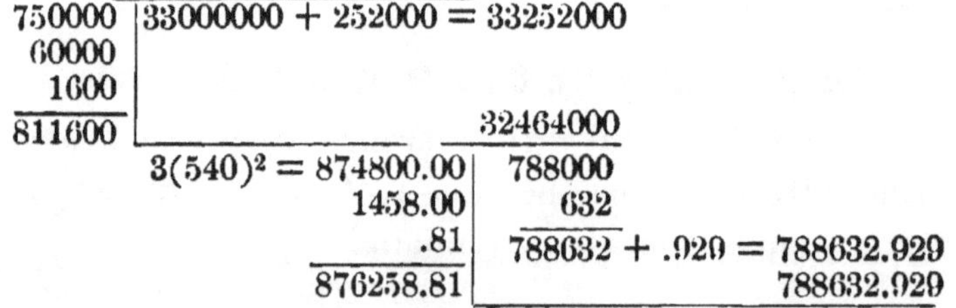

```
         158000000 + 252000 + 632. +. 929 |500 + 40 + 0 + .9
         125000000
750000 |33000000 + 252000 = 33252000
 60000 |
  1600 |
811600 |                     32464000
        3(540)² = 874800.00|  788000
                    1458.00|     632
                        .81|  788632 + .929 = 788632.929
                  876258.81|                  788632.929
```

Explanation.—Starting from the decimal point the number is separated into parts of three figures each annexing ciphers to fill out omitted orders. It is plain that one cipher in the root will give three in the power, or conversely, three in the power will give one in the root, six in the power will give two in the root, and so on. Seeking the greatest cube root in the first term, 500 is found, which is cubed and subtracted from that term. Squaring the root found and multiplying by 3 we get 750000 as a trial divisor. Dividing, the quotient 40 is set down as the second term of the root. As in the polynomial rule, taking three times the product of the first term of the root by the second, plus the square of the second, the results are 60000 and 1600, which added to the trial divisor make 811600, *the complete divisor.* Multiplying by the second term of the root and subtracting, the remainder is 788000 to which the next term is added. This process is now repeated for the third term of the root; but the trial divisor not being contained in the remainder a cipher is placed in the root and *two* at the right of the trial divisor.

2. The example abridged.

```
                158'252', 632', 929' (540.9
                125
     7500      |33252
      600      |
       16      |
     ----      |
     8116      |32464
                -----
      87480000  788632929
        145800 |
            81 |
      -------- |
      87625881 |788632929.
                ----------
```

277. Rule for Extracting the Cube Root of Numbers.

1. Starting at the decimal point, separate the number into periods of three figures each.

2. Find the greatest cube in the left hand period, place the root at the right, subtract, and bring down the next period.

3. Square the root figure found, and multiply the result by 300 for a trial divisor. Find how many times it is contained in the dividend, and write the quotient as the second figure of the root.

4. To the trial divisor add 30 times the product of the second by the first figure of the root and the square of the second figure for the complete divisor. Multiply the complete divisor by the second root figure and subtract the product.

5. Regard the figures already found as one number and proceed as before to find the third or remaining figures.

NOTE.—See again the remarks made at the end of 271.

278. Exercise in Extracting the Cube Root of Numbers.

1. 12167; 12812904; 1481.544; 167.284151.

2. .127263527; .008741816; $\frac{216}{2744}$

3. Extract the root to third decimal of 517; of 20.911; of 5; of .2; of 37.

4. Divide the cube root of $\frac{2515.456}{32768}$ by the square root of the square root of 8.3521

5. $\sqrt[3]{\frac{5}{9}}$; $\sqrt[3]{2456}$; $\sqrt[3]{999700.029999}$; $\sqrt[3]{\frac{29791}{68921}}$

SECTION IV.

EXTRACTION OF OTHER ROOTS OF POLYNOMIALS.

279. Other Roots Derived by Use of the Square and Cube Root Process.

1. The *fourth* root of a polynomial may be extracted by taking the square root twice.

2. The *sixth* root by taking in turn the square and cube roots. Either operation may be performed first.

3. The *eighth* by extracting the square root three times.

4. The *ninth* by extracting the cube root twice, and so on.

280. Prime Roots other than the Square and Cube Roots. These may be found by means of rules derived from formulæ in the same way that the rules for square and cube root were found.

For instance, the fifth root.

$$(A + B)^5 = A^5 + (5\,A^4 + 10\,A^3B + 10\,A^2B^2 + 5\,AB^3 + B^4)B$$

Here $5\,A^4$, i.e., 5 times the fourth power of the first term of the root, is the *trial* divisor; and to it must be added, for the complete divisor, $10\,A^3B + 10\,A^2B^2 + 5\,AB^3 + B^4$; i.e., 10 times the cube of the first times the second, etc., etc.

281. Exercise in Extracting Roots other than the Square and Cube Roots.

1. Find the fourth root of $16\,a^4 - 96\,a^3x + 216\,a^2x^2 - 216\,ax^3 + 81\,x^4$.
2. Find the fourth root of $1 - 4\,x + 10\,x^2 - 16\,x^3 + 19\,x^4 - 16\,x^5 + 10\,x^6 - 4\,x^7 + x^8$.
3. Find the fourth root of $625\,x^4 + 9600\,x^2y^2 + 4096\,y^4 - 10240\,xy^3 - 4000\,x^3y$.
4. Find the sixth root of $64 - 192\,x + 240\,x^2 - 160\,x^3 + 60\,x^4 - 12\,x^5 + x^6$.
5. Find the sixth root of $x^6 - 6\,x^5 + 15\,x^4 - 20\,x^3 + 15\,x^2 - 6\,x + 1$.
6. Find the eighth root of $x^8 + 8\,x^7 + 28\,x^6 + 56\,x^5 + 70\,x^4 + 56\,x^3 + 28\,x^2 + 8\,x + 1$.
7. Find the seventh root of $128\,x^7 - 448\,x^6 + 672\,x^5 - 560\,x^4 + 280\,x^3 - 84\,x^2 + 14\,x - 1$.

CHAPTER XIX.

OF FRACTIONAL EXPONENTS.

282. Fractional Exponents result naturally from the law of exponents in the extraction of roots. By **129** each exponent is divided by the index of the root to be extracted. When therefore this index is not contained exactly in any exponent, a *fractional power results.*

a. To illustrate. (See Ex. 266.) $\sqrt{a^6b^4} = a^{\frac{6}{2}}b^{\frac{4}{2}} = a^3b^2$; $\sqrt[5]{c^{20}} = c^{\frac{20}{5}} = c^4$; etc. If now instead of $\sqrt{a^6b^4}$ we had $\sqrt{a^7b^5}$, according to the same rule of division by the index of the root we should get $\sqrt{a^7b^5} = a^{\frac{7}{2}}b^{\frac{5}{2}}$.

283. Meaning of the Terms in a Fractional Exponent The numerator of any fractional exponent evidently denotes the *power* to which the quantity is to be raised while the denominator indicates the *root* to be taken.

a. The expression $\sqrt[m]{a^n}$ means the mth root of the nth power of a. Hence we see that a number affected with a fractional exponent has a perfectly definite meaning, and performing both operations gives rise to a resulting number, the value of the expression.

284. Fundamental Principle Governing the use of Fractional Exponents.

Any quantity affected with a fractional exponent may be separated into its factors, each factor taking the original fractional exponent.

This principle holds for integral exponents, thus, $(abc)^m = a^mb^mc^m$.

The question arises does it hold for fractional exponents as well. To show this let us prove that the expression

$$(1)\ (ab)^{\frac{m}{n}} = a^{\frac{m}{n}} \times b^{\frac{m}{n}}.$$

Raising both members of (1) to the nth power, since by definition the denominator n means the nth root in each case,

$$(2)\quad (ab)^m = a^m \times b^m = a^m b^m \qquad \text{(Ax. 5).}$$

The resulting equation (2) is plainly true; and consequently the members of equation (1) are equal. Moreover, it is easy to see that the same reasoning would apply generally to any fractional exponents and to any number of factors. Hence the theorem:—*Any root of a product is equal to the product of the like roots of the factors: and conversely, the product of like roots of two or more factors is equal to the same root of their product.*

a. If a number affected with a fractional exponent is to be evaluated *it is immaterial whether it be first raised to the power and then the root taken* or *vice versa.*

This amounts to proving, for example in $a^{\frac{m}{n}}$, that $(a^m)^{\frac{1}{n}} = \left(a^{\frac{1}{n}}\right)^m$.

We have

$(a^m)^{\frac{1}{n}} = a^{\frac{1}{n}} \times a^{\frac{1}{n}} \times a^{\frac{1}{n}}$ . . . to m factors. (By the theorem of this article) $= (a^{\frac{1}{n}})^m$. *Q. E. D.* (106, 3 (2.)

As a particular case, $(8^2)^{\frac{1}{3}} = (8^{\frac{1}{3}})^2 = 4$.

b. A quantity with a fractional exponent is not changed in value if the fraction be reduced to other terms.

Thus $a^{\frac{1}{2}} = a^{\frac{2}{4}}$, $m^{\frac{1}{4}}n^{\frac{1}{3}} = m^{\frac{3}{12}}n^{\frac{4}{12}}$.

For, when the numerator is doubled the quantity is squared; but when the denominator is doubled a square root must be taken in addition to the root denoted by the old index, and the two operations cancel each other. We may reason in like manner for any other factor by which both terms of the fractional index may be multiplied.

285. Fractional Exponents and Radical Signs. — Radical signs are used for the same purpose as fractional exponents, and the two notations are employed interchangeably. (See next chapter.)

It would be far preferable if the radical sign notation were entirely displaced by fractional exponents. As both are in constant use, both have to be taught.

286. Object of Treatment of fractional exponents in the present chapter. — This object is twofold:

1. To show that the same rules apply to fractional exponents as held for integral exponents.

2. To furnish the student exercise in the use of fractional exponents in all the simple operations.

287. Fractional Exponents and the Fundamental Operations.

1. Addition and subtraction of quantities involving fractional exponents.

Here it is plain that we can always add or subtract *similar* quantities whether affected by integral or fractional exponents.

$$\text{Thus. } 5\,a^{\frac{2}{3}}b^{\frac{1}{2}} + 6\,a^{\frac{2}{3}}b^{\frac{1}{2}} - 4\,a^{\frac{2}{3}}b^{\frac{1}{2}} = 7\,a^{\frac{2}{3}}b^{\frac{1}{2}}.$$

2. Multiplication and division of quantities involving fractional exponents.

Let us seek, e. g., the product of $x^{\frac{2}{3}}$ by $x^{\frac{5}{6}}$.

$$x^{\frac{2}{3}} \times x^{\frac{5}{6}} = x^{\frac{4}{6}} \times x^{\frac{5}{6}} \qquad \textbf{(284, } b\textbf{)}$$

$$x^{\frac{4}{6}} \times x^{\frac{5}{6}} = (x^4)^{\frac{1}{6}} \times (x^5)^{\frac{1}{6}} \qquad \textbf{(284, } a\textbf{)}$$

$$(x^4 \times x^5)^{\frac{1}{6}} = (x^9)^{\frac{1}{6}} = x^{\frac{9}{6}} \qquad \textbf{(284)}$$

Now $\frac{9}{6}$, the new exponent, is the *sum* of the old exponents $\frac{2}{3}$ and $\frac{5}{6}$, the very process of determining it giving their sum. In other words. the old rule of *adding the exponents* in multiplication holds for *fractional* as well as for *integral* exponents.

Since division is the reverse of multiplication, the exponent of the divisor is *subtracted* from that of the dividend.

3. Fractional powers and roots of quantities affected with fractional exponents.

Let us take, e. g., $(a^{\frac{2}{3}})^{\frac{4}{5}}$.

$$(a^{\frac{2}{3}})^{\frac{4}{5}} = \{[(a^{\frac{1}{3}})^2]^4\}^{\frac{1}{5}} = \{(a^{\frac{1}{3}})^8\}^{\frac{1}{5}} = \{a^{\frac{8}{3}}\}^{\frac{1}{5}} = \{(a^8)^{\frac{1}{3}}\}^{\frac{1}{5}} = (a^8)^{\frac{1}{15}}$$
$= a^{\frac{8}{15}}$, (**284**, *a*)

since the fifth root of the third root is, by definition, the fifteenth. Now $\frac{8}{15}$ the new exponent is the product of the old exponents $\frac{2}{3}$ and $\frac{4}{5}$. Hence the old rule of *multiplying* the exponent of the quantity by the exponent of the power holds for *fractional* as well as for *integral* exponents.

Thus, generally, $(a^{\frac{m}{n}})^{\frac{p}{q}} = a^{\frac{mp}{nq}}$; and conversely,

$$a^{\frac{mp}{nq}} = (a^{\frac{m}{n}})^{\frac{p}{q}} = (a^{\frac{p}{q}})^{\frac{m}{n}} = (a^{\frac{m}{q}})^{\frac{p}{n}} = (a^{\frac{p}{n}})^{\frac{m}{q}}.$$

a. Since fractional exponents are governed by precisely the same rules as held for integral, they might have been used from the beginning. There are weighty reasons, however, for deferring their introduction until now.

288. Exercise in the Use of Fractional Exponents.

1. Add $8^{\frac{1}{3}}$, $49^{\frac{1}{2}}$, and $27^{\frac{1}{3}}$.

2. $3a^{\frac{1}{2}}bc^2 + 9a^{\frac{1}{2}}bc^2 - 13a^{\frac{1}{2}}bc^2$.

3. $15m^{\frac{2}{3}}n^{\frac{3}{4}} + m^{\frac{2}{3}}n^{\frac{3}{4}} - m^{\frac{2}{3}}n^{\frac{3}{4}}$.

4. Find the sum of $2ax^{-m}y^{\frac{1}{2}} + 3bc - \frac{1}{2}a^{-2}x^{-\frac{1}{3}} + 3b$; $3ax^{-m}y^{\frac{1}{2}} + 2bc + \frac{3}{2a^2x^{\frac{1}{3}}} - 2b$; and $\frac{ay^{\frac{1}{2}}}{x^m} - b + 5\frac{1}{a^2x^{\frac{1}{3}}}$.

5. Find the sum of $6a^{\frac{1}{2}}b^{\frac{2}{3}} - 9c^{\frac{2}{3}}d + 10a^{\frac{1}{3}}b^{\frac{1}{3}}$; $-6a^{\frac{1}{2}}b^{\frac{2}{3}} - a^{\frac{1}{3}}b^{\frac{1}{3}} + 6c^{\frac{2}{3}}d$; $2c^{\frac{2}{3}}d - 3a^{\frac{1}{2}}b^{\frac{2}{3}} - 3a^{\frac{1}{3}}b^{\frac{1}{3}}$; and $-2a^{\frac{1}{2}}b^{\frac{2}{3}} + c^{\frac{2}{3}}d - a^{\frac{1}{3}}b^{\frac{1}{3}}$.

6. From the sum of $5ax^{\frac{1}{2}} - (x+y)^{\frac{1}{2}} + (a-b)^{\frac{1}{2}}$, $-7ax^{\frac{1}{2}} + 2(x+y)^{\frac{1}{2}} - 3(a-b)^{\frac{1}{2}}$, and $12ax^{\frac{1}{2}} - 3(x+y)^{\frac{1}{2}} + 12(a-b)^{\frac{1}{2}}$ take the sum of $3ax^{\frac{1}{2}} + 4(x+y)^{\frac{1}{2}} + (a-b)^{\frac{1}{2}}$ and $ax^{\frac{1}{2}} - (x+y)^{\frac{1}{2}} + 3(a-b)^{\frac{1}{2}}$.

7. To $5ax^2 - 7bc + 8m^{\frac{1}{2}} - 2c^{-n}$, add $3bc - 4c^{-n} + 2ax^2$, and then subtract $10m^{\frac{1}{2}} - 5xy + 3bc - 12c^{-n}$.

8. From $9c^{\frac56}d^{\frac34}+10c^{\frac23}d^{\frac13}-17c^{\frac12}d^{\frac12}$ take $c^{\frac56}d^{\frac34}-20c^{\frac13}d^{\frac12}$ $-11c^{\frac76}d^{\frac23}$.

9. Taking the even roots positively, simplify $16^{\frac14}+8^{\frac23}$ $+16^{\frac34}+125^{\frac13}-512^{\frac59}+100^{0.5}-81^{0.75}$.

10. Calculate, $36^{1\frac12}$, $49^{3\frac12}$, $4^{-3\frac12}$, $8^{-2\frac13}$, $9^{-0.5}$

11. Multiply

$$-3y^{\frac1n} \text{ and } 6y^2; \text{ also } 9a^{\frac1m}b^m \text{ and } 2a^{\frac2m}b^n.$$

12. Multiply together $7^{\frac34}$, $7^{\frac23}$, $7^{\frac78}$.

13. $a^{\frac12}\cdot a^{\frac34}\cdot a^{\frac78}\cdot a^{\frac{9}{16}}=?$ $a^2b^{\frac12}\cdot a^{\frac12}b\cdot a^{-1}b^2=?$

14. $x^{\frac23}\cdot x^{-\frac12}=?$ $m^{\frac23}\cdot m^{\frac32}=?$ $n\times n^{\frac12}\times n^{\frac13}\times n^{-\frac14}=?$

15. Multiply $a^{\frac56}+a^{-\frac76}+a$ by $a^{-\frac12}-a^{\frac34}$.

16. $(a^{\frac12}b^{-\frac23}c^{\frac34}d^{-\frac58})\div(a^{\frac78}b^{-\frac{9}{10}}c^{-\frac{11}{12}}d^{1\frac12})=?$

17. Multiply $2a^{-\frac23}-7x^{-\frac12}y^4-11c^{\frac16}$ by $axy^{-3}c^{\frac56}$.

18. Multiply $x^{\frac14}+y^{\frac13}$ by $x^{\frac14}-y^{\frac13}$.

19. Multiply $m^{\frac23}+m^{\frac13}n^{\frac13}+n^{\frac23}$ by $m^{\frac13}-n^{\frac13}$.

20. Multiply $a^{\frac14}-b^{-\frac23}$ by a^2-b.

21. Simplify $(x^{\frac23}\times x^{\frac47})^{\frac{11}{13}}$.

22. Find the product of $\left(\frac{ay}{x}\right)^{\frac12}$, $\left(\frac{bx}{y^2}\right)^{\frac14}$ and $\left(\frac{y^2}{a^2b^2}\right)^{\frac14}$

23. Multiply $x^{\frac32}-xy^{\frac12}+x^{\frac12}y-y^{\frac32}$ by $x+x^{\frac12}y^{\frac12}+y$.

24. Multiply $a^{\frac12}+b^{\frac12}+a^{-\frac12}b$ by $ab^{-\frac12}-a^{\frac12}+b^{\frac12}$.

25. Divide $11a^{\frac53}-33a^{\frac43}$ by $11a^{\frac23}$.

26. Divide $a^{\frac{3n}{2}}-b^{-\frac{3n}{2}}$ by $a^{\frac n2}-b^{-\frac n2}$.

27. Divide $x^{\frac43}+x^{\frac23}y^{\frac23}+y^{\frac43}$ by $x^{\frac23}+x^{\frac13}y^{\frac13}+y^{\frac23}$.

28. Divide $5x^{\frac mn}-10x^{-m}+15x^{\frac1n}y$ by $5x^p$.

29. Divide $x^{\frac52}-x^2-x^{\frac32}+6x-2x^{\frac12}$ by $x^{\frac32}-4x^{\frac12}+2$.

30. Divide $a-b$ by $a^{\frac14}-b^{\frac14}$.

31. Square the following:

$$-a^{\frac m2}b^{\frac m3};\quad -ab^{\frac12}c^{\frac m2};\quad \frac{(x+y)^{-\frac1p}}{(x-y)^{\frac mq}}$$

32. Cube the following: $a^{\frac{2}{3}}$, $a^{\frac{1}{n}} - a^{-\frac{1}{n}}b^{\frac{1}{n}}$, $a^{\frac{p}{q}}b^{-\frac{q}{p}}$.

33. Simplify $\left(\dfrac{x^{\frac{1}{2}}}{y^{\frac{1}{3}}}\right)^{\frac{3}{4}}$, $(a^{\frac{1}{2}}b^{\frac{1}{3}})^{-6}$, $(a^{\frac{1}{3}}b^{\frac{1}{2}}c^{-\frac{1}{2}})^{6}$.

34. Raise a^2 to the $\frac{3}{4}$ power.

35. Square $a^{\frac{1}{2}} + b^{\frac{1}{2}} + c^{\frac{1}{2}}$.

36. Extract the square root of $1 + 4x^{-\frac{1}{3}} - 2x^{-\frac{2}{3}} - 4x^{-1} + 25x^{-\frac{4}{3}} - 24x^{-\frac{5}{3}} + 16x^{-2}$.

37. Factor $a^{\frac{3}{2}} - 1$; $x^{\frac{3}{4}} - 27$; $a^2 + ab + b^2$. (135, 4)

38. $\dfrac{x - 7x^{\frac{1}{2}}}{x - 5x^{\frac{1}{2}} - 14} \div \dfrac{x^{\frac{1}{2}}}{x^{\frac{1}{2}} + 2} = ?$ $\dfrac{a^{\frac{3}{2}} + ab}{ab - b^3} - \dfrac{a^{\frac{1}{2}}}{a^{\frac{1}{2}} - b} = ?$

CHAPTER XX.

RADICALS.

289. Radical Quantities or simply Radicals are quantities of which some root is to be extracted. They are expressed sometimes with radical signs, sometimes with fractional exponents. (**285.**)

a. When a root of a quantity can be found exactly the radical sign or fractional exponent disappears in the simplified form (**272**). It is only when the root cannot be found exactly that these signs have to be retained. As a consequence the subject of radicals deals almost exclusively with the reduction and use of quantities whose roots cannot be exactly found.

290. Radicals whose Roots cannot be exactly extracted are termed **Irrational Quantities, or Surds,** (the English term).

Thus, $\sqrt{5}$, $\sqrt[3]{7a^2}$, etc.

a. A fraction always expresses the ratio of one number to another, viz., that of the numerator to the denominator. Upon reduction to the decimal form it gives rise to either a finite or circulating decimal (*i.e.*, one in which, sooner or later, sets of figures are repeated.) Now, the process of extracting the roots of numbers never gives rise to a circulating or repeating decimal, as the student can easily see with a little reflection. Consequently the latter can never exactly equal a fraction, and thus express a *ratio*. Hence the propriety in the use of the word *irrational*.

291. The Treatment of Radicals embraces:

1. Reduction of radical quantities.

2. Addition, subtraction, multiplication, and division of radical quantities.

3. Equations containing radicals.

a. It will often be convenient to solve, *pari passu*, the same exercise in the two notations. To indicate this brackets will be written at the left.

SECTION I. — Reduction of Radicals.

292. Kinds of Reduction. — The treatment will be under three heads. 1. Simplification. 2. Reduction to surd form. 3. Reduction to a common index.

I. — SIMPLIFICATIONS.

293. Reduction of Radicals to equivalent ones having a Lower Index, or to Rational Quantities.

1. Simplify $(4\,a^2)^{\frac{1}{4}} \equiv \sqrt[4]{4\,a^2}$ (**187**, *b*)

$$\begin{cases} (4\,a^2)^{\frac{1}{4}} = ((4\,a^2)^{\frac{1}{2}})^{\frac{1}{2}} = (2\,a)^{\frac{1}{2}} & \text{(287, 3, and 129)} \\ \sqrt[4]{4\,a^2} = \sqrt{\sqrt{4\,a^2}} = \sqrt{2\,a} & \text{(See 59. } a\text{)} \end{cases}$$

2. Simplify $(9\,a^4b^2)^{\frac{1}{6}} \equiv \sqrt[6]{9\,a^4b^2}$

$$\begin{cases} (9\,a^4b^2)^{\frac{1}{6}} = ((9\,a^4b^2)^{\frac{1}{2}})^{\frac{1}{3}} = (3\,a^2b)^{\frac{1}{3}} & \text{(287, 3, and 129)} \\ \sqrt[6]{9\,a^4b^2} = \sqrt[3]{\sqrt{9\,a^4b^2}} = \sqrt[3]{3\,a^2b} & \end{cases}$$

3. Simplify $(-8\,m^3n^6)^{\frac{1}{6}} \equiv \sqrt[6]{-8\,m^3n^6}$

$$\begin{cases} (-8\,m^3n^6)^{\frac{1}{6}} = ((-8\,m^3n^6)^{\frac{1}{3}})^{\frac{1}{2}} = (-2\,mn^2)^{\frac{1}{2}} \\ \sqrt[6]{-8\,m^3n^6} = \sqrt{\sqrt[3]{-8\,m^3n^6}} = \sqrt{-2\,mn^2} \end{cases}$$

294. Rule for reducing Radicals to Lower Indices. — Take an exact root of the quantity (whose index must be a factor of the given index). The result is still affected by a radical sign with the other factor as its index.

295. Exercise in the Simplification of Radicals by reduction to a lower index.

1. Simplify $\sqrt[8]{16\,a^4b^4c^4}$, $\sqrt[9]{27\,a^6d^6}$, $\sqrt[4]{2500}$

2. Find the fourth root of $81\,a^4x^2y^6$, $100\,x^2y^6$, $\frac{4}{25}\,x^4y^8z^2$

3. Simplify

$\sqrt[4]{36\,x^{4n+2}y^{2m}}$, $\sqrt[6]{-64\,a^3}$, $\sqrt[4]{81\,m^4n^4}$, $\sqrt[4]{4\,x^2-4\,xy+y^2}$

4. Simplify $\sqrt[mn]{a^m}$, $\sqrt[m]{a^{mn}}$, $\sqrt[p]{(x+y)^{2pq}}$, $\sqrt[37]{a^{703}}$, $\sqrt[105]{(x^3)^n}$

296. Simplification by removing a Factor from the Radical. This reduction depends upon **284**.

1. Simplify $(162)^{\frac{1}{2}} = \sqrt{162}$

$$\begin{cases}(162)^{\frac{1}{2}} = (81)^{\frac{1}{2}} \times (2)^{\frac{1}{2}} = \pm 9 \times (2)^{\frac{1}{2}} & (\textbf{284} \text{ and } \textbf{270})\\ \sqrt{162} = \sqrt{81} \times \sqrt{2} = \pm 9\sqrt{2} & (\textbf{52}, b)\end{cases}$$

2. Simplify $(25\,a^3b)^{\frac{1}{2}} = \sqrt{25\,a^3b}$

$$\begin{cases}(25\,a^3b)^{\frac{1}{2}} = (25\,a^2)^{\frac{1}{2}} \times (ab)^{\frac{1}{2}} = \pm 5\,a\,(ab)^{\frac{1}{2}}\\ \sqrt{25\,a^3b} = \sqrt{25\,a^2} \times \sqrt{ab} = \pm 5\,a\sqrt{ab}\end{cases}$$

3. Simplify $(24\,a^4b^3c^2)^{\frac{1}{3}} = \sqrt[3]{24\,a^4b^3c^2}$

$$\begin{cases}(24\,a^4b^3c^2)^{\frac{1}{3}} = (8\,a^3b^3)^{\frac{1}{3}} \times (3\,ac^2)^{\frac{1}{3}} = 2\,ab\,(3\,ac^2)^{\frac{1}{3}}\\ \sqrt[3]{24\,a^4b^3c^2} = \sqrt[3]{8\,a^3b^3} \times \sqrt[3]{3\,ac^2} = 2\,ab\sqrt[3]{3\,ac^2}\end{cases}$$

4. Simplify $7\,(625\,a^4b^4c)^{\frac{1}{4}} = 7\sqrt[4]{625\,a^4b^4c}$

$$7\sqrt[4]{625\,a^4b^4c} = 7\sqrt[4]{625\,a^4b^4 \times c} = \pm 7 \times 5\,ab\sqrt[4]{c} = 35\,ab\sqrt[4]{c}$$

5. Simplify $(3\,ax^2)^{\frac{3}{2}} = \sqrt{27\,a^3x^6}$

$$\sqrt{27\,a^3x^6} = \sqrt{9\,a^2x^6} \times \sqrt{3\,a} = \pm 3\,ax^3\sqrt{3\,a}$$

6. Simplify $\frac{a}{b}\sqrt{\frac{9\,ac^2}{4\,d}}$

$$\frac{a}{b}\sqrt{\frac{9\,ac^2}{4\,d}} = \frac{a}{b}\sqrt{\frac{9\,c^2}{4}} \times \sqrt{\frac{a}{d}} = \pm\frac{3\,ac}{2\,b}\sqrt{\frac{a}{d}}$$

297. Rule.

1. Separate the radical quantity into two factors one of which is the greatest perfect power of which the desired root can be taken.

2. Extract the root of the factor which is a perfect power and place the result in the coefficient of the other factor affected as before.

298. Exercise in removing Factors from Radicals.

1. $\sqrt{288},\ 2\sqrt{75},\ \sqrt[3]{256},\ \sqrt[3]{2187}$

2. $\sqrt{9a^4x},\ \sqrt{36a^8},\ 4\sqrt{27a^8b^5},\ \sqrt{50ab^2c^2},\ \sqrt{80a^8x^5}$

3. $\sqrt[3]{128a^7b^5},\ \sqrt[3]{-108x^4y^8},\ \sqrt[3]{25(5a^8+5a^4b)}$.
$\sqrt[3]{-512a^6y^9},\ \sqrt[4]{ax^4+bx^6},\ 5\sqrt[4]{80x^4y^7}$

4. $\sqrt{72x^{2n+1}},\ \sqrt[3]{96a^7x^9},\ \sqrt{3179a^5bx^7},\ \sqrt[3]{375x^4y^8}$,
$\sqrt{(a^2-b^2)(a+b)},\ \sqrt[p]{x^{a+p}y^{2p}}$

5. $3\sqrt[5]{32a^6b^2c^5},\ 5(a-b)\sqrt{a^2c+2abc+b^2c}$,
$\sqrt{x^3y^2-x^2y^8},\ \sqrt[5]{a^7b^{11}},\ \sqrt[3]{a^2b^3c^8-d^3b^8c^8}$

299. Simplification of Fractions under the Radical Sign.

a. There are good reasons for preferring to have quantities under the radical sign *integral.* This object can always be attained by means of the well-known principle in **156.** In order to remove the denominator from under the radical sign both terms of the fraction are multiplied by such a factor as will make the denominator a perfect power; then extracting its root the result is placed in the denominator *outside*, as in the previous case.

1. Simplify $(\frac{2}{3})^{\frac{1}{2}} = \sqrt{\frac{2}{3}}$

$$(\tfrac{2}{3})^{\frac{1}{2}} = (\tfrac{2}{3}\times\tfrac{3}{3})^{\frac{1}{2}} = (\tfrac{6}{9})^{\frac{1}{2}} = \pm\tfrac{1}{3}(6)^{\frac{1}{2}}$$

$$\sqrt{\tfrac{2}{3}} = \sqrt{\tfrac{2}{3}\times\tfrac{3}{3}} = \sqrt{\tfrac{6}{9}} = \pm\tfrac{1}{3}\sqrt{6}.$$

2. Simplify $\left(\frac{5ab}{2x^2y}\right)^{\frac{1}{3}} = \sqrt[3]{\frac{5ab}{2x^2y}}$

$$\left(\frac{5ab}{2x^2y}\right)^{\frac{1}{3}} = \left(\frac{5ab}{2x^2y}\times\frac{4xy^2}{4xy^2}\right)^{\frac{1}{3}} = \left(\frac{20abxy^2}{8x^3y^3}\right)^{\frac{1}{3}}$$

$$= \frac{1}{2xy}(20abxy^2)^{\frac{1}{3}}$$

REMARK. — The student may write this example in the radical notation. He should accustomhimself hereafter to use either at pleasure.

3. Reduce $m\left(\frac{a}{b}\right)^{\frac{1}{2}} = m\sqrt{\frac{a}{b}}$ to its simplest form

$$m\sqrt{\frac{a}{b}} = m\sqrt{\frac{a}{b}\times\frac{b}{b}} = m\sqrt{\frac{ab}{b^2}} = \frac{m}{b}\sqrt{ab}.\quad \textit{Ans.}$$

4. Reduce $\sqrt{\frac{a^3 x^2}{4c^2 y}}$ to its simplest form.

$$\sqrt{\frac{a^3 x^2}{4c^2 y}} = \frac{ax}{2c}\sqrt{\frac{a}{y}} = \frac{ax}{2c}\sqrt{\frac{a}{y} \times \frac{y}{y}} = \frac{ax}{2cy}\sqrt{ay} \quad Ans.$$

5. Simplify $\frac{2a}{b}\left(\frac{b^4}{8a^3}\right)^{\frac{1}{4}}$

SOLUTION.

$$\frac{2a}{b}\left(\frac{b^4}{8a^3} \times \frac{2a}{2a}\right)^{\frac{1}{4}} = \frac{2a}{b} \times \frac{b}{2a}(2a)^{\frac{1}{4}} = (2a)^{\frac{1}{4}} \quad Ans.$$

6. Reduce $\left(\frac{a-x}{a+x}\right)^{\frac{1}{2}} = \sqrt{\frac{a-x}{a+x}}$ to its simplest form.

$$\sqrt{\frac{a-x}{a+x}} = \sqrt{\frac{a-x}{a+x} \times \frac{a+x}{a+x}} = \frac{1}{a+x}\sqrt{a^2-x^2} \quad Ans.$$

7. Reduce $\sqrt[3]{\frac{5}{12}}$ to its simplest form.

$$\sqrt[3]{\frac{5}{12}} = \sqrt[3]{\frac{5}{2 \times 2 \times 3} \times \frac{2 \times 3 \times 3}{2 \times 3 \times 3}} = \frac{1}{6}\sqrt[3]{90}$$

300. Rule for Simplifying Radical Quantities by removing the denominator from under the radical sign.

1. Simplify the radical by the previous cases.

2. Multiply both terms of the fraction under the radical sign by such factors as will make the denominator a perfect power of the degree denoted by the index of the radical, and in so doing use no unnecessary factors.

3. Extract the root of the denominator and place it as a factor of the denominator of the coefficient of the radical.

301. Exercise in Simplifying Radicals in the Fractional Form.

1. $\sqrt{\frac{1}{3}}$, $\sqrt{\frac{2}{5}}$, $2\sqrt{\frac{3}{8}}$, $\frac{1}{3}\sqrt{\frac{3}{7}}$, $\frac{2}{5}\sqrt{\frac{11}{2}}$, $\left(\frac{18}{25}\right)^{\frac{1}{2}}$

2. $4\sqrt{\frac{3ac^3}{8b}}$, $\frac{3}{4}\sqrt{\frac{2ax^3}{3}}$, $\left(\frac{a^4b^3}{4}\right)^{\frac{1}{2}}$, $6\sqrt{\frac{a}{12b}}$, $18\sqrt{\frac{5}{72}}$

3. $\sqrt[3]{\frac{3a}{5b}}, \sqrt[3]{\frac{1}{2}}, \sqrt[3]{\frac{5}{9a}}, \sqrt[4]{\frac{3a^4}{4b^3}}, 7\sqrt[5]{\frac{16a^2}{m}}$

4. $\left(\frac{5a^5b}{8c^3}\right)^{\frac{1}{4}}, \left(\frac{ab^2}{4(a+x)}\right)^{\frac{1}{2}}, \frac{1}{x}\left(\frac{ax^3}{x-a}\right)^{\frac{1}{3}},$

$3am\sqrt[3]{\frac{(a-x)^2}{a+x}}$

II.—REDUCTION TO SURD FORM.—Converse Operation.

302. Reduction of Entire Quantities to the Form of Surds.

1. Reduce $3ax$ to the form of a square root.

$3ax = \sqrt{9a^2x^2}$, or $(9a^2x^2)^{\frac{1}{2}}$, by squaring.

2. Change $-\frac{1}{2}a^2b^3c$ to the form of a cube root.

$$-\frac{1}{2}a^2b^3c = \sqrt[3]{-\frac{a^6b^9c^3}{8}} = \left(-\frac{a^6b^9c^3}{8}\right)^{\frac{1}{3}}$$

3. Put $-3a^2x^{-2}$ under a radical whose index is 4.

$-3a^2x^{-2} = \sqrt[4]{81a^8x^{-8}}$

4. $\sqrt{\frac{2a}{3}} = \sqrt[6]{\frac{8a^3}{27}} = \left(\frac{8a^3}{27}\right)^{\frac{1}{6}}$

303. Rule for Reducing Rational Quantities to the form of surds, or radicals to equivalent ones of higher degrees. Involve the quantity to the proper power and place it under the sign.

304. Exercise in Reduction to the Surd Form.

1. Reduce $-5a^2b$ to the form of a cube root.

2. Place $6, 2a^2b, -\frac{3}{4}, -7m^2n$ under radicals whose indices are 2.

3. Reduce $\frac{3}{5}, a-x, 6a^2x^{-3}, \frac{3am}{2ny}$ to the form of cube roots.

4. Express $\sqrt[3]{x^2}, x^a, a^{\frac{1}{4}}$, and a as sixth roots.

5. Express $\sqrt[3]{2}$, $2\sqrt[9]{8}$, $\sqrt[6]{6}$, as surds of the same order, viz.. sixth roots.

6. Express $\sqrt{5}$, $\sqrt[3]{11}$, $\sqrt[6]{13}$ as surds having the same index.

7. Reduce $\dfrac{abx}{a+b+x}$ to the form of a cubic surd.

8. Reduce $\dfrac{\sqrt{2}}{5}$ to the form of a fourth root.

305. Reduction of Radical Qualities to the Form of Entire Surds: In Other Words, Placing Coefficients within the Sign.

1. Reduce $3\sqrt{5}$ to the form of an entire surd.

$$3\sqrt{5} = \sqrt{3 \times 3}\,\sqrt{5} = \sqrt{45}.$$

2. Reduce $2\,a\sqrt[5]{2\,a^2}$ to the form of an entire surd.

$$2\,a\sqrt[5]{2\,a^2} = \sqrt[5]{(2\,a)^5} \times \sqrt[5]{2\,a^2} = \sqrt{(2\,a)^5 \times 2\,a^2} = \sqrt{64\,a^7}$$

3. Put the coefficient of $-\dfrac{2\,a}{5\,y^2}\sqrt[3]{\dfrac{5\,y}{3\,a^2}}$ within the sign.

$$-\frac{2\,a}{5\,y^2}\sqrt[3]{\frac{5\,y}{3\,a^2}} = -\sqrt{\frac{8\,a^3}{125\,y^6} \times \frac{5\,y}{3\,a^2}} = -\sqrt{\frac{8\,a}{75\,y^5}}$$

306. Rule for Inserting Coefficients under the Sign.

1. Raise the coefficient to the power denoted by the index of the radical.

2. Multiply this result into the quantity already affected by the sign, and write the product under the sign.

307. Exercise in Placing Coefficients under the Sign.

1. $5\sqrt{14}$, $\frac{1}{3}\sqrt{3}$, $\dfrac{4}{11}\sqrt{\dfrac{77}{8}}$, $6\sqrt[3]{4}$, $2\sqrt[3]{5}$, $\frac{1}{2}\,(4)^{\frac{1}{3}}$.

2. $\dfrac{3\,ab}{2\,c}\sqrt{\dfrac{20\,c^2}{9\,a^2b}}$, $\dfrac{2\,a^3}{3\,x}\sqrt{\dfrac{27\,x^4}{a^3}}$, $a^4\,(b)^{\frac{1}{4}}$, $5\,x\,(25\,x^2)^{\frac{1}{m}}$

3. $\dfrac{a}{b}\sqrt{\dfrac{c}{a^2}}$, $\dfrac{ax}{a-x}\sqrt{\dfrac{a^2-x^2}{a^2x^2}}$, $\dfrac{a+b}{a-b}\left(\dfrac{a-b}{a+b}\right)^{\frac{1}{2}}$

4. Place the 2 in the coefficient of $2x^2\sqrt[4]{3ab}$ within the sign.

5. $(a-b)\sqrt{a^2+b^2+2ab}$, $\frac{2a}{b}\sqrt{\frac{b^4}{8a}}$, $\frac{2a}{3x}\sqrt[3]{\frac{27x^4}{a^2}}$

III. — REDUCTION OF RADICALS TO THE SAME INDEX.

308. **Reduction of Radicals to Equivalent Ones having a Common Index.** (See **284**, *b*.)

1. $\sqrt{3}, \sqrt[3]{6}, \sqrt[4]{10} = (3)^{\frac{1}{2}}, (6)^{\frac{1}{3}}, (10)^{\frac{1}{4}}$.

Evidently a common index can be found by reducing the exponents to equivalent ones having a common denominator.

$$(3)^{\frac{1}{2}}, (6)^{\frac{1}{3}}, (10)^{\frac{1}{4}} = (3)^{\frac{6}{12}}, (6)^{\frac{4}{12}}, (10)^{\frac{3}{12}} = (3^6)^{\frac{1}{12}}, (6^4)^{\frac{1}{12}}, (10^3)^{\frac{1}{12}},$$

or, $\sqrt{6}, \sqrt[3]{6}, \sqrt[4]{10} = \sqrt[12]{3^6}, \sqrt[12]{6^4}, \sqrt[12]{10^3} = \sqrt[12]{729}, \sqrt[12]{1296}, \sqrt[12]{1000}$.

2. Reduce $\sqrt{ax}, \sqrt[3]{bx^2}, \sqrt[6]{cx^3}$ to surds having a common index.

$$\begin{aligned}(ax)^{\frac{1}{2}}, (bx^2)^{\frac{1}{3}}, (cx^3)^{\frac{1}{6}} &= (ax)^{\frac{3}{6}}, (bx^2)^{\frac{2}{6}}, (cx^3)^{\frac{1}{6}}\\ &= (a^3x^3)^{\frac{1}{6}}, (b^2x^4)^{\frac{1}{6}}, (cx^3)^{\frac{1}{6}}\\ &= \sqrt[6]{a^3x^3}, \sqrt[6]{b^2x^4}, \sqrt[6]{cx^3}.\end{aligned}$$

3. Reduce $a\sqrt{x-y}$ and $\frac{b}{\sqrt[4]{x+y}}$ to a common index.

$$\begin{aligned}a(x-y)^{\frac{1}{2}}, b(x+y)^{-\frac{1}{4}} &= a(x-y)^{\frac{2}{4}}, b(x+y)^{-\frac{1}{4}}\\ &= a(x^2-2xy+y^2)^{\frac{1}{4}}, b((x+y)^{-1})^{\frac{1}{4}}\\ &= a\sqrt[4]{x^2-2xy+y^2}, b\sqrt[4]{\frac{1}{x+y}}.\end{aligned}$$

4. Reduce $a+c$ and $(a-c)^{\frac{1}{2}}$ to a common index.

$$\begin{aligned}(a+c), (a-c)^{\frac{1}{2}} &= (a+c)^{\frac{2}{2}}, (a-c)^{\frac{1}{2}}\\ &= \sqrt{a^2+2ac+c^2}, \sqrt{a-c}.\end{aligned}$$

309. Rule for Reducing Radicals to a Common Index.

1. If not already written with fractional exponents place them in that form.

2. Reduce these exponents to equivalent ones having a least common denominator.

3. Develop the radical quantities to the powers denoted by the numerators.

310. Exercise in Reducing Radicals to a Common Index.

1. $\sqrt[4]{8}, \sqrt{3}, \sqrt[8]{6}$.
2. $\sqrt[5]{a^8}, \sqrt{a}$.
3. $\sqrt[3]{5}, \sqrt{4}$.
4. a^2 and $b^{\frac{1}{3}}$.
5. $3^{\frac{2}{3}}, 2^{\frac{3}{4}}, 5^{\frac{1}{2}}$.
6. $\sqrt{a^8}, \sqrt[3]{a^2}, \sqrt[4]{a^8}$.
7. $\sqrt[8]{x^8}, \sqrt[9]{x^6}, \sqrt[12]{x^5}$.
8. $a\sqrt{a-x}, b\sqrt[3]{a^2-x^2}$.
9. $4(5x^2y)^{\frac{1}{3}}, 3(4xy)^{\frac{1}{6}}$. and $15a(3bx^2)^{\frac{1}{4}}$.
10. $x^{\frac{1}{m}}$, and $y^{\frac{1}{n}}$.
11. $a^{-\frac{1}{4}}, (8b^2)^{-\frac{1}{6}}. (3acy)^{\frac{1}{4}}$.
12. $(a+x)^{\frac{1}{2}}, (a-x)^{\frac{1}{3}}$.

SECTION II.

Fundamental Operations with Radicals.

311. Addition and Subtraction of Radicals. (**287**, 1.)

a. Radical qualities *seemingly unlike* can often be made similar by reducing them to their simplest forms.

1. Add together $3\sqrt{45}$, $\sqrt{20}$, and $7\sqrt{5}$.

$3\sqrt{45} = 9\sqrt{5}$; (**296**): $\sqrt{20} = 2\sqrt{5}$; Adding the coefficients, we have $18\sqrt{5}$. *Ans.*

2. Simplify $3\sqrt{28a^3x} - 13\sqrt{252a^3x} + 15a\sqrt{63ax}$.

$3\sqrt{28a^3x} = 6a\sqrt{7ax}$; $13\sqrt{252a^3x} = 78a\sqrt{7ax}$; $15a\sqrt{63ax} = 45a\sqrt{7ax}$. Adding, $-27a\sqrt{7ax}$. *Ans.*

3. $2\sqrt{\frac{5}{3}} + \frac{1}{6}\sqrt{60} + \sqrt{15} + 2\sqrt{\frac{3}{5}} = ?$

$2\sqrt{\frac{5}{3}} = \frac{2}{3}\sqrt{15}$ (**299**); $\frac{1}{6}\sqrt{60} = \frac{1}{3}\sqrt{15}$; $2\sqrt{\frac{3}{5}} = \frac{2}{5}\sqrt{15}$ Sum $\frac{12}{5}\sqrt{15}$.

4. $2\sqrt[4]{16a^8} + \sqrt[3]{81} - \sqrt[3]{-512a^6} + \sqrt[3]{192} - 7\sqrt[6]{9}.$

$2\sqrt[4]{16a^8} = 4a^2$; $\sqrt[3]{81} = 3\sqrt[3]{3}$; $\sqrt[3]{-512a^6} = -8a^2$; $\sqrt[3]{192} = 4\sqrt[3]{3}$; $7\sqrt[6]{9} = 7\sqrt[3]{3}$.

Adding, the sum is found to be $12a^2$.

312. Rule for Addition and Subtraction of Radicals.

1. Reduce each radical to its simplest form by one of the rules for simplification.

2. Add the coefficients of similar terms as in simple addition.

a. Similar terms, as will be remembered, are those which have the same letters affected by the same exponents fractional or integral. In the language of radicals, those which have the *same index* and the *same quantity* within the sign.

313. Exercise in Addition and Subtraction of Radicals.

1. Add $8\sqrt{125}$ and $2\sqrt{80}$; $9\sqrt{192}$ and $7\sqrt{75}$.

2. From $5\sqrt{363a^3y^2}$ take $3\sqrt{242a^3y^2}$.

3. Add $10a\sqrt{28a^2x}$, $5b\sqrt{63b^2x}$, and $c\sqrt{112c^2x}$.

4. $\sqrt{44} + 5\sqrt{176} - 2\sqrt{99}$; $4\sqrt{128} + 4\sqrt{75} - 5\sqrt{162}$.

5. $2\sqrt{3} - \frac{1}{2}\sqrt{12} + 4\sqrt{27} - 2\sqrt{\frac{3}{16}}$.

6. $\sqrt{48ab^2} + b\sqrt{75a} + \sqrt{3a(a-9b)^2}$.

7. $2\sqrt[3]{\frac{1}{4}} + 8\sqrt[3]{\frac{1}{32}}$.

8. $6\sqrt[6]{4a^2} + 2\sqrt[3]{2a} + \sqrt[9]{8a^3}$.

9. $\sqrt{2ax^2 - 4ax + 2a} - \sqrt{2ax^2 + 4ax + 2a}$.

10. From $(a-x)\sqrt{a^2 - x^2}$ take $a(a-x)\sqrt{\frac{a+x}{a-x}}$.

11. $\sqrt{\frac{ab^3}{c^2}}+\frac{1}{2c}\sqrt{(a^3b-4a^2b^2+4ab^3)}$.

12. $3\sqrt{147}-\frac{7}{3}\sqrt{\frac{1}{3}}-\sqrt{\frac{1}{27}}$.

13. $\sqrt[3]{16a^3b}+\sqrt[4]{4a^2b}+\sqrt[3]{54a^3b}+\sqrt{a^2b}$.

14. $\sqrt[3]{8a^3b+16a^4}-\sqrt[3]{b^4+2ab^3}$.

15. $3b^2(a^3c)^{\frac{1}{2}}+\frac{2}{c}(a^5c^3)^{\frac{1}{2}}-c^4\left(\frac{ac}{b^2}\right)^{\frac{1}{2}}$.

16. $(54a^{m+6}b^3)^{\frac{1}{3}}-(16a^{m-3}b^6)^{\frac{1}{3}}+(2a^{4m+9})^{\frac{1}{3}}+(2c^8a^m)^{\frac{1}{3}}$.

17. $7\frac{1}{9}\sqrt[3]{7\frac{19}{32}}+4\sqrt[3]{0.21875}-5\sqrt[4]{0.0256}$.

18. $c\sqrt[5]{a^6b^7c^3}-a\sqrt[5]{ab^7c^8}+b\sqrt[5]{a^6b^2c^8}$.

314. Multiplication and Division of Radicals. (See **287**, 2)

a. By **284** two radical quantities can be multiplied together provided they have the same index. We readily see that they cannot be multiplied until reduced to the same index.

Thus $\sqrt{6}\times\sqrt[3]{5}\neq\sqrt{30}\neq\sqrt[3]{30}$.

By **309** radicals can always be reduced to a common index, and therefore they can always be multiplied or divided as desired.

1. Multiply $\sqrt{6}$ by $\sqrt{3}$. *Ans.* $\sqrt{18}=3\sqrt{2}$.

2. Divide $\sqrt[3]{a^4b^2x^2y^4}$ by $\sqrt[3]{a^2by^4}$. *Ans.* $\sqrt[3]{a^2bx^2}$.

3. Multiply $2\sqrt[3]{3}$ by $3\sqrt{2}$.

$$2(3)^{\frac{1}{3}}\times3(2)^{\frac{1}{2}}=2(3^2)^{\frac{1}{6}}\times3(2^3)^{\frac{1}{6}}=6(9\times8)^{\frac{1}{6}}=6\sqrt[6]{72}.$$

(287, 2)

4. Divide $\sqrt[3]{2a^2x^3}$ by $\sqrt[5]{2ax^4}$.

$(2a^2x^3)^{\frac{1}{3}}=(32a^{10}x^{15})^{\frac{1}{15}}$; $(2ax^4)^{\frac{1}{5}}=(8a^3x^{12})^{\frac{1}{15}}$ Dividing, we have $(4a^7x^3)^{\frac{1}{15}}$ *Ans.*

5. Multiply $11\sqrt{2} - 4\sqrt{15}$ by $\sqrt{6} + \sqrt{5}$.

$$
\begin{array}{l}
11\sqrt{2} - 4\sqrt{15} \\
\quad \sqrt{6} + \quad \sqrt{5} \\
11\sqrt{12} - 4\sqrt{90} + 11\sqrt{10} - 4\sqrt{75},
\end{array}
$$

Or, $22\sqrt{3} - 12\sqrt{10} + 11\sqrt{10} - 20\sqrt{3} = 2\sqrt{3}$
$- \sqrt{10}$ *Ans.*

6. Divide $a + 2\sqrt{ab} + b - c$ by $\sqrt{a} + \sqrt{b} + \sqrt{c}$.

$$
\begin{array}{llll}
a + 2\sqrt{ab} + b - c & (\sqrt{a} + \sqrt{b} + \sqrt{c}. & & \\
\underline{a + \sqrt{ab} + \sqrt{ac}} & (\sqrt{a} + \sqrt{b} - \sqrt{c} & & \\
\quad \sqrt{ab} - \sqrt{ac} + b & & & \\
\quad \underline{\sqrt{ab} \qquad\qquad + b + \sqrt{bc}} & & & \\
\qquad - \sqrt{ac} \qquad - \sqrt{bc} - c & & & \\
\qquad \underline{- \sqrt{ac} \qquad - \sqrt{bc} - c.} & & &
\end{array}
$$

315. Rule for Multiplying and Dividing Radicals.

1. If the radicals have the same index, multiply and divide the quantities under the radical signs as desired, placing the result under the common sign.

2. If the radicals do not have a common index, reduce them as in **310**.

3. When polynomials involving radicals are to be multiplied or divided, proceed as with rational quantities, observing the rules for the multiplication and division of monomials.

a. The coefficients are to be multiplied or divided as indicated to form the coefficient of the result.

316. Exercise in Multiplying and Dividing Radicals.

1. $2\sqrt{14} \times \sqrt{21}$; $3\sqrt{8} \times \sqrt{6}$; $5\sqrt{a} \times 2\sqrt{3a}$; $\sqrt[3]{168} \times \sqrt[3]{147}$.

2. $a\sqrt{b^3} \times b^2\sqrt{a}$; $\sqrt{\frac{2}{3}} \times \sqrt{\frac{3}{4}}$; $2\sqrt{3xy} \times 5\sqrt{3xy^3}$.

3. $\sqrt{\frac{2\,xy}{3\,a}} \times \sqrt{\frac{3\,bx}{2\,y}}$; $\sqrt[3]{2\,a} \times \sqrt[3]{4\,a^5}$; $5\sqrt{3} \times 7\sqrt{\frac{8}{3}} \times \sqrt{2}$.

4. Divide $\sqrt{40}$ by $\sqrt{2}$; $6\sqrt{54} \div 3\sqrt{2}$; $77\sqrt[3]{9} \div 7\sqrt[3]{18}$.

5. Multiply $\sqrt[3]{10}$ by $4\sqrt[4]{2}$; $\sqrt{2} \times \sqrt[3]{3} \times \sqrt[4]{5}$.

6. $a\sqrt[m]{x} \times b\sqrt[n]{y}$; $2\sqrt[3]{\frac{1}{2}} \times 3\sqrt{\frac{5}{6}}$.

7. $\sqrt[3]{6\,a^{\frac{1}{2}}bc^{-1}} \times \sqrt[3]{3^{-1}a^{-\frac{1}{2}}bc^2}$; $\frac{1}{2}\sqrt[3]{\frac{1}{2}} \div \frac{1}{3}\sqrt[3]{\frac{1}{3}}$.

8. $\sqrt[3]{2} \cdot \sqrt[6]{\frac{1}{3}} \div \sqrt[8]{3}$; $(\sqrt{5} + 2\sqrt{7} + 3\sqrt{10})\,2\sqrt{5}$.

9. Multiply $5\,a^{\frac{1}{2}}$ by $3^{\frac{1}{2}}$; $4\,a^{\frac{1}{2}}b^{\frac{1}{2}} \times 5\,a^{\frac{1}{3}}b^{\frac{2}{3}}$; $\sqrt[4]{3\,ac} \times \sqrt[3]{2ac}$.

10. Multiply $3\,x^{\frac{1}{2}}y^{\frac{1}{3}}$ by $2\,x^{\frac{1}{4}}y^{\frac{1}{6}}$ and express the product without fractional exponents.

11. Divide $\sqrt[m]{x^p}$ by $\sqrt[n]{x^q}$; $\sqrt[3]{y^5} \div \sqrt[4]{y^8}$.

12. Multiply together $\frac{ax}{bc}\sqrt{ax}$, $\frac{by}{cd}\sqrt[3]{by}$ and $\frac{c^2d}{a}\sqrt[4]{cz}$.

13. $\left(\sqrt[4]{\frac{a}{b}} \times \sqrt[3]{\frac{a^2}{b^2}}\right) \div \sqrt{\frac{b}{a}}$; $\sqrt[5]{64} \div 2$; $\sqrt{a^2 - x^2} \div (a - x)$.

14. $(3 + \sqrt{5})(3 - \sqrt{5})$; $(\frac{3}{2} + \frac{5}{2}\sqrt{\frac{1}{2}})(\frac{1}{2} - 7\sqrt{\frac{1}{2}}.)$

15. $(\sqrt[3]{5} - 2\sqrt[3]{6})(3\sqrt[3]{4} - \sqrt[3]{36})$; $\left(\sqrt{\frac{ad^2}{c^3}} + \sqrt{\frac{a^2}{b}}\right)$
$(\sqrt{ac} + \sqrt{b^3})$.

16. $(\sqrt{2} + \sqrt{3} - \sqrt{5})(\sqrt{2} + \sqrt{3} + \sqrt{5})$;
$(a^2 - a\sqrt{b} - 6\,b) \div (a - 3\sqrt{b})$.

17. $(a^2 + 2\,a^{\frac{1}{2}}b^{\frac{2}{3}} - 4\,a^{\frac{3}{2}}b^{\frac{1}{2}} - 8\,b^{\frac{7}{6}}) \div (a^{\frac{1}{2}} - 4\,b^{\frac{1}{2}})$.

18. $(\sqrt{x} + \sqrt{y} + \sqrt{z})(\sqrt{x} - \sqrt{y} - \sqrt{z})(\sqrt{x} - \sqrt{y} + \sqrt{z})$.

19. $(\sqrt{6} + 4\sqrt{18} - 3 - 8\sqrt{2}) \div \sqrt{3}$; $(\sqrt{72} + \sqrt{32} - 4) \div \sqrt{8}$.

20. $(2\sqrt{8} + 3\sqrt{5} - 7\sqrt{2})(\sqrt{72} - 5\sqrt{20} - 2\sqrt{2})$.

21. $(2\sqrt{3} - \sqrt[3]{2})(2 + \sqrt[3]{9})$; $(5 + \sqrt[3]{4} + 2\sqrt[4]{5})(\sqrt{6} + \sqrt{5})$.

22. Divide 20 by $\sqrt{40}$; $m\sqrt{\frac{x-1}{x+1}} \div n\sqrt{\frac{x+1}{x-1}}$.

23. $4\sqrt{2} \times (3\sqrt{8} \div \frac{1}{2}\sqrt{2}) \times \frac{3}{8}\sqrt{2}$; $(a+b\sqrt{c})$
$(d-e\sqrt{f})$.

317. Rationalization of Denominators. — There is one case of reduction of radicals (**292**) which it is desirable to treat at this point. It is that of rationalizing denominators; i.e., transforming fractions with radical quantities in their denominators into equivalent ones whose denominators are rational. Many fractional expressions containing radicals in their denominators can have them removed by multiplying both terms of the fraction by the proper factor (**156**).

1. Rationalize the denominator of $\frac{2}{\sqrt{3}}$.

$$\frac{2}{\sqrt{3}} = \frac{2}{\sqrt{3}} \times \frac{\sqrt{3}}{\sqrt{3}} = \frac{2\sqrt{3}}{3} \quad \textit{Ans.} \qquad (\mathbf{156})$$

2. Rationalize the denominator of $\frac{4}{2+\sqrt{3}}$.

$$\frac{4}{2+\sqrt{3}} = \frac{4}{2+\sqrt{3}} \times \frac{2-\sqrt{3}}{2-\sqrt{3}} = \frac{4(2-\sqrt{3})}{4-3}$$
$$= 8 - 4\sqrt{3} \quad \textit{Ans.}$$

3. Rationalize the denominator of

$$\frac{\sqrt{3}-4\sqrt{5}-2\sqrt{7}}{2\sqrt{3}-\sqrt{5}+\sqrt{7}}.$$

By actual multiplication and addition, we have

$$\frac{\sqrt{3}-4\sqrt{5}-2\sqrt{7}}{2\sqrt{3}-\sqrt{5}+\sqrt{7}} = \frac{\sqrt{3}-4\sqrt{5}-2\sqrt{7}}{(2\sqrt{3}-\sqrt{5})+\sqrt{7}}$$
$$\times \frac{2\sqrt{3}-\sqrt{5}-\sqrt{7}}{(2\sqrt{3}-\sqrt{5})-\sqrt{7}} = \frac{40-9\sqrt{15}-5\sqrt{21}+6\sqrt{35}}{2(5-2\sqrt{15})}$$

To rationalize this result both terms are multiplied by $5+2\sqrt{15}$, giving (the student should verify the result),

$$\frac{35\sqrt{15}+35\sqrt{21}-70}{2\times-35}=\frac{2-\sqrt{15}-\sqrt{21}}{2}\quad Ans.$$

4. * To rationalize a fraction of the form $\dfrac{K}{a^{\frac{1}{m}}\pm b^{\frac{1}{n}}}$.

Let p be the l. c. m. of m and n. Then

$$\frac{\left(a^{\frac{1}{m}}\right)^p-\left(b^{\frac{1}{n}}\right)^p}{a^{\frac{1}{m}}\pm b^{\frac{1}{n}}}=\left(a^{\frac{1}{m}}\right)^{p-1}\mp\left(a^{\frac{1}{m}}\right)^{p-2}\left(b^{\frac{1}{n}}\right)+\left(a^{\frac{1}{m}}\right)^{p-3}\left(b^{\frac{1}{n}}\right)^2\mp$$

$$\ldots\mp\left(b^{\frac{1}{n}}\right)^{p-1}\qquad\text{(By 130. 1 and 2)}$$

$$=Q\ \text{(say)}$$

If now both terms of $\dfrac{K}{a^{\frac{1}{m}}+b^{\frac{1}{n}}}$ be multiplied by Q, we have $\dfrac{K}{a^{\frac{1}{m}}\pm b^{\frac{1}{n}}}=\dfrac{KQ}{\left(a^{\frac{1}{m}}\right)^p-\left(b^{\frac{1}{n}}\right)^p}$ since the quotient Q multiplied by the divisor $a^{\frac{1}{m}}\pm b^{\frac{1}{n}}$ ought to equal the dividend. But $\left(a^{\frac{1}{m}}\right)^p$ and $\left(b^{\frac{1}{n}}\right)^p$ or $a^{\frac{p}{m}}$ and $b^{\frac{p}{n}}$ are rational, since p was taken as the l. c. m. of m and n. Therefore $\dfrac{KQ}{a^{\frac{p}{m}}-b^{\frac{p}{n}}}$ is the equivalent fraction with a rational denominator.

In Examples 2—4 the multiplier in each instance is termed *the complementary* radical. In the case where the radical or radicals are square roots, as in Ex's 2 and 3, when the denominator is a *sum* the complementary factor is a *difference*, and vice versa.

318. Rule for Rationalizing the Denominators of Fractions.

1. For Monomial Denominators.—Multiply both terms of the fraction by such a factor as will rationalize the denominator.

2. For Binomial Denominators. — Multiply both terms of the fraction by the factor complementary to the denominator.

3. For Polynomial Denominators. — Regard one part of the denominator as one term and the others as the second term of a binomial, and proceed as in **2.**

a. The advantage gained by rationalizing denominators may be shown by an example. Thus, $\dfrac{3\sqrt{5}}{4\sqrt{7}}$ if calculated just as it stands would require the extraction of two roots and the division of one long decimal by another. Whereas if the denominator be rationalized, the extraction of only one root is necessary, and the division is a decimal divided by an integer (usually not a large number). So, likewise, with binomial denominators.

319. Exercise in Rationalizing the Denominators of Expressed Divisions.

1. $\dfrac{m}{\sqrt{5}}, \quad \dfrac{2}{\sqrt[3]{9}}, \quad \dfrac{\sqrt{3}}{\sqrt{6a^3}}, \quad \dfrac{3+\sqrt{8}}{2\sqrt{18}}, \quad \dfrac{a+\sqrt[4]{5}}{a\sqrt[4]{8b^3}}.$

SUGGESTION. — Multiply these fractions respectively by

$\sqrt{5}, \quad \sqrt[3]{3}, \quad \sqrt{6a}, \quad \sqrt{2}, \quad \sqrt[4]{2b}.$

2. $\dfrac{3}{\frac{1}{2}\sqrt[4]{5}}, \quad \dfrac{6\sqrt[5]{2}}{\sqrt[5]{4^3}}, \quad \dfrac{\sqrt[3]{5}}{\sqrt{10}}, \quad \dfrac{\sqrt{a}}{\sqrt[3]{b}}, \quad \dfrac{5\sqrt{4\frac{1}{2}}+3\sqrt{12.5}}{\sqrt{2}}.$

3. $\dfrac{\sqrt{3}+\sqrt{2}}{\sqrt{3}-\sqrt{2}}, \quad \dfrac{3\sqrt{5}-2\sqrt{2}}{2\sqrt{5}-\sqrt{18}}, \quad \dfrac{1}{\sqrt{2}-1}, \quad \dfrac{5}{\sqrt{7}-\sqrt{2}}.$

4. $\dfrac{3}{\sqrt[3]{3}-\sqrt[3]{2}}.$

5. $\dfrac{1+2\sqrt{3}}{5-\sqrt{3}}, \quad \dfrac{\sqrt{1+x^2}-\sqrt{1-x^2}}{\sqrt{1+x^2}+\sqrt{1-x^2}}, \quad \dfrac{a^2}{\sqrt{x^2+a^2}+x}.$

6. $\dfrac{1}{a^{\frac{1}{3}}+b^{\frac{1}{3}}}.$

7. $\dfrac{2\sqrt{15}+8}{5+\sqrt{15}} \div \dfrac{8\sqrt{3}-6\sqrt{5}}{5\sqrt{3}-3\sqrt{5}}; \quad [\sqrt[4]{x}-\sqrt[4]{y}] \div [\sqrt[4]{x}+\sqrt[4]{y}].$

8. $\dfrac{1}{\sqrt{x}-\sqrt{y}}, \quad \dfrac{1}{x+\sqrt{x^2-1}}+\dfrac{1}{x-\sqrt{x^2-1}}.$

9. $\dfrac{\sqrt{2}+\sqrt[3]{3}}{\sqrt{2}-\sqrt[3]{3}}.$

10. $\dfrac{1+\sqrt{3}+\sqrt{6}}{1+\sqrt{2}-\sqrt{3}}; \quad \dfrac{\sqrt{2}}{\sqrt{2}+\sqrt{3}-\sqrt{5}}.$

320. Powers and Roots of Radicals. For Principles and Rules see **287**, 3.

a. Cases arise, however, in the extraction of roots for which the rules heretofore given do not apply. These problems will be treated separately in the next article.

1. Square $3\sqrt[3]{3}$.

$$(3\sqrt[3]{3})^2 = (3\,(3)^{\frac{1}{3}})^2 = 9\,(3)^{\frac{2}{3}} = 9\,(3^2)^{\frac{1}{3}} = 9\sqrt[3]{9}.$$

Or, $(3\sqrt[3]{3})^2 = 3\sqrt[3]{3} \times 3\sqrt[3]{3} = 9\sqrt[3]{9}.$ **(315)**

Cube $\sqrt{ax^2}$; raise $\sqrt[3]{-10}$ to the fourth power.

2. Raise $\frac{1}{3}\sqrt{6}$ to the fourth power; $2\sqrt{\frac{1}{3}}$ to the fifth power.

3. Raise $\frac{1}{2}\sqrt[3]{-4x^2y}$ to the fifth power; $-3\sqrt{\frac{2}{7}}$ to the third power.

4. $(-\sqrt[6]{3c^2})^8, \sqrt{\left(\frac{16}{25}\right)^7}, \sqrt{\left(\frac{25}{64}\right)^6}. \left(\sqrt[3]{\sqrt[7]{8a^3}}\right)^7$

5. $(\sqrt[3]{25})^5$, $(\sqrt[5]{2})^2$; raise $\sqrt[m]{2a}$ to the nth power.

6. Square $\sqrt{3}+x\sqrt{3}$; cube $2-\sqrt{3}$.

7. Square $x^{\frac{1}{2}} - y^{-\frac{2}{3}}$; raise $\sqrt{3}-\sqrt{2}$ to the fourth power.

8. Raise $\sqrt{x}-y$ to the third power; cube $\dfrac{\sqrt{a^2-x^2}}{\sqrt{a}\,\sqrt[3]{a+x}}$

9. Cube $\sqrt{x}+3\sqrt{y}$; cube $-\sqrt[3]{\sqrt{a}-\sqrt{bc}}$.

10. Extract the square root of $\sqrt{a^8}$; of $(\sqrt{a^8})^{\frac{1}{3}}$.

11. Extract the fourth root of $16\,a^8\,\sqrt[3]{2\,c}$; of $\sqrt{\sqrt{a^4b^2}}$.

12. Extract the cube root of $32\,\sqrt[3]{27\,a^9x^2}$, of $(\sqrt[3]{49\,x^2})^{\frac{1}{2}}$.

13. $\sqrt[2]{64\,a^8\sqrt[3]{2\,c}}$, $\sqrt{9\,\sqrt[4]{5}}$, $\sqrt[3]{125\,\sqrt[4]{y}}$; $\sqrt{\overline{\sqrt[3]{2}}}$.

14. Required the cube root of $125\,x^{\frac{1}{2}}$, of $64\,a^6b^4\,\sqrt{2\,cd}$.

15. Extract the fifth root of $486\,a\,\sqrt[3]{4\,a^2}$.

16. Extract the square root of $x^{\frac{2}{3}} + 6\,x^{\frac{1}{3}}y^{\frac{1}{2}} + 9\,y$; the cube root of $(a + x)\,\sqrt{a + x}$.

17. Extract the fifth root $\sqrt{32\,x^{10}}$; the eighteenth root of $\sqrt[5]{a^{90}b^{120}}$.

18. Extract the cube root of $8\,\sqrt{x^3} + 36\,xy + 54\,y^2\,\sqrt{x} + 27\,y^3$.

19. Extract the square root of $9\,x - 6\sqrt{xy} + y - 6\sqrt{x} + 2\sqrt{y} + 1$.

321. Roots of Radical Quantities not in the Original Product form. — Binomial Surds.

a. If x and y be replaced by 2 and 3 respectively in Ex. 18 of the last article, we have $16\,\sqrt{2} + 36 \times 2 \times 3 + 54 \times 9\,\sqrt{2} + 27 \times 27 = 945 + 502\,\sqrt{2}$, which has only two terms, one being a surd, and may therefore be appropriately called a *binomial surd*. In order to extract the cube root of $945 + 502\,\sqrt{2}$ by the usual cube root rule, it becomes necessary to arrange it in the form first given. This is generally difficult to do, when one does not know how it was obtained. Such roots, when they exist, may often be discovered by special processes.

Only the simplest case, viz., the square root of binomial surds, is presented in most treatises on elementary algebra. The method of solution for other cases is very similar. Two or three lemmas are a necessary preliminary to the derivation of the root formula.

222. Theorems Relating to Equations containing Radicals.

1. *The square root of a rational quantity cannot be partly rational and partly surd.*

Let n be any number, and suppose it has for its square root $a+\sqrt{m}$. Then

$$\sqrt{n} = a + \sqrt{m}$$
$$n = a^2 + 2a\sqrt{m} + m \qquad \text{(Ax. 5)}$$
$$-2a\sqrt{m} = a^2 + m - n$$
$$\sqrt{m} = -\frac{a^2 + m - n}{2a}$$

Thus a surd equals a rational fraction or a whole number (dependent on whether $2a$ is contained in the numerator exactly) which is impossible. (See **290**, *a*.) *Q. E. D.*

2. *In any equation containing radicals, the rational part on one side equals the rational part on the other, and the surd part on one side equals the surd part on the other.*

Thus in the equation $a+\sqrt{b}=x+\sqrt{y}$, if a represent all the rational quantities on the left side, and x those on the right side, they must be equal; also $\sqrt{b}=\sqrt{y}$.

$$a + \sqrt{b} = x + \sqrt{y} \qquad \text{(Hyp.)}$$
$$\sqrt{b} = x - a + \sqrt{y} \qquad \text{(Ax. 2)}$$

Now, if x is not equal to a, $\sqrt{b}$ is partly rational and partly surd, which is impossible by 1 of this article. Whence x must equal a, and if $x=a$, $\sqrt{b}=\sqrt{y}$. *Q. E. D.*

3. If $\sqrt{a+\sqrt{b}} = \sqrt{x} \pm \sqrt{y}$, then will $\sqrt{a-\sqrt{b}} = \sqrt{x} \mp \sqrt{y}$

$$\sqrt{a+\sqrt{b}} = \sqrt{x} + \sqrt{y} \qquad \text{(Hyp.)}$$
$$a + \sqrt{b} = x + 2\sqrt{xy} + y \qquad \text{(Ax. 5)}$$
$$\therefore a = x + y, \text{ and } \sqrt{b} = 2\sqrt{xy} = \sqrt{4xy} \qquad \text{(2, above)}$$
$$\therefore a - \sqrt{b} = x - 2\sqrt{xy} + y = (\sqrt{x} - \sqrt{y})^2$$
$$\therefore \sqrt{a - \sqrt{b}} = \sqrt{x} - \sqrt{y}. \quad Q.\ E.\ D.$$

323. The Square Root of Binomial Surds. — Demonstration, Rule and Examples.

1. Let A represent the rational part and $\sqrt{B}$ the radical part of any binomial surd. Suppose $\sqrt{x} + \sqrt{y}$ to represent its square root. Then

(1) $\sqrt{A + \sqrt{B}} = \sqrt{x} + \sqrt{y}$

(2) $A = x + y$ and (3) $B = 4xy$ (**322**, 3)

Now, $(x - y)^2 = (x + y)^2 - 4xy = A^2 - B$ (Identities)

(4) $\therefore\ x - y = \sqrt{A^2 - B}$ (Ax. 6)

(2) $x + y = A$

$2x = A + \sqrt{A^2 - B}$ (Ax. 1)

$\sqrt{x} = \sqrt{\dfrac{A + \sqrt{A^2 - B}}{2}}$ (Ax. 6)

$2y = A - \sqrt{A^2 - B}$ (Ax. 2)

$\sqrt{y} = \sqrt{\dfrac{A - \sqrt{A^2 - B}}{2}}$ (Ax. 6)

Hence,

$$\sqrt{y} + \sqrt{y} = \sqrt{\frac{A + \sqrt{A^2 - B}}{2}} + \sqrt{\frac{A - \sqrt{A^2 - B}}{2}}$$

Also,

$$\sqrt{x} - \sqrt{y} = \sqrt{\frac{A + \sqrt{A^2 - B}}{2}} - \sqrt{\frac{A - \sqrt{A^2 - B}}{2}}$$

(**322**, 3)

2. Rule. — The rule is written from the formula just found.

(1. Square the rational quantity, subtract the quantity under the radical sign from the result, and extract the square root of the remainder. If this cannot be done there is no root of the form sought.

(2. Add the root just obtained to the rational quantity and also subtract it from the rational quantity, and divide

the results by two. The sum or difference (according as the binomial surd is a sum or difference) of the square roots of these quotients is the root sought.

a. Instead of following the rule, *inspection* will often enable one to determine the root.

3. Examples.

(1. Extract the square root of $31 + 10\sqrt{6}$.

Here $A = 31$, and $\sqrt{B} = 10\sqrt{6} = \sqrt{600}$, or $B = 600$.

$A^2 - B = 961 - 600 = 361$; and $\sqrt{A^2 - B} = 19$

$$\therefore \sqrt{31 + \sqrt{600}} = \sqrt{\frac{31 + 19}{2}} + \sqrt{\frac{31 - 19}{2}} =$$

$\pm (5 + \sqrt{6})$ *Ans.*

2. Reduce $\sqrt{np + 2m^2 - 2m\sqrt{np + m^2}}$ to its simplest form.

Here $\sqrt{A^2 - B} = \sqrt{n^2p^2} = np$. Hence, the answer is $\sqrt{np + m^2} - \sqrt{m^2} = \pm (\sqrt{np + m^2} - m)$.

3. $\sqrt{11 + 2\sqrt{30}} = ?$ $\sqrt{11 + 2\sqrt{30}} = \sqrt{6 + 2\sqrt{30} + 5}$ $= \sqrt{6} + \sqrt{5}$ *Ans.*

324. Exercise in Extracting the Square Roots of Binomial Surds.

1. $3 - 2\sqrt{2}$, $49 - 20\sqrt{6}$, $87 - 12\sqrt{42}$, $10 - \sqrt{96}$, $42 + 3\sqrt{174\frac{2}{3}}$.

2. $75 + 12\sqrt{21}$, $4\frac{1}{3} - \frac{4}{3}\sqrt{3}$, $\frac{3}{2} + \sqrt{2}$.

3. $x - 2\sqrt{x - 1}$, $x + xy - 2x\sqrt{y}$, $2 - \sqrt{4 - 4a^2}$, $ax - 2a\sqrt{ax - a^2}$, $x + y + z + 2\sqrt{xz + yz}$.

4. Solve by inspection $4 + 2\sqrt{3}$, $6 - 2\sqrt{5}$, $9 - 2\sqrt{14}$, $23 - 8\sqrt{7}$, $11 + \sqrt{72}$, $28 - 5\sqrt{12}$.

3. Multiply $c\sqrt{-a}$ by $d\sqrt{-b}$

$c\sqrt{-a} = c\sqrt{a}\sqrt{-1}$; $d\sqrt{-b} = d\sqrt{b}\sqrt{-1}$;
now since $\sqrt{-1} \times \sqrt{-1} = -1$, by definition,
$c\sqrt{-a} \times d\sqrt{-b} = cd\sqrt{ab} \times -1 = -cd\sqrt{ab}$.

NOTE. — It should be remembered that the sign $\pm$ is only written in case of ambiguity (See **264**). When it is known by what factors a product has been formed the appropriate sign must be prefixed.

Thus, $\sqrt{+1} \times \sqrt{+1} = +1$; $\sqrt{-1} \times \sqrt{-1} = -1$
Here $c\sqrt{-a} \times d\sqrt{-b}$ equals not $\pm cd\sqrt{+ab}$ but $-cd\sqrt{+ab}$.

4. Divide $3\sqrt{-4} - 2\sqrt{-12} + \sqrt{6} - 9$ by $3\sqrt{-2}$.

$$(3 \times 2\sqrt{-1} - 4\sqrt{3}\sqrt{-1} + \sqrt{6} - 9) \div 3\sqrt{2}\sqrt{-1}$$

$$= \sqrt{2} - \tfrac{4}{3}\sqrt{\tfrac{3}{2}} + \frac{\sqrt{3}}{3\sqrt{-1}} - \frac{3}{\sqrt{2}\sqrt{-1}}$$

Simplifying these terms, we have

$$\tfrac{4}{3}\sqrt{\tfrac{3}{2}} = \tfrac{2}{3}\sqrt{6}; \quad \frac{\sqrt{3}}{3\sqrt{-1}} = \frac{\sqrt{3}}{3\sqrt{-1}} \times \frac{\sqrt{-1}}{\sqrt{-1}}$$

$$= \frac{\sqrt{-3}}{-3}; \quad \frac{3}{\sqrt{2}\sqrt{-1}} = \frac{3}{\sqrt{2}\sqrt{-1}} \times \frac{\sqrt{-1}}{\sqrt{-1}} \cdot \frac{\sqrt{2}}{\sqrt{2}}$$

$$= \frac{3\sqrt{-2}}{-2}$$

$$\text{Hence, } \sqrt{2} - \tfrac{4}{3}\sqrt{\tfrac{3}{2}} + \frac{\sqrt{3}}{3\sqrt{-1}} - \frac{3}{\sqrt{2}\sqrt{-1}}$$

$$= \sqrt{2} - \tfrac{2}{3}\sqrt{6} - \frac{\sqrt{-3}}{3} + \tfrac{3}{2}\sqrt{-2} \quad \textit{Ans.}$$

5. Square $(\sqrt{-a} + \sqrt{-b})$

$$[(\sqrt{a} + \sqrt{b})\sqrt{-1}]^2 = -1(a + 2\sqrt{ab} + b).$$

5. Extract the fourth root of $17 + 12\sqrt{2}$, $56 + 24\sqrt{}$ $\frac{3}{2}\sqrt{5} + 3\frac{1}{2}$, $248 + 32\sqrt{60}$.

6. $\sqrt{3\sqrt{5} + \sqrt{40}} = ?$ $\sqrt{\sqrt{1573} - 4\sqrt{78}}$, $\sqrt{2\frac{1}{4} + \sqrt{5}}$ $\sqrt{\frac{a^2}{4} + \frac{c}{2}\sqrt{a^2 - c^2}}$, $\sqrt{28 + 10\sqrt{3}} + \sqrt{67 - 16\sqrt{3}}$.

325. Fundamental Operations with Imaginary Radical Quantities.

Any pure imaginary quantity, as $\sqrt{-a}$, can be reduced to $\pm\sqrt{a}\cdot\sqrt{-1}$ (**284**), of which $\sqrt{a}$ may be regarded as the coefficient of the imaginary $\sqrt{-1}$. Thus, all pure imaginaries can be expressed in terms of $\sqrt{-1}$. Let us examine the powers of $\sqrt{-1}$.

$(\sqrt{-1})^2 = -1$ by *definition* of a square root.

$(\sqrt{-1})^3 = (\sqrt{-1})^2\sqrt{-1} = -1\sqrt{-1}$

$(\sqrt{-1})^4 = (\sqrt{-1})^2(\sqrt{-1})^2 = -1 \times -1 = +1$

$(\sqrt{-1})^{4n} = ((\sqrt{-1})^4)^n = +1$, where n is any integral number.

$(\sqrt{-1})^{4n+1} = (\sqrt{-1})^{4n}\sqrt{-1} = \sqrt{-1}$

$(\sqrt{-1})^{4n+2} = (\sqrt{-1})^{4n}(\sqrt{-1})^2 = -1$

$(\sqrt{-1})^{4n+3} = (\sqrt{-1})^{4n}(\sqrt{-1})^3 = -\sqrt{-1}$.

1. Add $\sqrt{-a^2}$ and $\sqrt{-b^2}$.

$\sqrt{-a^2} = \sqrt{a^2} \times \sqrt{-1} = \pm a\sqrt{-1}$; $\sqrt{-b^2} = \pm b\sqrt{-1}$.

$\therefore \sqrt{-a^2} + \sqrt{-b^2} = (\pm a \pm b)\sqrt{-1}$. Or, taking positive roots only, $= (a + b)\sqrt{-1}$.

2. From $4\sqrt{-27}$ take $2\sqrt{-12}$

$4\sqrt{-27} = 12\sqrt{-3}$; $2\sqrt{-12} = 4\sqrt{-3}$;

$\therefore 4\sqrt{-27} - 2\sqrt{-12} = 8\sqrt{-3}$.

6. Extract the square root of $4\sqrt{-6} - 2$.

Evidently the product under the radical sign must be -24 (since $2\sqrt{-6} = \sqrt{-24}$). This number suggests the root $\sqrt{-6} + \sqrt{+4} = \sqrt{-6} + 2$. Squaring $\sqrt{-6} + 2$, we get $-6 + 4\sqrt{-6} + 4 = 4\sqrt{-6} - 2$.

326. Exercise in the Application of the Fundamental Rules to Imaginaries.

1. Add $2\sqrt{-48}$, $3\sqrt{-12}$, $5\sqrt{-8}$, and $-7\sqrt{-32}$.

2. Find the sum of $2 + \sqrt{-1}$ and $3 - \sqrt{-64}$.

3. Add $\sqrt[4]{-1}$ and $\sqrt[4]{-16}$; $\sqrt[4]{-4}$ and $\sqrt[4]{-9}$.

4. From $\sqrt{-18}$ subtract $\sqrt{-8}$; from $a + \sqrt{-b}$ take $a + \sqrt{-c}$.

5. Multiply $4\sqrt{-3}$ by $2\sqrt{-2}$; $4\sqrt{-3} \times 9\sqrt{-12}$.

6. $2\sqrt{-6} \times 5\sqrt{-4} \times 3\sqrt{-7}$; $2\sqrt[6]{-4} \times 3\sqrt[6]{-16}$.

7. Find the third and fourth powers of $a\sqrt{-1}$.

8. Divide $6\sqrt{-3}$ by $2\sqrt{-4}$; $(4 + \sqrt{-2}) \div 2 - \sqrt{-2}$.

9. Divide $2\sqrt{8} - \sqrt{-10}$ by $-\sqrt{-2}$; $-\sqrt{-1} \div -6\sqrt{-3})$.

10. Simplify $\dfrac{1 + \sqrt{-1}}{1 - \sqrt{-1}}$; $\dfrac{8}{-1 + \sqrt{-1}\sqrt{3}}$.

11. $(7 - \sqrt{-5})(10 - 3\sqrt{-6})$; $(2 - 5\sqrt{3}\sqrt{-1})(7 - 4\sqrt{3}\sqrt{-1})$.

12. Of what number are $24 + 7\sqrt{-1}$ and $24 - 7\sqrt{-1}$ the factors?

13. Find the continued product of $x + a$, $x + a\sqrt{-1}$, $x - a$, and $x - a\sqrt{-1}$.

14. Multiply, $(x - p - q\sqrt{-1})(x - p + q\sqrt{-1})$.

15. Raise $a + b\sqrt{-1}$ to the third power.

16. Extract the square root of $3 + 4\sqrt{-22}$.

17. Extract the square root of $8\sqrt{-1}$ $(=0 + 8\sqrt{-1}$.

18. Extract the square root of $-2 + 4\sqrt{-6}$; of $31 + 42\sqrt{-2}$.

19. $(\sqrt{-17} + \sqrt{-19})(\sqrt{-119} - \sqrt{-133}) = ?$

20. Prove that an imaginary quantity of the form $\sqrt{-a}$ cannot be partly real and partly imaginary, as $\sqrt{-a} = m + \sqrt{-n}$. (See **322**, 1.)

21. Prove that in any equation containing imaginaries of the form $a + b\sqrt{-1} = m + n\sqrt{-1}$, the real part on one side equals the real part on the other, and the imaginary part on the one side, the imaginary part on the other. In particular, if $a + b\sqrt{-1} = 0$, then $a = 0$, and $b = 0$. (See **322**, 2.)

Remark on Imaginaries. — It was not till the beginning of this century that Argand invented his diagram, since which time great progress has been made in this and allied subjects. An extended presentation of imaginaries would be out of place in any except an advanced algebra; but some little idea of the graphical representation of imaginaries will probably be very suggestive and instructive even here.

It was early shown that multiplying a positive number by -1 reverses its character and makes it negative, and multiplying a negative number by -1 reverses its character and makes it positive. Now we may conceive of -1 as an *operator* which revolves the whole positive series around into the negative, and the negative around into the positive, zero being the pivot.

But we saw that $\sqrt{-1} \times \sqrt{-1} = -1$, or $\sqrt{-1} \times \sqrt{-1} \times a = -a$; i.e. the operator $\sqrt{-}$ acting on a *twice* revolves it from the positive to the negative position. We are thus led to the idea that $\sqrt{-1}$ acting on a *once* would revolve it half way, or into the position at right angles. Generalizing this result we have the imaginaries as represented in the figure.

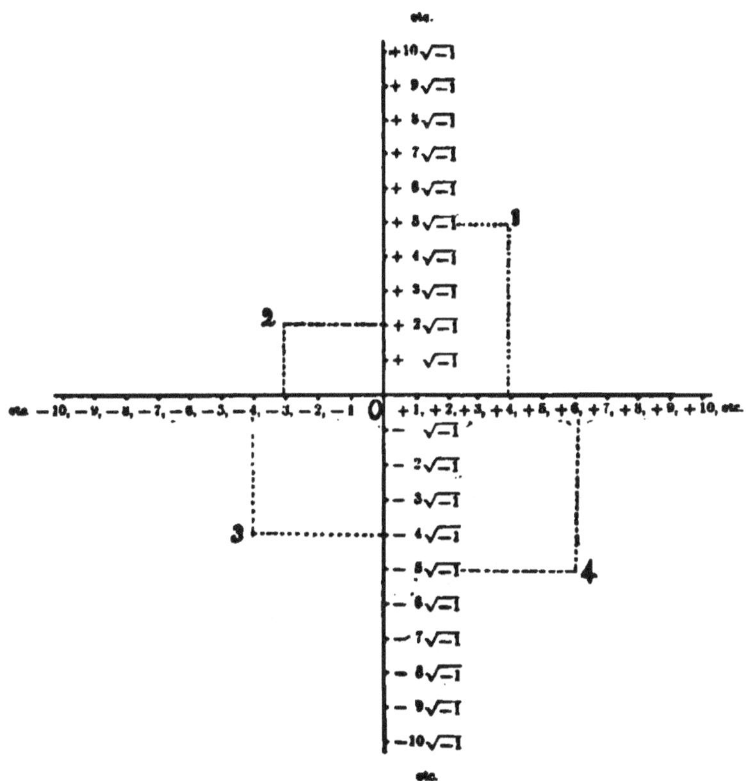

In Chapter II., we regarded the numbers of the double series as fixing lengths and directions on the line containing it. Thus, 0 fixed the origin, $+3$, the distance to the point 3 units to the right from the origin, and so on. Similarly, here, $+2\sqrt{-1}$ fixes the line from the origin to the point 2 units *above* it, and $-4\sqrt{-1}$, the distance from the origin to the point 4 units *below* it. In Argand's diagram the distance and direction of any point in the whole plane may be fixed by a *complex number*, i. e., one partly real and partly imaginary, as $\pm a \pm b\sqrt{-1}$, by letting a denote the perpendicular distance to the vertical line (measured parallel to the real series), and b the perpendicular distance to the horizontal line (measured parallel to the imaginary series). Thus, the line joining **0** and **1** in the figure would be represented by $4+5\sqrt{-1}$; the line joining **0** and **2** by $-3+2\sqrt{-1}$; the line joining **0** and **3** by $-4-4\sqrt{-1}$; the line joining **0** and **4** by $+6-5\sqrt{-1}$.

The meaning of sums, differences, products, and quotients of complex numbers may now be interpreted on the diagram. But the examination of these operations would lead us too far astray from the subject of radicals.

SECTION III.

Equations Involving Radicals.

327. Equations containing Radicals. — Since there is no reason why such equations should be of one degree rather than another, this subject might be postponed until the solution of quadratic equations is learned. Still, because the preliminary transformations are much the same in all cases, they can be treated here. Only such as reduce to simple equations are given in the following set.

1. Solve $\sqrt{x+5}+3=8-\sqrt{x}$

$$\sqrt{x+5}+3=8-\sqrt{x}$$

$$\sqrt{x+5}=5-\sqrt{x} \qquad \text{(Ax. 2)}$$

$$x+5=25-10\sqrt{x}+x \qquad \text{(Ax. 5, \textbf{114}, 2)}$$

$$10\sqrt{x}=20 \qquad \text{(Ax. } a\text{)}$$

$$\sqrt{x}=2 \qquad \text{(Ax. 4)}$$

$$x=4 \qquad \text{(Ax. 5)}$$

2. Solve $2\sqrt[3]{5x-35}=5\sqrt[3]{2x-7}$

$$8(5x-35)=125(2x-7) \qquad \text{(Ax. 5)}$$

$$x=2\tfrac{5}{6} \qquad \text{(\textbf{219}, 2)}$$

3. Solve $\sqrt{12x-5}+\sqrt{3x-1}=\sqrt{27x-2}$

$$12x-5+2\sqrt{36x^2-27x+5}+3x-1$$
$$=27x-2 \qquad \text{(Ax. 5, } \mathbf{114}, 1)$$
$$\sqrt{36x^2-27x+5}=6x+2 \qquad \text{(Axs. } a \text{ and } 4)$$
$$36x^2-27x+5=36x^2+24x+4 \qquad \text{(Ax. 5)}$$
$$\therefore\ x=\tfrac{1}{51}\quad \textit{Ans.} \qquad (\mathbf{219}, 2)$$

But upon substituting this value in the given equation we find it is not verified.

Thus, $\sqrt{-\frac{243}{51}}+\sqrt{-\frac{48}{51}}\neq\pm\sqrt{-\frac{75}{51}}$

i.e., $9\sqrt{-\frac{1}{17}}+4\sqrt{-\frac{1}{17}}\neq\pm 5\sqrt{-\frac{1}{17}}$

It is sufficient for our present purposes to know that values of the unknown obtained by such processes as these *may not satisfy the original equation*, and are therefore not roots of it unless the equation is taken in a certain way. In the present case $\frac{1}{51}$ is a root if the sign of the first *or* second terms be taken negatively.

To avoid mistakes, roots obtained for radical equations *should be verified by substitution* in the original form of the equation. (See **360**.)

4. Given $x+a=\sqrt{a^2+x\sqrt{b^2+x^2}}$ to find x

$$x^2+2ax+a^2=a^2+x\sqrt{b^2+x^2} \qquad \text{(Ax. 5)}$$
$$x^2+2ax=x\sqrt{b^2+x^2} \qquad \text{(Ax. 2)}$$
$$x+2a=\sqrt{b^2+x^2} \qquad \text{(Ax. 4)}$$
$$x^2+4ax+4a^2=b^2+x^2 \qquad \text{(Ax. 5)}$$
$$\therefore\ x=\frac{b^2-4a^2}{4a}.$$

VERIFICATION.

$$\frac{b^2-4a^2}{4a}+a=\sqrt{a^2+\left(\frac{b^2-4a^2}{4a}\right)\sqrt{b^2+\left(\frac{b^2-4a^2}{4a}\right)^2}}$$
$$\frac{b^2}{4a}=\sqrt{a^2+\frac{b^2-4a^2}{4a}\sqrt{\frac{16a^2b^2+b^4-8a^2b^2+16a^4}{16a^2}}}$$
$$=\sqrt{a^2+\left(\frac{b^2-4a^2}{4a}\right)\left(\frac{b^2+4a^2}{4a}\right)}$$
$$=\sqrt{\frac{16a^4+b^4-16a^4}{16a^2}}=\frac{b^2}{4a}$$

5. Solve $\sqrt{2x} - \sqrt{x-8} = \dfrac{2}{\sqrt{x-8}}$

$$\sqrt{2x^2 - 16x} - (x-8) = 2$$
$$\sqrt{2x^2 - 16x} = x - 6$$
$$2x^2 - 16x = x^2 - 12x + 36$$
$$2x^2 - 4x = 36$$

which is a quadratic equation, whose solution we are not now in position to obtain.

328. Rules and Directions for solving Equations involving Radicals.

No general rule can be given. The advantage of deviations from the normal process is here even more marked than in simple equations. (See **217.**)

1. When but one radical appears in an equation, transpose all the other quantities to the opposite side of the equation, and then raise both sides to the power indicated by the index of the radical.

2. When two or more radicals occur, it is usually best to place one alone on one side and then raise both members to the desired power.

3. The process just described will often have to be repeated one or more times in the same problem before all the radicals disappear.

a. When fractions occur, it is generally best to clear of fractions before squaring or cubing or raising to other powers as the case may be. Sometimes an advantage is gained by reducing a fraction to its lowest terms, or by rationalizing a denominator, or by transposing and uniting, etc.

b. Care should be taken to combine all similar terms before raising to a power.

c. By example 3 the student is admonished that the value of the unknown quantity should be verified in the original equation.

329. Exercise in the Solution of Equations containing Radicals.

1. $\sqrt{x-5}=3$.
2. $x+2-\sqrt{16+x^2}=0$
3. $x=9-\sqrt{x^2+9}$.
4. $8+\sqrt{3x+6}=14$.
5. $\sqrt[4]{2x+11}=\sqrt{5}$.
6. $\sqrt{x+25}=1+\sqrt{x}$.
7. $\sqrt{x+x^2}-x-\frac{1}{3}=0$.
8. $\sqrt{x-4}+3=\sqrt{x+11}$.
9. $\sqrt{4x+5}-\sqrt{x}=\sqrt{x+3}$.
10. $\sqrt{ax+2ab}-a=b$.
11. $\sqrt{3-x}-\sqrt[4]{3+x^2}=0$.
12. $3+\sqrt[3]{x^3-9x^2}=x$.
13. $\sqrt{4a+x}=2\sqrt{b+x}-\sqrt{x}$.
14. $\sqrt{4-\sqrt{x^4-x^2}}=x-2$.
15. $\sqrt[5]{9x-4}+6=8$.
16. $\sqrt{4x^2-7x+1}=2x-1\frac{1}{3}$.
17. $\sqrt{x}+\sqrt{4+x}=\dfrac{2}{\sqrt{x}}$.
18. $\sqrt{x}+\sqrt{9+x}=\dfrac{45}{\sqrt{9+x}}$.
19. $\sqrt{x+3}+\sqrt{x+8}-\sqrt{4x+21}=0$.
20. $\sqrt{x}+\sqrt{a}-\sqrt{ax+x^2}=\sqrt{a}$.
21. $\dfrac{x-ax}{\sqrt{x}}=\dfrac{\sqrt{x}}{x}$.
22. $\sqrt{\sqrt{x}+3}-\sqrt{\sqrt{x}-3}=\sqrt{2\sqrt{x}}$.

23. $\dfrac{x-4}{\sqrt{x}+2}=5\sqrt{x}-8+\dfrac{3\sqrt{x}}{2}.$

24. $\dfrac{1}{x}+\dfrac{1}{a}=\sqrt{\dfrac{1}{a^2}+\sqrt{\dfrac{4}{a^2x^2}+\dfrac{9}{x^4}}}.$

25. $\dfrac{a}{x}+\dfrac{\sqrt{a^2-x^2}}{x}=\dfrac{x}{b}.$

26. $\sqrt[3]{64+x^2-8x}=\dfrac{4+x}{\sqrt[3]{4+x}}.$

27. $\dfrac{3x-1}{\sqrt{3x}+1}=1+\dfrac{\sqrt{3x}-1}{2}.$

FOURTH GENERAL SUBJECT.—QUADRATIC EQUATIONS.

CHAPTER XXI.

QUADRATIC EQUATIONS CONTAINING ONE UNKNOWN QUANTITY.

330. Quadratic Equations are equations of the *second degree*. There are two kinds, **Complete** and **Incomplete**.

Complete quadratic equations (also called *adfected* quadratic equations) contain both the second and the first powers of the unknown. Incomplete quadratic equations (also called *pure* quadratic equations) contain only the second power.

Thus, $ax^2 + bx = c$ is a complete quadratic equation, while $mx^2 = n$ is an incomplete quadratic equation.

a. These are the only cases to be considered. For, the first equation contains the square and the first power of the unknown and a term which does not contain it. There can be no other. This equation reduces to the second when b (the coefficient of x) equals zero, giving $ax^2 = c$, a pure quadratic; if $a = 0$, the equation reduces to $2bx = c$ which is no longer a quadratic; while if $c = 0$, it reduces to $ax^2 + 2bx = 0$, and x divides out leaving $ax + 2b = 0$, which is also a simple equation.

SECTION I.

Solution of Incomplete Quadratic Equations.

331. Method of Solution of Incomplete Quadratic Equations.

The type form of these equations is $mx^2 = n$, i.e., a term containing x^2 and a term independent of x.

The solution is very simple.

$$mx^2 = n \qquad \text{(Hyp.)}$$

$$x^2 = \frac{n}{m} \qquad \text{(Ax. 4)}$$

$$x = \pm\sqrt{\frac{n}{m}} \qquad \text{(Ax. 6)}$$

a. It is not necessary to prefix $\pm$ to both members of the equation, for doing so *merely duplicates the roots.*

Thus, while $\pm x = \pm\sqrt{\frac{n}{m}}$ gives rise to four equations, viz.,

$$(1)\ +x = +\sqrt{\frac{n}{m}} \quad (2)\ +x = -\sqrt{\frac{n}{m}} \quad (3)\ -x = +\sqrt{\frac{n}{m}}$$

$$(4)\ -x = -\sqrt{\frac{n}{m}}.$$

two of these equations are identical with two others. Eqs. (1) and (4) are the same, as may be seen by multiplying (4) through by -1. So also (2) and (3) are identical.

1. Given $\frac{1}{3}(x^2 - 10) + \frac{1}{10}(6x^2 - 100) = 3x^2 - 65$ to find x.

$$10(x^2 - 10) + 3(6x^2 - 100) = 90x^2 - 1950 \qquad \text{(Ax. 3)}$$

$$10x^2 - 100 + 18x^2 - 300 = 90x^2 - 1950$$

$$-62x^2 = -1550 \qquad \text{(Ax. a, 86)}$$

$$x^2 = 25 \qquad \text{(Ax. 4)}$$

$$x = \pm\sqrt{5}.\ \textit{Ans.} \qquad \text{(Ax. 6)}$$

Verifications. Substituting either $+5$ or -5 in the given equation, $\frac{1}{3}(25 - 10) + \frac{1}{10}(150 - 100) = 75 - 65$ since whether we take $+5$ or -5, when squared, it gives the same number.

2. $\frac{x}{a} + \frac{b}{x} = \frac{c}{x} - \frac{x}{d}$

$$dx^2 + abd = acd - ax^2 \qquad \text{(Ax. 3)}$$
$$ax^2 + dx^2 = acd - abd$$
$$(a + d)\,x^2 = ad\,(c - b)$$
$$x^2 = \frac{ad\,(c - b)}{a + d}$$
$$x = \pm\sqrt{\frac{ad\,(c - b)}{a + d}}.$$

3. $\sqrt{x + a} = \sqrt{x + \sqrt{b^2 + x^2}}$

$$x + a = x + \sqrt{b^2 + x^2}$$
$$a = \sqrt{b^2 + x^2}$$
$$a^2 = b^2 + x^2$$
$$-x^2 = b^2 - a^2$$
$$x^2 = a^2 - b^2$$
$$x = \pm\sqrt{a^2 - b^2}.$$

332. Exercise in the Solution of Incomplete Quadratic Equations.

1. $3x^2 - 2 = 2x^2 + 2.$
2. $x^2 - 36 = \frac{x^2}{4} + 12.$
3. $x^2 - 3 = \frac{4x^2 + 18}{9}.$
4. $(9 + x)(9 - x) = 19.$
5. $(x + 1)^2 = 2x + 17.$
6. $4x - 150x^{-1} = x - 3x^{-1}.$
7. $a^2x^2 - b^2 + cx^2 = 0.$
8. $\frac{3}{1 + x} + \frac{3}{1 - x} = 8.$
9. $\frac{a^2 - x^2}{x^2 - b^2} = \frac{a}{b}.$

10. $\frac{4(x^2-5)}{3}-\frac{1}{12}=20+\frac{3(25-x^2)+10}{4}$.

11. $\frac{1+x}{1-x}=\frac{x+25}{x-25}$.

12. $x=\frac{a+bx}{b+cx}$.

13. $\frac{x+a}{x-a}+\frac{x-a}{x+a}-2=\frac{4a^2}{2a+1}$.

14. $(4+x)(5-7x)+23=(9+2x)(2-3x)$.

15. $x+\sqrt{x^2-17}=\frac{4}{\sqrt{x^2-17}}$.

16. $\frac{x-m}{x+m}=\frac{n-x}{n+x}$.

17. $x^2+4:x^2-11::100:40$.

18. $\frac{1}{2}x^2-\frac{1}{4}x^2-3:\frac{1}{4}x^2-\frac{1}{8}x^2+3::9:3$.

SECTION II.

Solution of Complete Quadratic Equations.

333. Complete Quadratic Equations are solved by a method called completing the square, which makes their solution different from anything we have yet had.

1. The equation $(ax-b)^2=c^2$ is a complete quadratic, as may be seen by squaring the left member. Moreover, it can be solved somewhat like an incomplete quadratic, viz., by extracting the square root of both members.

$$(ax-b)^2=c^2 \qquad \text{(Hyp.)}$$
$$ax-b=\pm c \qquad \text{(Ax. 6)}$$
$$ax=b\pm c \qquad \text{(Ax. 1)}$$
$$x=\frac{b\pm c}{a} \qquad \text{(Ax. 4)}$$

2. This suggests the question, can all complete quadratics be put in the same form? A little consideration will show that they can. Thus, every complete quadratic equation can be reduced to the type form.

$$ax^2 + 2\,bx = c.$$

This means that, if necessary, the equation is cleared of fractions, the terms transposed, collected, and arranged so that the term containing x^2 is first, that containing x is next, and the known terms form the right member.

Thus, in the equation

$$\frac{3}{5}(x-6)(x+2) = \frac{2}{3}\left(62\frac{1}{10} + \frac{18\,x}{5}\right)$$

$$18\,x^2 - 72\,x - 216 = 1242 + 72\,x \qquad \textbf{(111)}$$

$$18\,x^2 - 144\,x = 1458,$$

$$a = 18,\ 2\,b = -144, \text{ or } b = -72,\ c = 1458.$$

and so with any other equation.

3. To show that the type equation $ax^2 + 2\,bx = c$ can be reduced to essentially the same form as $(ax - b)^2 = c^2$, and can then be solved, we proceed as follows: —

$$ax^2 + 2\,bx = c$$

$$a^2x^2 + 2\,abx = ac. \qquad \text{(Ax. 3)}$$

Now the left member consists of the first two terms of the trinomial square of $ax + b$, and all it needs is the third term, b^2, to make it a perfect square as desired. But there is no reason why we may not add b^2 to the left member provided we also add it to the right member.

$$a^2x^2 + 2\,abx + b^2 = b^2 + ac \qquad \text{(Ax. 1)}$$

$$ax + b = \pm\sqrt{b^2 + ac} \qquad \text{(Ax. 6)}$$

$$x = \frac{-b \pm \sqrt{b^2 + ac}}{a}. \qquad \text{(Axs. 2 and 4)}$$

We conclude that every complete quadratic can be reduced to the form $(ax \pm b)^2 = b^2 \pm ac$, and can then be solved (after extracting the square root of both sides,) as *two simple equations.*

Two points in the reduction from the type form deserve especial attention, viz., making the coefficient of x^2 a perfect square, and completing the square.

334. The Coefficient of x^2 must be a perfect square *and positive.* If the coefficient of x^2 is not already a perfect square, as 1, a^2, 64, $9\,b^2$, or the like, the equation must be multiplied or divided through by such a quantity that the coefficient of x^2 is made a perfect square and positive. The extremes in a trinomial square, as is evident, must be positive.

335. Completing the Square. The terms added must be so chosen as to unite with the other two to form a trinomial square. (See **135, 1.**) Let us investigate how this term may with certainty be found.

If in the expression

$$A^2 + 2\,AB + B^2$$

we have $A^2 + 2\,AB$ given to find B^2, the problem is easily solved. For, dividing $2\,AB$, the second term by $2\,A$, *twice the square root of the first term*, the quotient is B, which must be squared for the third term. To illustrate this let us take the 5th example in **135, 1.**

Given $9\,x^2 + 30\,x$ to find 25.

$$2\,\sqrt{9\,x^2} = 6\,x;\ 30\,x \div 6\,x = 5;\ 5^2 = 25.$$

Given $4a^2x^2 - 12\,abxy$ to find $9\,b^2y^2$.

$$2\,\sqrt{4\,a^2x^2} = 4\,ax;\ 12\,abxy \div 4\,ax = 3\,by;\ (3\,by)^2 = 9\,b^2y^2.$$

As an exercise in this process, the student should solve all the problems of **135, 1** in the same way.

It is easily seen that *when the coefficient of x^2 is* 1, the process gives the square of half the coefficient of x, as the third term. See Ex's 1–4, 11–13, 16, 19 in **135, 1.**

336. Examples of the Solution of Complete Quadratic Equations.

1. Given $2x = 4 + \frac{6}{x}$

$$2x^2 = 4x + 6 \qquad \text{(Ax. 3)}$$
$$x^2 - 2x = 3 \qquad \textbf{(334)}$$
$$2\sqrt{x^2} = 2x;\ \ 2x \div 2x = 1;\ \ 1^2 = 1 \qquad \textbf{(335)}$$
$$x^2 - 2x + 1 = 3 + 1 = 4$$
$$x - 1 = \pm 2 \qquad \text{(Ax. 6)}$$
$$\text{i.e. } x - 1 = +2, \text{ or } x - 1 = -2$$
$$\therefore x = 3, \text{ or } x = -1. \quad \textit{Ans.}$$

VERIFICATION.

1st. $x = 3$ in the given equation, $2 \times 3 = 4 + \frac{6}{3}$ i.e. $6 = 6$.

2d. $x = -1$ in the given equation, $2 \times -1 = 4 + \frac{6}{-1}$, i.e., $-2 = -2$.

2. Solve $50x^2 - 15x = 27$

$$100x^2 - 30x = 54 \qquad \textbf{(334)}$$
$$2\sqrt{100x^2} = 20x;\ 30x \div 20x = \tfrac{3}{2};\ (\tfrac{3}{2})^2 = \tfrac{9}{4}$$
$$100x^2 - 30x + \tfrac{9}{4} = 54 + 2\tfrac{1}{4} = 56\tfrac{1}{4} = 56.25$$
$$10x - \tfrac{3}{2} = \pm 7\tfrac{1}{2} \text{ i.e. } +7\tfrac{1}{2} \text{ and } -7\tfrac{1}{2}.$$
$$\therefore \quad 10x = \tfrac{3}{2} + 7\tfrac{1}{2}, \text{ or } \tfrac{3}{2} - 7\tfrac{1}{2}$$
$$x = \tfrac{9}{10}, \text{ or } -\tfrac{6}{10} = -\tfrac{3}{5}. \quad \textit{Ans.}$$

3. Solve $(x + 1)(2x + 3) = 4x^2 - 22$

$$2x^2 + 5x + 3 = 4x^2 - 22$$
$$-2x^2 + 5x = -25$$
$$x^2 - \tfrac{5}{2}x = +\tfrac{25}{2}$$
$$\tfrac{5}{2} \div 2 = \tfrac{5}{4};\ (\tfrac{5}{4})^2 = \tfrac{25}{16}$$
$$x^2 - \tfrac{5}{2}x + \tfrac{25}{16} = \tfrac{25}{2} + \tfrac{25}{16} = \tfrac{225}{16}$$
$$x - \tfrac{5}{4} = \pm \tfrac{15}{4}$$
$$x = 5, \text{ or } -2\tfrac{1}{2}.$$

4. Solve $3x^2 + 10x = 57$

$$36x^2 + 120x = 684 \qquad \textbf{(334)}$$
$$2\sqrt{36x^2} = 12x;\ 120x \div 12x = 10;\ 10^2 = 100$$
$$36x^2 + 120x + 100 = 784$$
$$6x + 10 = \pm 28$$
$$x = 3 \text{ or } -\tfrac{38}{6}.$$

5. Solve $nx^2 + px = q$

$$4\,n^2x^2 + 4\,npx = 4\,nq \qquad (334)$$

$$2\sqrt{4\,n^2x^2} = 4\,nx;\ 4\,npx \div 4\,nx = p;\ (p)^2 = p^2$$

$$4\,n^2x^2 + 4\,npx + p^2 = 4\,nq + p^2$$

$$2\,nx + p = \pm\sqrt{4\,nq + p^2}$$

$$x = \frac{-p \pm \sqrt{4\,nq + p^2}}{2\,n}.$$

OBSERVATION. — This example shows that if an equation be multiplied through by 4 times the coefficient of x^2, the square of the old coefficient of x, viz., p, is the quantity to be added to complete the square. See also Ex. 4.

337. Rules for Solving Complete Quadratic Equations. — One rule only is really needed, the first and second rules as given below being special cases of the general process explained in the third rule.

1. First rule.

(1. Reduce the given equation to the type form $ax^2 + bx = c$, and divide through by the coefficient of x^2, if it is not already 1.

(2. Add the square of one half the coefficient of x to both sides of the equation and extract the square root of the two members.

(3. Solve the two resulting simple equations for the two values of x. (See Examples 1 and 3 of the last article.)

2. Second rule.

(1. Reduce the equation to the type form (removing any monomial factors which may exist in it, Ax. 4); then multiply through by 4 times the coefficient of x^2, merely indicating the multiplications in the left member.

(2. Add to both sides of the equation the square of the old coefficient of x, and complete the solution as in rule **1**. (See Exs. 4 and 5.)

3. Third rule.

(1. Reduce the equation to the type form and multiply or divide through by such a quantity as will make the coefficient of x^2 a perfect square, and at the same time shorten to the greatest extent the subsequent reductions.

(2. Add to both sides of the equation such a quantity as will make the left member a perfect trinomial square. To do this divide the second term by twice the square root of the first term, and square the quotient for the third term. Complete the solution as in the first rule.

a. The first and second rules are easier for beginners. The rule just given is really the best because it is most flexible. However, considerable experience is needed to be able to use it to the best advantage. Suggestions will be made from time to time to show the student its superiority over the others, and therefore the desirability of mastering it as well as the others. The second, which ensures the avoidance of fractions, is sometimes called the Hindoo Rule.

4. Fourth rule.

(1. Reduce the equation to the type form.

(2. Substitute the values of n, p, and q from the given problem in the answer of example 5 of the last article. (See **255**.)

Thus, given $\frac{x^2}{2} - \frac{x}{3} + 7\frac{3}{8} = 8$

$12x^2 - 8x = 15$ (to type form)

$nx^2 + px = q$ (Eq. of Ex. 5)

So here, $n = 12$; $p = -8$; and $q = 15$.

Substituting these values in the value of x found in Ex. 5.

$$x = \frac{-(-8) + \sqrt{64 + 4 \times 12 \times 15}}{2 \times 12}$$

$$\text{or } \frac{-(-8) - \sqrt{64 + 4 \times 12 \times 15}}{2 \times 12}$$

$$= \frac{8 + 28}{24} = \frac{3}{2}, \text{ or } \frac{8 - 28}{24} = -\frac{5}{6} \qquad \textit{Ans.}$$

338. Exercise in the Solution of Complete Quadratic Equations.

1. $x^2 - 15 = 45 - 4x$.
2. $x^2 + 22x = 75$.
3. $x^2 = x + 72$.
4. $3y^2 + 48 = 30y$.
5. $5x^2 + 20x = 25$.
6. $x^2 - 6x = 0$.
7. $x^2 - 341 = 20x$.
8. $23x = 120 + x^2$.
9. $x^2 - 6x = 6x + 28$.
10. $4x^2 + 4x = -1$.
11. $x^2 - \frac{2}{3}x = 32$.
12. $\frac{x^2}{10} + 350 - 12x = 0$.
13. $\frac{x + 22}{3} - \frac{4}{x} = \frac{9x - 6}{2}$.
14. $(x - 1)(x - 2) = 1$.
15. $\frac{19}{5}x = \frac{11}{5} - x^2$
16. $8x^2 - 20x = 21$.

Suggestion. — Multiply through by 2.

17. $\frac{17x^2}{2} + \frac{14}{5}x = 20x + 25\frac{1}{5} - \frac{13}{2}x^2$.

Suggestion. — After clearing divide through by 6.

18. $252x^2 + 360x = -125$.

Suggestion. — $252 = 7 \times 36$. Therefore if an extra factor, 7, is introduced, the product is a perfect square. Multiplying the equation through by 7, we get

$$7 \times 252x^2 + 7 \times 360x = -875$$

$$2\sqrt{7^2 \times 36x^2} = 2 \times 42x;\ 7 \times 360x \div 2 \times 42x = 30;\ (30)^2 = 900.$$

$$(42x)^2 + (\quad) + (30)^2 = 900 - 875 = 25$$
$$42x + 30 = \pm 5$$
$$x = -\tfrac{25}{42}, \text{ or, } -\tfrac{5}{6}.$$

Verification. — $\begin{cases} \frac{625}{7} - \frac{1500}{7} = -\frac{875}{7} = -125 \\ \text{or, } 175 - 300 = -125. \end{cases}$

REMARK. — To see the advantage gained, the student should solve this exercise by the first and second rules.

19. $72x^2 + 408x = 1222.$

SUGGESTION. — Divide through by 2.

20. $84x^2 + 45 = 124x.$

SUGGESTION. — $84 = 7 \times 3 \times 4$. Hence multiply through by 7×3.

21. $96x^2 = 4x + 15.$

SUGGESTION. — $96 = 6 \times 16$. Hence multiply through by 6.

22. $622x = 15x^2 + 6384.$

SUGGESTION. — Multiply through by 15.

23. $\frac{3}{5}(x+6)(x-2) = \frac{3}{4}\left(2\frac{1}{2} + \frac{36x}{5}\right).$

24. $3x + 4 = 39x^{-1}.$

25. $\frac{x}{a} + \frac{a}{x} = \frac{2}{a}.$

26. $\frac{a^2 - b^2}{c} = 2ax - cx^2.$

27. $16x^{-1} - 4 = 12x^{-2}.$

28. $x + \frac{1}{x} = \frac{4}{\sqrt{3}}.$

SUGGESTION. — Multiply through by x, and not by $x\sqrt{3}$

29. $986x - 145080 = x^2.$

30. $7x^2 - 7x = 1.$

31. $(\frac{1}{3}x)^2 + 1 = (\frac{5}{13})^2 - 1\frac{8}{3}x - (\frac{1}{4}x)^2$

SUGGESTION. — Transpose and unite, without clearing of fractions.

32. $x^2 - (a+b)x + ab = 0.$

33. $\frac{1}{x + \frac{1}{x}} = 1.$

34. $(m-n)x^2 - nx = m.$

35. $\frac{1}{a-x} - \frac{1}{a+x} = \frac{-3+x^2}{a^2-x^2}.$

36. $\frac{1}{a-b+x} = \frac{1}{a} - \frac{1}{b} + \frac{1}{x}.$

37. $\sqrt{x-1} = x - 1$.

38. $x - \sqrt{x} = 20$.

39. $\sqrt{2x+2} + \sqrt{7+6x} = \sqrt{7x+72}$.

40. $7(x+7) + \frac{7(3x+50)}{x} = 0$.

41. $9a^4b^4x^2 - 6a^3b^3x = b^2$.

42. $\sqrt{x+a} - \sqrt{x-a} = \sqrt{2x}$.

43. $cx^2 - 2cx\sqrt{d} = ax^2 - cd$.

SECTION III.

PROBLEMS.

339. Problems Involving the Solution of Quadratic Equations, Complete and Incomplete.

a. Algebra is a formal science made to cover all cases, and without any reference to particular problems. Some problems by their nature admit of negative numbers: in such a negative answer has its proper significance; while in others negative values for the answer are inadmissible. Moreover, in algebra, imaginary values for the unknown denote that the problem is *impossible*.

1. Find two numbers one of which is 5 times as great as the other, and the difference of whose squares is 96.

SUGGESTION. — Let x = the less, then $5x$ = the greater. Then the equation is $(5x)^2 - x^2 = 96$.

2. The square of a certain number diminished by 17 is equal to 130 diminished by twice the square of the number. Required the number.

3. A person bought a quantity of cloth for \$120; and if he had bought 6 yards more for the same sum the price per yard would have been \$1 less. What was the price?

Let $x =$ the price per yard, then the equation is

$$\frac{120}{\frac{120}{x}+6} = x - 1. \quad \text{Therefore } x = 5 \text{ or } -4. \qquad (219, 2)$$

But -4 can have no meaning in this problem. (253). We can, however, so modify the statement of the problem as to be able to make use of the second answer. Thus, — The exchange account of a banker amounted in a certain number of days to $120, during which time exchange remained the same. Had the period differed from what it was by 6 days, and *in his favor* (i.e. 6 more if premiums and 6 less if discounts) it would have made a difference of $1 per day. What was his daily premium or daily discount?

Let $x =$ the daily premium or discount.

Then referring to the equation above, we see that if x is positive the statement is satisfied; also if x is negative, $\frac{120}{x}$ is negative and 6 added diminishes the number of days *numerically*. The two answers 5 and -4 signify a daily *premium* of $5 or a daily *discount* of $4.

4. Find a number such that if you subtract it from 10 and multiply this number by the number itself the product shall be 21.

5. Divide the number 346 into two such parts that the sum of their square roots shall be 26.

Let $x =$ the square root of one part, and $26 - x$, that of the other.

6. A merchant bought a piece of cloth for $324, and the number of dollars he paid for a yard was to the number of yards as $4:9$. How many yards did he buy?

7. If a certain number be added to 94 and then the same number be taken from 94, the product of these derived numbers is 8512. What are the numbers?

8. A man traveled 60 miles and if he had traveled 1 mile an hour more he would have required 3 hours less to perform the journey. At what rate did he travel?

9. Find three consecutive numbers whose sum is equal to the product of the first two.

10. A rectangular field whose length is 3367 and whose breadth is 37 yards has by it another field of an equal number of acres whose length is to its breadth as 13 : 7. What are the dimensions of the latter?

11. An individual bought a certain number of kilograms of salt, 4 times as many of sugar, and 8 times as many of coffee, and paid for each of them 40 times as many cents as there were kilos of that material. How many kilos of coffee did he buy if he paid altogether $32.40?

12. If a certain number is diminished by 3 and also increased by 3, then is the sum of the quotients which we get by dividing the greater by the less, and the less by the greater equal to $3\frac{1}{3}$. What is the number?

13. A capital stands at 4% interest. If the number of dollars of capital be multiplied by the number of dollars in 5 months' interest, the product is 1053375. How many dollars are there in the capital?

14. The hypothenuse of a certain right-angled triangle is 2 ft. greater than the base, and 9 ft. greater than the perpendicular. Find the sides of the triangle.

15. A girl bought a number of oranges for 40 cts. Had she bought at another place she would have received 3 more oranges for the same money, each orange costing $\frac{2}{3}$ of a ct. less. How many did she buy?

16. Two peasant women together brought 260 eggs to market, and both lost the same amount. "Had I sold your eggs," said the first to the second, "and had they brought my price, I should have lost on them 7.20 marks." "That may be," responded the other, "but if I had sold your eggs at the price mine brought, I should have lost 9.80 marks." How many eggs did each bring to market?

SUGGESTION. — The equation is $x \cdot \frac{7.20}{260 - x} = \frac{9.80}{x}(260 - x)$, which, after clearing, may have the square root of each member extracted *as it stands*.

17. The perimeter of a rectangular field is 500 yds., and its area 14400 sq. yds. Find the length of the sides.

18. What are eggs a dozen when two more in a shilling's worth lowers the price a penny per dozen?

19. Find two numbers whose difference is d, and whose product is p.

20. Divide a straight line a inches in length into two parts such that the longer part may be a mean proportional between the whole line and the shorter part.

This is called in geometry dividing the line in "golden section," or into mean and extreme ratio. (See **378**, **1**)

21. What number is that whose square increased by 5 is greater by 23 than $7\frac{1}{2}$ times the number?

22. The product of two numbers is p and their quotient is q; required the numbers.

23. The sum of the squares of two numbers is c, and their difference is d. Required the numbers.

What happens if $2c$ is less than d^2? Thus, e.g., take $d = 9$ and $2c = 80$. (See Art. **265**.)

24. Two points start out together from the vertex of a right angle along its respective sides, the one moving m ft. per second, the other n ft. per second. How long will they require to be c ft. apart?

25. It is required to find three numbers such that the product of the first and second equals a, the product of the first and third equals b, and the sum of the squares of the second and third equals c.

26. A set out from C towards D and travelled 7 miles per day. After he had gone 32 miles, B set out from D towards C and went every day $\frac{1}{15}$ of the whole journey, and after he had traveled as many days as he went miles in a day, he met A. Required the distance from C to D.

27. A reservoir can be filled by two pipes, by one 2 hours sooner than by the other. If both pipes are open at the same time the reservoir is filled in $1\frac{7}{8}$ hours. In how many hours can it be filled by the smaller pipe alone?

28. A farmer sowed one year a hektoliter of wheat; the next year he sowed what he harvested the first year, less b hektoliters, and reaped c fold of what he sowed and d hektoliters beside. Assuming a like fruitfulness both years how much did he reap the first year?

29. It is required to divide each of the numbers 21 and 30 into two parts, so that the first part of 21 may be 3 times as great as the first part of 30; and that the sum of the squares of the remaining parts may be 585.

30. A looking glass, in size $a \times b$ inches, has a frame of uniform width and of the same area as the glass. What is the width of the frame? Suppose $a = 12$, and $b = 18$, i.e., a glass 12×18, and substitute in the answer as by the fourth rule.

31. A square courtyard has a rectangular walk around it (on the inside). The side of the court is 2 yards less than six times the breadth of the walk; and the number of square yards in the walk exceeds the number of yards in the perimeter of the court by 92. Find the area of the court.

32. The driving wheels of a locomotive are two feet greater in diameter than the running wheels; the running wheels make 140 turns more than the driving wheels in a mile. What are the diameters? (The circumference of a wheel $= 3\frac{1}{7}$ times the diameter, nearly.)

33. Generalize the preceding problem.

34. A grazier bought a certain number of oxen for \$240, and after losing 3 sold the remainder at \$8 per head more than they cost him, thus gaining \$59. What number did he buy?

35. Generalize the 34th problem.

36. A wall is built by two masons of which the first begins to work $1\frac{1}{2}$ days later than the other. It takes $5\frac{1}{2}$ days from the time the first began. How long would it take to finish the wall should the first work 3 days less than the second?

SOLUTION. — Let x = the number of days the first requires in which to build the wall, and y = the number of days the second requires.

Then $\frac{7}{y} + \frac{5\frac{1}{2}}{x} = 1$ whence $\frac{1}{y} = \frac{2x - 11}{14x}$

There is need for still another unknown which we may denote by z. Let z = the number of days required when the first works 3 days less than the second. Then,

$$\frac{z-3}{x} + z\left(\frac{2x-11}{14x}\right) = 1$$

$$14z - 42 + 2xz - 11z = 14x.$$

Thus it appears that we have to find the values of x y, z, a *single* quadratic equation in two unknowns, which is *indeterminate.* The method of solution will be like that of **247**. Solving for x,

$$x = \frac{3z - 42}{14 - 2z} = -1 - \frac{z - 28}{2z - 14}$$

To make x positive the signs of the two terms of the *fraction* must be unlike. For this we can assume $z = 8$, and 10, both of which give positive integral values for x, but only $z = 8$ gives y positive and integral. Thus we have $z = 8$, $x = 9$, $y = 18$. Of course in the present problem there is nothing to interfere with z having *fractional* values within the limits determined by the above equation.

37. A man has \$1300, which he divides into two portions and loans at different rates of interest. If the first portion had been loaned at the second rate, it would have produced \$36; and if the second portion had been loaned at the first rate it would have produced \$49. Required the rates of interest.

38. Required the rates in the preceding if the two portions produce equal returns. How does this problem differ from the preceding?

39. There is a certain fraction to each of whose terms if 1 be added the product of the resulting fraction and original fraction is $\frac{2}{5}$. Required the fraction.

CHAPTER XXII.

SIMULTANEOUS QUADRATIC EQUATIONS.

340. Simultaneous Equations in Quadratics. See **225**. Two cases may be distinguished. First, when two or more of the given equations are of the second degree; and second, when there is but one quadratic equation given. Perhaps the latter ought to be called simultaneous quadratic and simple equations. See **195**.

Thus, $\begin{cases} 2x^2 + 3y^2 = 15 \\ 5x^2 + y = 10 \end{cases}$ is an example of the first case; while, $\begin{cases} 11xy + 9x + 13y = 19 \\ x + 7y = 14 \end{cases}$ is an example of the second case.

SECTION I.

Simultaneous Equations, One of which is of the Second Degree, and the Others Simple Equations.

341. Solution of Simultaneous Equations where One is Quadratic.—To solve such, substitute in the quadratic the values obtained from the other equations, so as to get a single quadratic equation containing one unknown quantity. Solve this and substitute the roots in the other equations. Two sets of answers will generally result. Examples will make this plain.

1. Given (1) $x^2 + 3xy = 54$, and (2) $x + 4y = 23$

(2₁) $x = 23 - 4y$.

(3) $(23 - 4y)^2 + 3(23 - 4y)y = 54$ (**228**)

$(3_1)\ 529 - 184\,y + 16\,y^2 + 69\,y - 12\,y^2 = 54$

$(3_2)\ 4\,y^2 - 115\,y = -475$ (334)

$$4\,y^2 - 115\,y + \left(\frac{115}{4}\right)^2 = \frac{13225}{16} - \frac{7600}{16} \qquad (335)$$

$$2\,y - \frac{115}{4} = \pm\frac{75}{4}$$

$$\begin{cases} y = 23\tfrac{3}{4}, \text{ or} \\ \therefore x = -72, \end{cases} \begin{cases} y = 5 \\ x = 3. \end{cases} \qquad (229)$$

2. Given $\begin{cases} (1)\ x^2 + y^2 + z^2 = 50 \\ (2)\ 2\,x + 3\,y + z = 23 \\ (3)\ x + 2\,y + 3\,z = 23 \end{cases}$

Eliminating x and y (so as to have z only in the quadratic equation) from equations (2) and (3), and substituting their values, —

$(2_1)\ 2\,x + 3\,y = 23 - z$

$(3_1)\ 2\,x + 4\,y = 46 - 6\,z$

$(4)\ \qquad y = 23 - 5\,z$ (Ax. 2)

$(2_2)\ 4\,x + 6\,y = 46 - 2\,z$

$(3_2)\ 3\,x + 6\,y = 69 - 9\,z$

$\qquad x = 7\,z - 23$ (Ax. 2)

$(1)\ (7\,z - 23)^2 + (23 - 5\,z)^2 + z^2 = 50$ (228)

$49\,z^2 - 322\,z + 529 + 529 - 230\,z + 25\,z^2 + z^2 = 50$

$75\,z^2 - 552\,z = -1008$

$25\,z^2 - 184\,z = -336$

$25\,z^2 - (\quad) + (18.4)^2 = 2.56$

$5\,z - 18.4 = \pm 1.6$

$$\begin{cases} \quad z = 4, \text{ or } z = 3.36 \\ (4)\ y = 3 \qquad y = 6.20 \\ (5)\ x = 5 \qquad x = .52 \end{cases} \quad \textit{Ans.} \qquad \begin{matrix} \\ (229) \\ (229) \end{matrix}$$

342. Special Forms for which there are more elegant solutions than by substitution. — If the two given equations enable us to calculate quickly the values of the expressions $x^2 \pm 2xy + y^2$, we have immediately, by extracting their square roots, the values of $x + y$ and $x - y$, from which the values of x and y are readily found by adding and subtracting these equations. **(255, 4.)**

1. Given (1) $x^2 + y^2 = 74$ and (2) $x + y = 12$

$$
\begin{array}{lrcll}
(2_1) & x^2 + 2xy + y^2 & = & 144 & \text{(Ax. 5)} \\
(1) & x^2 \qquad\quad + y^2 & = & 74 & \\ \hline
(3) & 2xy & = & 70 & \\
(3_1) & 4xy & = & 140 & \\
(4) & x^2 - 2xy + y^2 & = & 4 & [(2_1) - (3_1), \text{ Ax. } 2] \\
(4_1) & x - y & = & \pm 2 & \text{(Ax. 6)} \\
(2) & x + y & = & 12 & \\ \hline
 & 2x & = & 14 \text{ or } 10 & \text{(Ax. 1)} \\
 & x & = & 7 \text{ or } 5 & \\
 & 2y & = & 10 \text{ or } 14 & \text{(Ax. 2)} \\
 & y & = & 5 \text{ or } 7. &
\end{array}
$$

Ans. $x = 7$, $y = 5$; or, $x = 5$ and $y = 7$.

2. (1) $x - y = 12$ (2) $xy = 85$

$$
\begin{array}{lrcll}
(1_1) & x^2 - 2xy + y^2 & = & 144 & \text{(Ax. 5)} \\
(2_1) & 4xy & = & 340 & \\ \hline
(3) & x^2 + 2xy + y^2 & = & 484 & \text{(Ax. 1)} \\
(3_1) & x + y & = & \pm 22 & \text{(Ax. 6)} \\
(1) & x - y & = & 12 & \\
 & x & = & 17, \text{ or } -5 & \\
 & y & = & 5, \quad -17 &
\end{array}
$$

a. A verification will show in this as in all the examples of this and the preceding article that the answers are obtained in sets. Thus + 17 and + 5 go together, and − 5 and − 17. But + 17 and − 17 as values of x *and* y *would not satisfy the equations.* So with all quadratic solutions.

343. Exercise in the Solution of Simultaneous Equations. — Those which come under **342** should be solved by the method there explained.

1. $\begin{cases} x + y = 7 \\ x^2 + 2y^2 = 34. \end{cases}$

2. $\begin{cases} x + 4y = 23 \\ x^2 + 3xy = 54. \end{cases}$

3. $\begin{cases} 5x - y = 17 \\ xy = 12. \end{cases}$

4. $\begin{cases} 3x - y = 11 \\ 3x^2 - y^2 = 47. \end{cases}$

5. $\begin{cases} x + y = 15 \\ xy = 36. \end{cases}$

6. $\begin{cases} 3x - 2y = 2 \\ 9x^2 + 4y^2 = 394. \end{cases}$

7. $\begin{cases} xy = 923 \\ x + y = 84. \end{cases}$

8. $\begin{cases} xy = -2193 \\ x + y = -8. \end{cases}$

9. $\begin{cases} \frac{1}{x} + \frac{1}{y} = 2 \\ x + y = 2. \end{cases}$

10. $\begin{cases} 2x + y = 7 \\ 4x^2 + y^2 = 25. \end{cases}$

11. $\begin{cases} x^2 + y^2 + xy = 208 \\ x + y = 16. \end{cases}$

SUGGESTION. — First find the value of xy and then solve as in the last article.

12. $\begin{cases} x - y = 3 \\ x^2 - 3xy + y^2 = -19. \end{cases}$

SUGGESTION. — Divide the first equation by the second in Ex. 13.

13. $\begin{cases} x^2 - y^2 = 16 \\ x - y = 2. \end{cases}$

14. $\begin{cases} 4(x^2y^2) = 13xy \\ x - y = 6. \end{cases}$

15. $\begin{cases} \frac{x}{y} + \frac{y}{x} = 2\frac{1}{2} \\ x + y = 6. \end{cases}$

16. $\begin{cases} x + \frac{1}{3} = \frac{2x + y}{3} \\ \frac{x + y}{x} = \frac{4x - y}{2}. \end{cases}$

17. $\begin{cases} \frac{5y}{x} + \frac{2y + 3}{x + 3} = 0 \\ 4x + 3y = 1. \end{cases}$

18. $\begin{cases} \frac{x}{3} + \frac{y}{2} = 1 \\ \frac{3}{x} + \frac{2}{y} = 4. \end{cases}$

19. $\begin{cases} xy = a \\ \dfrac{x}{y} = b. \end{cases}$

21. $\begin{cases} x + y = a \\ xy = \frac{1}{4}(a^2 - b^2). \end{cases}$

20. $\begin{cases} (x - 2)(y - 3) = a \\ (x - 2) \div (y - 3) = b. \end{cases}$

22. $\begin{cases} ax + by = p \\ cx^2 + dy^2 = q. \end{cases}$

23. $\begin{cases} 12 : x :: 16 : 3y \\ \sqrt{x} + \sqrt{y} = 5. \end{cases}$

SECTION II.

Simultaneous Equations, two or more of which are of the Second Degree.

344. Solution of Simultaneous Quadratic Equations. — It is not possible to give a general solution. Many particular examples, however, can be solved by special methods.

1. Given (1) $x^2 - 2pxy = a^2$ and (2) $y^2 + 2qxy = b$.

(1_1) $x^2 - 2pxy + p^2y^2 = a^2 + p^2y^2$ **(333,** 3**)**

(1_2) $x = py \pm \sqrt{a^2 + p^2y^2}$ (Ax. 6)

(2_1) $y^2 + 2q(py \pm \sqrt{a^2 + p^2y^2})\, y = b$ **(228)**

$\pm 2qy\sqrt{a^2 + p^2y^2} = b - y^2(1 + 2pq)$ **(328)**

$4q^2y^2(a^2 + p^2y^2) = b^2 - 2by^2(1 + 2pq) + y^4(1 + 2pq)^2$ (Ax. 5)

which is an equation of the fourth degree and cannot in general be solved.

a. The general equation of the second degree in two unknowns, i.e., one which contains every possible term, is

$$ax^2 + bxy + cy^2 + dx + ey + f = 0.$$

If, now, two such equations as this be combined the process of solution is quite like the example just reduced (only very much more complex), and therefore is not solvable by direct algebraic methods.

2. Given (1) $x^2 + y^2 = a$ and (2) $xy = b$.

(3) $x^2 + 2xy + y^2 = a + 2b$ (Ax. 1)

(3_1) $x + y = \pm\sqrt{a + 2b}$ (Ax. 6)

(4) $x^2 - 2xy + y^2 = a - 2b$ (Ax. 2)

(4_1) $x - y = \pm\sqrt{a - 2b}$ (Ax. 6)

$$\therefore x = \frac{\pm\sqrt{a+2b} \pm \sqrt{a-2b}}{2}$$

$$y = \frac{\pm\sqrt{a+2b} \mp \sqrt{a-2b}}{2}.$$

REMARK. — The sign $\pm$ (read "minus or plus") means minus *first.*

3. (1) $xy + 6x + 7y = 50$. (2) $3xy + 2x + 5y = 72$

(1_1) $3xy + 18x + 21y = 150$

(2) $3xy + \ \ 2x + \ \ 5y = \ \ 72$

(3) $16x + 16y = \ \ 78$

(3_1) $\therefore x = \dfrac{39 - 8y}{8}$

(1) $\dfrac{39 - 8y}{8} \cdot y + \dfrac{234 - 48y}{8} + 7y = 50$ **(228)**

$$39y - 8y^2 + 234 - 48y + 56y = 400$$

$$8y^2 - 47y = -166$$

$$y^2 - \frac{47}{8}y = -\frac{166}{8}$$

$$y^2 - \frac{47}{8}y + \left(\frac{47}{16}\right)^2 = \frac{2209}{256} - \frac{5312}{256}$$

$$y = \frac{47}{16} \pm \frac{1}{16}\sqrt{-3103}.$$

(3) $x = +\dfrac{31}{16} \mp \dfrac{1}{16}\sqrt{-3103}.$ **(229)**

4. (1) $2x^2\ 3xy + y^2 = 63 +$ (2) $x^2 + 2xy + 6y^2 = 174$.

Put $y = rx$, r being unknown.

Then substituting the value of y in each equation.

$$(1)\quad 2x^2 + 3rx^2 + r^2x^2 = 63$$

$$(2)\quad x^2 + 2rx^2 + 6r^2x^2 = 174$$

$$(1_1)\quad (2 + 3r + r^2)\,x^2 = 63 \qquad (97)$$

$$(1_2)\quad x^2 = \frac{63}{2 + 3r + r^2} \qquad (\text{Ax. } 4)$$

$$(2_1)\quad (1 + 2r + 6r^2)\,x^2 = 174 \qquad (97)$$

$$(2_2)\quad x^2 = \frac{174}{1 + 2r + 6r^2} \qquad (\text{Ax. } 4)$$

$$\frac{63}{2 + 3r + r^2} = \frac{174}{1 + 2r + 6r^2} \qquad (231, \text{ Ax. } 7)$$

$$63 + 126r + 378r^2 = 348 + 522r + 174r^2 \qquad (\text{Ax. } 3)$$

$$204r^2 - 396r = 285$$

$$68r^2 - 132r = 95 \qquad (\text{Ax. } 4)$$

$68 = 4 \times 17$; $\therefore 4 \times 17 \times 17$ is a perfect square $= 34^2$.

$$17 \times 68r^2 - 17 \times 132r = 1615.$$

$$(34r)^2 - (\quad) + (33)^2 = 1615 + 1089.$$

$$34r - 33 = \pm 52$$

$$r = +\tfrac{5}{2}, \text{ or } -\tfrac{19}{34}$$

$$(1_2)\quad x^2 = \frac{63}{2 + 3\times\frac{5}{2} + \frac{25}{4}} = 4; \text{ or } \frac{63}{2 - 3\times\frac{19}{34} + \frac{361}{1156}} = \frac{3468}{35} \qquad (229)$$

$$\therefore\ x = \pm 2, \text{ or } \pm\sqrt{\tfrac{3468}{35}} \quad \textit{Ans.}$$

$$y = rx = \tfrac{5}{2} \times (\pm 2) = \pm 5, \text{ and } \mp \tfrac{19}{34}\sqrt{\tfrac{3468}{35}} \quad \textit{Ans.}$$

5. Given (1) $3xy - 4x - 4y = 0$ (2) $x^2 + y^2 + x + y - 26 = 0$

$$(1_1)\quad 2xy - \tfrac{8}{3}(x + y) = 0$$

$$(2_1)\quad x^2 + y^2 + (x + y) = 26$$

$$x^2 + 2xy + y^2 - \tfrac{5}{3}(x + y) = 26 \qquad (\text{Ax. } 1)$$

$$(x + y)^2 - \tfrac{5}{3}(x + y) = 26$$

$$(x + y)^2 - \tfrac{5}{3}(x + y) + (\tfrac{5}{6})^2 = 26 + \tfrac{25}{36} = \tfrac{961}{36} \qquad (335)$$

$$(x+y)-\tfrac{5}{6}=\pm\tfrac{31}{6}$$

$$x+y=6, \text{ or } -\tfrac{13}{3}$$

$$x=6-y, \text{ or } -\tfrac{13}{3}-y$$

(1) $3(6-y)y+4y-24-4y=0$ **(228)**

$$18y-3y^2-24=0$$

$$y^2-6y=-8$$

$$y^2-6y+9=1$$

$$y-3=\pm 1$$

(4) $\left.\begin{array}{l} y=4 \text{ or } 2 \\ x=2 \text{ or } 4 \end{array}\right\}$ *Ans.* **(229)**

(1) $3\left(-\tfrac{13}{3}-y\right)y+\tfrac{52}{3}+4y-4y=0$ **(228)**

$$-13y-3y^2+\tfrac{52}{3}=0$$

$36y^2+12\times 13y=208$ (**334**, Ax. 3)

$$36y^2+(\ )+(13)^2=169+208=377$$

$$6y+13=\pm\sqrt{377}$$

(4) $\left.\begin{array}{l} y=\dfrac{-13\pm\sqrt{377}}{6} \\ x=\dfrac{-13\mp\sqrt{377}}{6} \end{array}\right\}$ *Ans.* **(229)**

345. Classes of Simultaneous Equations, two or more being Quadratic, and the others, if any, simple (**344**). Methods of solution.

1. Special forms akin to those explained in **342** can be solved by a similar process. See Ex. 2, **344**.

2. When each equation contains only one term of the second degree, and that term has the same product or square of the unknowns in all of the equations. See Ex. 3 of the preceding article.

3. When neither of two given equations contains the first power of either x or y. This is usually spoken of as the homogeneous case, and to solve it y is put equal to rx, or x equal to ry. See Ex. 4.

4. When an equation can be framed which can be solved for a *function* of the unknowns. See Ex. 5 where the function regarded as unknown was $x + y$.

5. Miscellaneous methods. — Various substitutions, and expedients of other kinds, are often of use in effecting solutions in particular problems. Suggestions will be made in the proper places to meet such cases.

a. The methods just explained are not mutually exclusive. Sometimes an exercise can be solved in a number of different ways. In some instances particular methods have decided advantages over others.

346. Exercise in the Solution of Simultaneous Quadratic Equations, two or more of which are of the second degree.

1. $\begin{cases} x^2 + y^2 = 170 \\ xy = 13. \end{cases}$

2. $\begin{cases} x^2 + 3xy + y^2 = 5 \\ 2x^2 + xy + 2y^2 = 5. \end{cases}$

3. $\begin{cases} x^2 - 3xy + y^2 + 1 = 0 \\ 3x^2 - xy + 3y^2 = 13. \end{cases}$

4. $\begin{cases} 4x^2 + 9y^2 = 585 \\ 8xy = 336. \end{cases}$

5. $\begin{cases} x\left(\dfrac{a^2}{x} + x\right) = y\left(\dfrac{4a^2}{y} - y\right). \\ xy = a^2 \end{cases}$

6. $\begin{cases} a^2x^2 + 25b^2y^2 = 14ab \\ xy = 1. \end{cases}$

7. $\begin{cases} xy + x = 66 \\ xy - y = 50. \end{cases}$

8. $\begin{cases} 2y^2 + y = 28 \\ y^2 + 3x - 4y = 26. \end{cases}$

9. $\begin{cases} x - \dfrac{1}{y} = a \\ y - \dfrac{1}{x} = \dfrac{1}{a}. \end{cases}$

10. $\begin{cases} x^2 - xy + y^2 = 3 \\ x^2 - 2xy + 4y^2 = 4. \end{cases}$

11. $\begin{cases} 2x^2 - 5xy + 3y^2 = 1 \\ 3x^2 - 5xy + 2y^2 = 4. \end{cases}$

12. $\begin{cases} x^2 + xy + y^2 - 3\frac{1}{4} = 0 \\ 2x^2 - 3xy + 2y^2 - 2\frac{3}{4} = 0. \end{cases}$

13. $\begin{cases} x^2 - 2xy = 21 \\ xy + y^2 = 18. \end{cases}$

14. $\begin{cases} \frac{5}{3} x^2 - 4y^2 = a \\ 4xy - 3y^2 = 3a. \end{cases}$

15. $\begin{cases} 2y^2 - 4xy + 3x^2 = 17 \\ y^2 - x^2 = 16. \end{cases}$

16. $\begin{cases} (1)\ x^2 + 4y^2 + 80 = 15x + 30y \\ (2)\ xy = 6. \end{cases}$

Suggestion. — Multiply (2) by 4 and add it to (1). Then arrange in the form $(x + 2y)^2 - 15(x + 2y) = -56$. Solve this quadratic for $x + 2y$, getting two values for it.

17. $\begin{cases} x^2 + 3xy = 54 \\ xy + 4y^2 = 115. \end{cases}$

Suggestion. — Add the two equations and solve for $x + 2y$.

18. $\begin{cases} 9x^2 + y^2 - 63x + 21y + 86 = 0 \\ xy = 4. \end{cases}$

19. $\begin{cases} \dfrac{1}{x^2} + \dfrac{1}{y^2} = \dfrac{485}{576} \\ \dfrac{1}{x} + \dfrac{1}{y} = \dfrac{23}{24}. \end{cases}$

Suggestion. — Solve such exercises in the *reciprocal* form, i.e., without clearing.

20. $\begin{cases} x^2 + y = 4x \\ y^2 + x = 4y. \end{cases}$

Suggestion. — Subtract one equation from the other and divide through by $x - y$.

21. $\begin{cases} (1)\ x^2 + y^2 + z^2 = 30 \\ (2)\ xy + yz + xz = 17 \\ (3)\ x - y - z = 2. \end{cases}$

Suggestion. — Add 2 times (2) to (1) and extract the square root of the result.

22. $\begin{cases} x^2 + y^2 + 4\sqrt{x^2 + y^2} = 45 \\ xy = 12. \end{cases}$

23. $\begin{cases} x^2 + 2xy + y + 3x = 73 \\ y^2 + 3y + x = 44. \end{cases}$

24. $\begin{cases} x + y + a\sqrt{x + y} = 6a^2 \\ x^2 + y^2 = b. \end{cases}$

25. $\begin{cases} (1) & x^2 + y^2 + z^2 = 84 \\ (2) & x + y + z = 14 \\ (3) & xy = 8. \end{cases}$

SUGGESTION. — Add $2xy = 16$ to (1), and substitute $z = 14 - (x + y)$ in the new equation, and after arranging solve for $x + y$.

26. $\begin{cases} x^2 + xy + y^2 = 91 \\ x + \sqrt{xy} + y = 13. \end{cases}$

SUGGESTION. — Divide the first equation by the second.

27. $\begin{cases} (1) & x - y = 8(\sqrt{x} - \sqrt{y}) \\ (2) & \sqrt{xy} = 15. \end{cases}$

SUGGESTION. — Divide (1) through by $\sqrt{x} - \sqrt{y}$.

28. $\begin{cases} x^2 + y^2 = a \\ x^2 + z^2 = b \\ y^2 + z^2 = c. \end{cases}$

SUGGESTION. — Add the three equations and divide through by 2. From this equation subtract each of the equations in turn.

29. $\begin{cases} x = a\sqrt{x + y} \\ y = b\sqrt{x + y}. \end{cases}$

SUGGESTION. — Square each equation and then subtract one from the other. The resulting equation is divisible by $x + y$. Also divide one of the equations by the other.

SECTION III. — Problems.

347. Problems Involving the Solution of Simultaneous Quadratic Equations.

1. Find two numbers which multiplied give 576 and divided the one by the other give $2\frac{1}{4}$.

2. Two numbers are to each other as 11 to 13, and the sum of their squares is 14210. What are the numbers?

3. Find two numbers whose sum, product, and difference of their squares are all equal to each other.

4. What number being divided by the product of its digits gives the quotient 2, and if 27 be added to the number the digits will be inverted?

5. What numbers are there whose sum is 100 and the sum of whose square roots is 14?

6. A and B have each a small field in the shape of a square, and it requires 200 rods of fence to enclose both. The sum of the contents of these fields is 1300 sq. rods. What is the value of each at $2.25 a square rod?

7. What two numbers are those whose sum multiplied by the greater is 120, and whose difference multiplied by the less is 16?

8. Two farmers together drove to market 100 sheep, and returned with equal sums. If each of them had sold his sheep at the price the other actually did, the one would have returned with $180 and the other with $80. At what price per sheep did they sell respectively and how many sheep had each?

9. The small wheel of an ordinary bicycle makes 135 revolutions more than the large wheel in a distance of 260 yds.; if the circumference of each were one foot more the small wheel would make 27 revolutions more than the large wheel in a distance of 70 yds. Find the circumference of each wheel.

10. A tailor has noticed that broadcloth on being wet shrinks up $\frac{1}{8}$ in its length and $\frac{1}{16}$ in its breadth. If the surface of a piece of broadcloth is $5\frac{3}{4}$ square yds. less, and the distance round it $4\frac{1}{4}$ yds. less than before it was wet. what was the length and width of the broadcloth originally?

11. A cistern which is half full can be filled by one of two pipes in a certain time and emptied by another in a dif-

ferent time. If both pipes be left open, the cistern is emptied in 12 hours. But now if the opening of both pipes be made smaller so that the one needs an hour longer in filling and the other also an additional hour in emptying, then if both pipes are left open, 15¾ hours are needed to empty the cistern. In what time can the empty cistern be filled by the first pipe alone, and in what time can the full cistern be emptied by the other acting alone?

12. A certain number of laborers remove a heap of stones from one place to another in 8 hours. Were the number of laborers 8 more, and did each carry 2½ kilograms less at a time, then the heap would be transported in 7 hours. But if the number of laborers were 8 less, each carrying 5½ kilograms more, then the heap would be removed in 9 hours. How many laborers were there, and how many kilos did each carry at a time?

13. Find two numbers whose sum added to the sum of their squares is 42, and whose product is 15.

14. Find two numbers such that their product added to their sum shall be 47, and their sum taken from the sum of their squares shall leave 62.

15. Find three numbers such that when the sum of the first and second is multiplied by the third, the product is 63; when the sum of the second and third is multiplied by the first, the product is 28; and when the sum of the third and first is multiplied by the second, the product is 55.

SUGGESTION. — Solve in a manner similar to Ex. 28, **346**.

16. The diagonal of a box is 125 inches, the area of the lid is 4500 sq. in., and the sum of the conterminous edges is 215 inches. Find the lengths of these edges.

CHAPTER XXIII.

QUADRATIC EQUATIONS AND EQUATIONS IN GENERAL.

348. Properties and Solutions of Quadratic Equations and Equations of Higher Degrees. Some of the properties of quadratic equations obviously belong to equations of other degrees. It will be found also that many equations of other degrees can be solved like quadratics. It will be convenient besides to introduce into this chapter the general discussion of problems and the validity of processes of solution.

SECTION I.

PROPERTIES.

349. The Sum of the Roots of a Quadratic Equation.— When the coefficient of x^2 is unity, the coefficient of x taken negatively is equal to the sum of the roots.

If in the type form of quadratic equation, $ax^2 + 2bx = c$, the right member be transposed, and the equation divided through by a, there results $x^2 - 2\frac{b}{a}x - \frac{c}{a} = 0$. To simplify the equation we replace $-2\frac{b}{a}$ by p, and $-\frac{c}{a}$ by q, and let us call the new equation the *normal form*.

$$x^2 + px + q = 0$$

$$4x^2 + 4px = -4q \qquad (334)$$

$$(2x)^2 + (\ ) + p^2 = p^2 - 4q \qquad (335)$$

$$2x + p = \pm\sqrt{p^2 - 4q}$$

$$x = -\tfrac{1}{2}p + \tfrac{1}{2}\sqrt{p^2 - 4q}$$

$$\text{or } x = -\tfrac{1}{2}p - \tfrac{1}{2}\sqrt{p^2 - 4q}$$

If now the two values of x, which are the roots of the equation, be added, the *radical parts cancel and the sum is* $-p$, i.e., *the sum of the roots is equal to the coefficient of* x *in the normal equation taken with opposite sign.* Q. E. D.

350 The Product of the Roots of a Quadratic Equation. — The *product* of the roots is equal to the known term in the normal equation.

Multiplying together the two roots of the last article,

$$-\frac{p}{2}+\frac{1}{2}\sqrt{p^2-4q}$$
$$-\frac{p}{2}-\frac{1}{2}\sqrt{p^2-4q}$$
$$\frac{p^2}{4}-\frac{1}{4}(p^2-4q)=q \quad Q.\ E.\ D.$$

351. Solving a Quadratic Equation by Factoring. — By comparing the theorems of the last two articles with **116**, 8, we learn that *factoring the normal form of any quadratic equation gives its roots.*

1. Given the equation $x^2+7x+12=0$ to find its roots. Factoring the left member.

$$(x+4)(x+3)=0. \qquad \textbf{(135, 2)}$$

By **349** and **350** the roots are -4 and -3. For, $(-4)+(-3)=-7$ which is the coefficient of x taken *negatively:* while $-4\times-3=+12$, which is the known term. Hence, after factoring, take the second terms of the factors *with their signs changed* for the roots of the equation.

2. $x^2-(a+b)x+ab=0$

$$(x-a)(x-b)=0. \qquad \textbf{(135, 2)}$$

Hence the roots are a and b.

Note. — Compare this solution with that given by the regular process, Ex. 32, **338**

3. Solve $x^2-7x+12=0$ by factoring, and also by completing the square.

4. $x^2 - 9x + 20 = 0$.
5. $z^2 + 11z + 30 = 0$.
6. $x^2 + 13x + 12 = 0$.
7. $x^2 + 21bx + 110b^2 = 0$.
8. $x^2 - 6x - 27 = 0$.
9. $x^2 - 20x - 300 = 0$.
10. $x^2 - 8x = -15$.
11. $x^2 - 11x - 50 = 160$.
12. $x^2 - \frac{13}{14}x + \frac{3}{14} = 0$.
13. $6x^2 - 17x + 12 = 0$.

REMARK. — Factoring as in **135**, 3 we get $(3x - 4)(2x - 3) = 0$. Or $(x - \frac{4}{3})(x - \frac{3}{2}) = 0$ by dividing the equation through by 3×2 (Ax. 4.) Hence the roots are $\frac{4}{3}$ and $\frac{3}{2}$.

14. $x + 4 + \dfrac{7x - 8}{x} = 13$
15. $\dfrac{x + 11}{x} + \dfrac{18 + x}{x^2} = 7$
16. $\dfrac{3x^2}{3} + 3\frac{1}{2} = \dfrac{3x}{2} + 8$.
17. $6x^2 - x - 12 = 0$.
18. $x^2 - 0.3x = 0.7$.

352. Solving Equations of Any Degree by Factoring.

1. Let us verify some of the examples of the last article, substituting the values not in the original but in the factored form.

Putting $x = -4$ in Ex. 1, we have

$$(-4 + 4)(-4 + 3) = 0, \text{ or, } 0(-1) = 0. \qquad \textbf{(111, } a\textbf{)}$$

Thus when *one* factor of a product is zero the product is zero, and so the equation is *verified.* Substituting $x = -3$ will make the second factor zero and so satisfy the equation. **(189,** *a.***)**

Likewise in the second example, either $x = a$, or $x = b$, will make one of the factors zero, and so satisfy the equation, and they are therefore roots of the equation.

2. But this reasoning will hold for any number of factors as well as for two. Thus the equation $(x - 5)(x + 2)(x - 1) = 0$ (or, $x^3 - 4x^2 - 7x + 10 = 0$, by multiplication) is satisfied by either $x - 5 = 0$, $x + 2 = 0$, or $x - 1 = 0$.

$$\therefore x = 5, \qquad x = -2, \qquad x = 1.$$

And these are the three roots of the equation.

Again, taking the equation $6x^2 + 5x - 4 = 0$, we may write it in the form $(2x - 1)(3x + 4) = 0$. (**135**, 3). Placing each factor in turn equal to zero,

$$2x - 1 = 0 \therefore x = \tfrac{1}{2}; \text{ and } 3x + 4 = 0 \therefore x = -\tfrac{4}{3}$$

3. When the unknown, x, divides out of an equation, one root of that equation is zero. Thus in the equation

$$2x^3 - 11x^2 + 12x = x(x - 4)(2x - 3) = 0,$$

placing each factor equal to zero,

$x = 0$; $x - 4 = 0, \therefore x = 4$; $2x - 3 = 0, \therefore x = \tfrac{3}{2}$.

Manifestly $x = 0$ satisfies the equation $2x^3 - 11x^2 + 12x = 0$, for it makes every term equal to zero.

353. Rule for Solving Equations of Any Degree when they can be Factored.

1. By transposition if necessary make the right member zero.

2. Factor the left member into binomials the first term of each of which is a multiple of the unknown.

3. Set each factor equal to zero, thus obtaining as many values of the unknown as there are binominal factors.

The following corollaries of this method are highly important in the general theory of equations. Their truth is merely suggested here.

a. **An equation has as many roots as there are units in its degree.**

b. **If $x - a$ is one factor of the left member of an equation whose right member is zero, *then a is a root*. Hence if we can find one or more factors whether we can find the others or not, these factors give roots.**

c. **If one or more factors are known, the equation divided by these leaves an equation which contains the other factors, or the other roots.**

354. Exercise in the Solution of Equations by Factoring.

1. $x^2 + 23\,ax + 90\,a^2 = 0.$

2. $6\,x^3 + 17\,x^2 + 12\,x = 0.$

3. $12\,x^3 - 9\,x^2 - 8\,x + 6 = 0.$

4. $x^3 + x^2 + x + 1 = 0.$

5. $x\,(x - 6)^2 + 10\,x - 60 = 0.$

SUGGESTION. — This equation is evidently divisible by $x - 6$ giving a quotient $x^2 - 6\,x + 10 = 0$. Hence 6 is one root and the others must be found from the equation $x^2 - 6\,x + 10 = 0$. It is not possible, however, to resolve this equation by factoring. Turning then to the old process of completing the square the roots are readily found to be $x = 3 \pm \sqrt{-1}$. So in all cases. If an equation can be reduced down to the second degree, the solution can always be perfected by the old process.

6. $x^3 + 4\,x^2 - 4\,x - 16 = 0.$

7. $x^4 - 2\,x^3 - 16\,x^2 + 32\,x = 0.$

8. $x^3 - 3\frac{1}{2}\,x^2 + 3\frac{1}{2}\,x - 1 = 0.$

9. $x^4 - 5\,x^3 - 6x^2 + 15\,x + 9 = 0.$

SUGGESTION. — Arrange into $x^4 - 6\,x^2 + 9 - 5\,x\,(x^2 - 3) = 0.$

10. $x^4 + 7\,ax^3 - 2\,a^2\,x^2 - 28\,a^3\,x - 8\,a^4.$

SUGGESTION. — Arrange as follows: —

$$x^4 - 2a^2\,x^2 - 8\,a^4 + 7\,ax\,(x^2 - 4\,a^2) = 0.$$

11. One root of the equation $x^3 - 3\,x^2 - 14\,x + 42 = 0$ is 3; find the other two.

12. Two roots of the equation $x^4 - 2\,x^3 + 4\,x^2 + 2\,x - 5 = 0$ are 1 and -1: find the other two.

13. Solve the equations $27\,x^3 - 1 = 0$; $x^3 + 1 = 0$.

355. Conversely. — To Construct an Equation when its Roots are given: subtract each root from x and multiply all the resulting binomial factors together.

a. Thus far we have always found irrational and imaginary roots occurring in pairs, one with the $-$ sign before its radical term and the other with the + sign before its radical term. They are sometimes described as conjugate surds or imaginaries.

Upon multiplying two such binomial factors a trinomial with real and rational coefficients results. Thus taking the two imaginary roots of example 5 of the preceding article,

$$\begin{array}{l} x-(3+\sqrt{-1}) \\ x-(3-\sqrt{-1}) \\ \hline x^2-3x-x\sqrt{-1} \\ \quad -3x+x\sqrt{-1}+9-(-1) \\ \hline x^2-6x+10. \end{array}$$

Furthermore it is easy to see that if the radical terms did not disappear from *this* product, they would certainly not disappear when multiplied by dissimilar radicals in the other roots, and so would not disappear in the final product, which is the equation. Hence we conclude that if an equation having real coefficients does have imaginary or irrational roots they occur in pairs, and should be *taken together* in reconstructing the equation.

1. Form the equation whose roots are $+3$ and -4.

   $(x-3)(x-(-4))=0$, or $x^2+x-12=0$.

2. Form an equation whose roots are 1, 2, and -3.

3. Construct the equation whose roots are $0, -1, 2$, and -5.

4. Find the equation whose roots are $7+\sqrt{3}$, $7-\sqrt{3}$, and 1.

5. What is the equation whose roots are, $1\pm\sqrt{2}$, $2\pm\sqrt{-3}$, i.e., $1+\sqrt{2}$, $1-\sqrt{2}$, $2+\sqrt{-3}$, $2-\sqrt{-3}$.

6. Form the equation with the roots $2\pm\sqrt{3}$, -2, and 1.

SECTION II.

Discussion of Problems. — Failure of Processes of Solution.

356. Discussion of the Equation of the Second Degree. — Greatest and least values of coefficients.

Let us take the equation as found in **349**,

$$x^2 + px + q = 0,$$

the roots of which are $x = -\frac{p}{2} \pm \frac{1}{2}\sqrt{p^2 - 4q}$.

There are two classes, real roots and imaginary roots.

1. Equations having real roots.

In order that the roots may be real the radical quantity $\sqrt{p^2 - 4q}$ must be zero or positive.

(1. If $p^2 - 4q = 0$, the radical in the roots disappears, and $x = -\frac{1}{2}p$, or $x = -\frac{1}{2}p$, i. e., the two roots are *equal*.

In this case, $x^2 + px + q = x^2 + px + \frac{p^2}{4} = \left(x + \frac{p}{2}\right)\left(x + \frac{p}{2}\right) = 0$, thus giving two roots each equal to $-\frac{p}{2}$. **(351)**

(2. If $p^2 - 4q > 0$, the roots are real. Let us suppose a and b to be the two roots. The signs of the roots depend upon the signs of p and q. Compare articles **116**, 8, **135**, 2, and **351**.

(1) $x^2 + px + q = (x + a)(x + b) = 0$,
or, when p and q are both positive, the roots are both negative.

(2) $x^2 - px + q = (x - a)(x - b) = 0$,
i.e., when p is negative and q positive, the roots are both positive.

(3) $x^2 + px - q = (x - a)(x + b) = 0$, where $a < b$. i.e., if p is positive and q negative, the less root is positive and the greater is negative.

(4) $x^2 - px - q = (x + a)(x - b) = 0$, where $a < b$, i.e., if p and q are both negative, the less root is negative and the other positive.

2. Equations having imaginary roots. Here the quantity $p^2 - 4q$ under the radical sign is negative. If we conceive either p or q to vary in value, say p, the quantity $p^2 - 4q$ will also change, and may pass from + to −, or from − to +. But in making such a change at one epoch it must have the value zero. Corresponding to this is the *critical* value of x: for, as long as the quantity under the radical sign is positive the value of the unknown is real, and as soon as it becomes negative the value of the unknown becomes *imaginary*. The value of q must be positive to give imaginary roots: for, so long as q remains negative $p^2 - 4q$ can never be anything else than positive, making $\sqrt{p^2 - 4q}$ real.

To illustrate. — Suppose $q = 4$, then $p^2 - 4q = 0$ gives $p = \pm 4$, and so p cannot be greater than 4 or less than -4 in the equation $x^2 + px + 4 = 0$; for the moment that it is made larger, $\sqrt{p^2 - 4q}$ becomes imaginary, i.e., x becomes imaginary. Such values of p may be called maximum or minimum values, because they are the greatest or least which yield real values of x.

357. * Maxima and Minima Values of Functions. — If y be a function of x, and if as x increases y increases for a time and then decreases, the greatest value that y attains is a maximum; but if as x increases y decreases for a time and then increases the least value that y attains is a minimum.

a. Functions are of two kinds, explicit and implicit. Thus, in $y = x^2 + 2x + 5$, y is an explicit function of x, its value being given directly: while in $x^2 + xy = y^2 + 3y + 7$, y is an implicit function of

x, since to find its value in terms of x it is necessary to solve the equation for y. This gives

$$y = \frac{x-3}{2} \pm \frac{1}{2}\sqrt{-19-6x+5x^2}$$

and y has now become an explicit function of x.

1. Given $y = x^2 + 6x - 17$ to find the minimum value of y.

To show that there is a minimum value of y, we substitute a series of values of x in the equation and find the corresponding values of y.

Now $x = -9$, or any larger negative number, makes $y = 10$, or some increasingly larger positive number.

$$\begin{cases}\text{When } x = -8 \\ \text{then } \;\; y = -1\end{cases}; \;\; \begin{matrix} x = -7 \\ y = -10 \end{matrix}; \;\; \begin{matrix} x = -6 \\ y = -17 \end{matrix}; \;\; \begin{matrix} x = -5 \\ y = -22 \end{matrix}; \;\; \begin{matrix} x = -4 \\ y = -25 \end{matrix};$$

$$\begin{matrix} x = -3 \\ y = -26 \end{matrix}; \;\; \begin{matrix} x = -2 \\ y = -25 \end{matrix}; \;\; \begin{matrix} x = -1 \\ y = -22 \end{matrix}; \;\; \begin{matrix} x = 0 \\ y = -17 \end{matrix};$$

$$\begin{matrix} x = 1 \\ y = -10 \end{matrix}; \;\; \begin{matrix} x = +2 \\ y = -1 \end{matrix};$$

and $x = 3$ or any larger positive number makes $y = +10$ or some increasingly larger number.

By examining we see that as x changed y reached its least value, -26, when x was -3. To find the *exact limit* we proceed as follows: *we solve the equation for x, and then see for what value of y the radical becomes imaginary.*

$$y = x^2 + 6x - 17$$
$$x^2 + 6x = y + 17$$
$$x^2 + 6x + 9 = y + 26$$
$$x + 3 = \pm\sqrt{y+26}$$
$$x = -3 \pm \sqrt{y+26}.$$

The radical passes from real to imaginary just before it changes to a minus quantity, i.e., at zero. Hence to find the limiting value of y, we place

$$y + 26 = 0$$

Whence, $y = -26$.

This value of y is the *minimum* value.

2. Given y as the implicit function of x as contained in the equation $y^2 + x^2 = 3x + 4y$ to find the maximum or minimum values.

By substituting we learn that there is both a maximum and a minimum.

If $x = -2$ or any greater negative number, y is imaginary.

$$\text{If } x = -1 \quad x = 0 \quad x = 1 \quad x = 2$$
$$\text{then } y = \begin{cases} +2 \\ \text{or, } +2 \end{cases}; \quad y = \begin{cases} 0. \\ 4. \end{cases}; \quad y = \begin{cases} -.45 \\ +4.45 \end{cases}; \quad y = \begin{cases} -.45 \\ +4.45 \end{cases}$$

$$x = 3 \quad x = 4$$
$$y = \begin{cases} 0 \\ 4 \end{cases}; \quad y = \begin{cases} +2 \\ +2 \end{cases};$$

and $x = 5$ or any greater positive number, y is imaginary.

To find the *exact value*

$$y^2 + x^2 = 3x + 4y$$
$$x^2 - 3x + \tfrac{9}{4} = 4y - y^2 + \tfrac{9}{4} \qquad \textbf{(335)}$$
$$x - \tfrac{3}{2} = \pm\sqrt{\tfrac{9}{4} + 4y - y^2}$$
$$\text{Then } \tfrac{9}{4} + 4y - y^2 = 0 \qquad \text{(See previous example)}$$
$$y^2 - 4y = \tfrac{9}{4}$$
$$y^2 - 4y + 4 = \tfrac{25}{4}$$
$$y - 2 = \pm\tfrac{5}{2}$$
$$y = 4.5 \text{ or } -.5.$$

Returning to the series of values given above we see that 4.5 is the maximum value between 4.45 and 4.45 (the function was increasing at the first and decreasing at the second, so that it must have reached a maximum at some point between), while $-.5$ is the minimum between the values $-.45$ and $-.45$.

Substituting (**229**), the corresponding value of x is found to be 1.5.

These values may be inserted in the above series,

$$\text{Thus,} \quad \begin{matrix} x = 1 \\ y = \begin{cases} -.45 \\ 4.45 \end{cases} \end{matrix} \quad \begin{matrix} x = 1.5 \\ y = \begin{cases} -.5 \\ 4.5 \end{cases} \end{matrix} \quad \begin{matrix} x = 2 \\ y = \begin{cases} -.45 \\ 4.45 \end{cases} \end{matrix}$$

3. Divide a given number a into two parts such that their product shall have the greatest possible value.

$$\text{Let } x = \text{one part}$$
$$a - x = \text{the other part}$$
$$\text{then } x\,(a - x) = y, \text{ is a maximum.}$$
$$x^2 - ax = -y$$
$$x^2 - ax + \frac{a^2}{4} = \frac{a^2}{4} - y$$
$$x - \frac{a}{2} = \sqrt{\frac{a^2}{4} - y}$$

So that for the maximum $\frac{a^2}{4} - y = 0$ whence $y = \frac{a^2}{4}$.

Corresponding to this, $x = \frac{a}{2}$, i.e., the maximum product is obtained when the *two halves* of the number are multiplied.

4. To inscribe in a triangle a rectangle of maximum area. There is such a rectangle; for if we conceive of a rectangle at the bottom of the figure having a base nearly as long as the base of the triangle, BC, (and consequently a very small altitude) have its base diminish in size and its altitude increase, it will finally come to have an altitude nearly equal to the altitude of the triangle, but with a very small base. Evidently if the rectangle first increase in size and then diminish, there must be some position for which it has a maximum area.

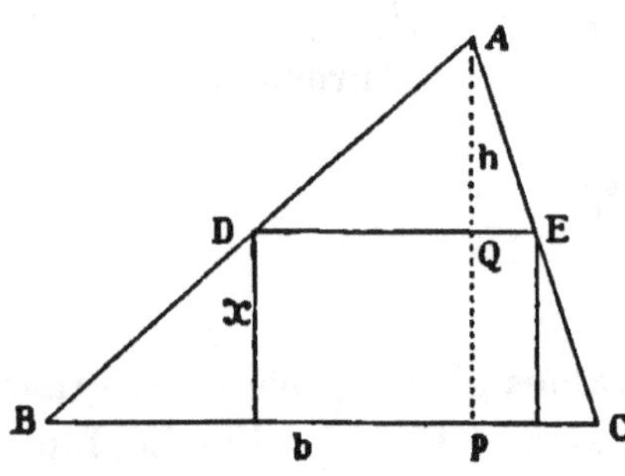

By a theorem in geometry, since the triangles ABC and ADE are similar,

$$AP : AQ :: BC : DE :$$

Call BC, b; AP, h; and y, the area of the rectangle. Required to find the maximum value of the last.

Substituting in the preceding proportion,

$$h : h - x :: b : DE.$$

$$\therefore DE = \frac{bh - bx}{h}$$

and since the area of a rectangle = base × altitude.

$$y = \frac{bh - bx}{h} . x$$

$$hy = bhx - bx^2 \qquad \text{(Ax. 3)}$$

$$4 b^2x^2 - 4 b^2hx = - 4 bhy \qquad \textbf{(334)}$$

$$4 b^2x^2 - 4 b^2hx + b^2h^2 = b^2h^2 - 4 bhy$$

$$2 bx - bh = \pm \sqrt{b^2h^2 - 4 bhy}.$$

Now $b^2h^2 - 4 bhy = 0$ gives the maximum.

Hence $2 bx - bh = 0$; or $x = \frac{h}{2}$: $b^2h^2 - 4 bhy = 0$; or $y = \frac{bh}{4}$.

Thus the maximum rectangle has for its altitude *half* the altitude of the triangle, and for its area $\frac{1}{4}$ of the base × the altitude, which is half the area of the whole triangle.

5. Determine the minimum and maximum values of the function $y = \dfrac{x^2 - 2x + 21}{6x - 14}$ for real values of x.

358. Discussion of the Equation $x = \frac{a}{b}$ for Limiting Values of a and b. Infinites and Infinitesimals.

a. As stated in 62, the symbol 0 is often used to denote any exceedingly small number less than any that can be named. Such quantities are called infinitesimals. Any number greater than any that can be named is called an infinite. A finite quantity is neither an infinite nor an infinitesimal. The student must understand at the outset that all infinitesimals are not of the same size, nor are all infinites of the same size, though 0 is used to denote all of the former, and ∞ all of the latter.

b. The sign $\doteq$ (read "approaches the limit") is used to signify that one quantity approximates to the value of another more and more, so that they differ by less than any assignable magnitude.

1. To find the value of $\frac{a}{b}$ when the denominator b is finite, but the numerator is infinitesimal.

The result may be arrived at perhaps better by an example than in any other way. Suppose b is any number, say 500, and a assumes the values 100000, 10000, 1000. 100, 10, 1, .1, .01, and so on without end,·

$$\frac{100000}{500} = 200,\ \frac{10000}{500} = 20,\ \frac{1000}{500} = 2;\ \frac{100}{500} = \frac{1}{5},$$
$$\frac{10}{500} = \frac{1}{50},\ \frac{1}{500} = \frac{1}{500},\ \frac{.1}{500} = \frac{1}{5000},\ \frac{.01}{500} = \frac{1}{50000},$$

and so on without end. Evidently the value of the fraction gets smaller and smaller, or what is the same thing approaches the limit zero. Stating this principle generally, if in the fraction $\frac{a}{b}$, the numerator approach the limit 0, the *fraction* approaches the limit 0. Stated in the form of an equation:

$$(1)\quad \text{When } a \doteq 0,\ x = \frac{a}{b} \doteq 0.$$

2. Similarly the value of $\frac{a}{b}$, when a is *finite* and b *infinite*, approaches the limit zero.

3. To find the value of $\frac{a}{b}$ when a is finite and b infinitesimal.

This is the reverse of the first case. Whatever finite value the numerator may have, as the denominator diminishes the value of the fraction increases, and ultimately when the denominator has become infinitesimally small, the value of the fraction has grown *infinitely large*.

$$(2)\quad \text{When } b \doteq 0,\ \frac{a}{b} \doteq \infty.$$

4. When b is finite and a is infinite the fraction approaches the same limit, ∞.

5. To find the value of $x = \frac{a}{b}$ when both a and b are infinitesimals.

In the first example given, however small the denominator might become, the value of the fraction was always $\frac{1}{500}$ of it. Thus of these two infinitesimals one was 500 times greater than the other. In general when b is finite and a infinitesimal the value of the fraction is an infinitesimal which is $\frac{1}{b}$ times the first. Now, if two quantities grow small together and we have no means of knowing their relative size, *the value of the fraction is indeterminate.*

(3) When $a = 0$, and $b = 0$, $x = \frac{a}{b}$ is *indeterminate.*

However, in such a fraction as $\frac{5x}{x}$, if x be infinitesimal, the value of the fraction does not become indeterminate, since the *same* infinitesimal is in both terms and so divides out, giving 5. Likewise the fraction $\frac{a^2 + b^2}{a - b}$ does not become indeterminate when

$$a = b \text{ since } \frac{a^2 - b^2}{a - b} = \frac{(a - b)(a + b)}{(a - b) \times 1} = a + b = 2a.$$

359. * Problem of the Lights. — This celebrated problem, due to Clairaut, furnishes an excellent exercise in the interpretation of algebraic results.

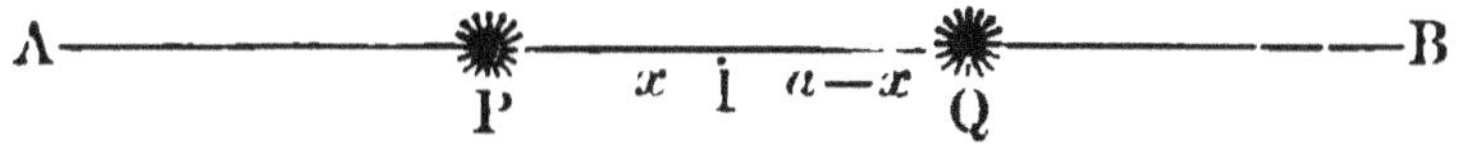

$$A \text{———} P \text{——} x \,|\, a - x \text{——} Q \text{———} B$$

Two lights are at P and Q. It is required to find the points on A B having equal illumination.

By a law of optics the intensity with which a light shines at any outside point is inversely proportional to the square of the distance from the point to the source of the light.

Let $a =$ PQ be the distance between the lights,
$m^2 =$ the intensity of illumination of the light P at the distance of one ft. from the source,
$n^2 =$ the similar intensity of the light Q.

Suppose I to be the point of equal illumination; and let $x =$ PI, then QI $= a - x$. If now the illumination from P is m^2, at 1 ft., it is $\frac{1}{4}m^2$ at 2 ft., $\frac{1}{9}m^2$, at 3 ft., $\frac{1}{x^2}m^2$, at x ft. according to the law of intensity. Likewise the illumination from Q at a distance of $a - x$ ft. is $\frac{n^2}{(a-x)^2}$. But by the supposition the two intensities are equal.

$$\frac{m^2}{x^2} = \frac{n^2}{(a-x)^2}$$

$$\frac{m}{x} = \pm \frac{n}{a-x} \qquad \text{(Ax. 6)}$$

$$ma - mx = \pm nx$$

$$(m \pm n)\,x = ma$$

$$x = \frac{ma}{m \pm n} \text{ i.e. } \frac{m}{m+n}a, \text{ or } \frac{m}{m-n}a.$$

From the values of x, we learn

1. That there are *two* points equally illuminated.

2. One position of the point of illumination may always be found *within* the line PQ. For, since by the nature of the problem m and n can only be positive numbers $\frac{m}{m+n}$ will always be a proper fraction, and therefore $\frac{m}{m+n}a$ is less than a, and positive. Now we started by assuming that a position of I would be found to the right of P, and the positive result verifies our assumption.

3. The other position of the point of equal illumination is either to the right of Q or to the left of P according

as m is greater or less than n. For, $\frac{ma}{m-n}$ *is positive and greater than a* when $m-n$ is positive; i.e., when $m>n$. In this case the second point of equal illumination lies *beyond* Q. And $\frac{ma}{m-n}$ is *negative* when $n>m$, and the second point of equal illumination lies to the *left* of P.

Moreover, all this is in accord with the conditions of the problem; for the two lights will illuminate equally a point between them situated close enough to the weaker light to make up for the stronger illumination of the other. Besides this point there will be another, situated on the side next the weaker light, so that here again the closer proximity of the one light is offset by the greater intensity of the other.

4. Certain special cases are worthy of attention for the analytical results they give.

(1. Suppose $m \doteq 0$; then $x \doteq +0$ or -0. (358, 1)

Here while the light at P is infinitesimally weak, the two points of equal illumination are infinitely close to it on either side, making up over the other light in closeness what is lost in intensity.

This can be made clearer by an example. Suppose at the distance of one foot Q is very bright and P is *very weak*. Then at $\frac{1}{2}$ ft. P is 4 times brighter; at $\frac{1}{4}$ ft., 16 times brighter; at $\frac{1}{100}$ ft., 10000 times brighter; and so on. Thus by going closer the illuminating power of the weak light may be made to increase indefinitely. Of course this reasoning assumes the source of the light to be a mathematical point.

(2. Suppose $m \doteq n$, then $x \doteq \frac{1}{2}a$, and ∞. (358, 3)

Here one position of I is midway; the outside position is at infinity, for it is only at such a distance that the advantage one light has over its equal is reduced to an infinitesimal.

(3. Suppose $a \doteq 0$, then both values of $x \doteq 0$.

This is the same problem as the general case, only *in miniature.* The value of a and the two values of x are by no means equal. What is conveyed by representing all three by the limit 0, is that all are infinitesimals, and that the two points of equal illumination, one between and the other outside of the infinitesimal distance between P and Q, are infinitely close to them.

(4. Suppose $a \doteq 0$, and $m \doteq n$, then $x = 0$, and $\frac{0}{0} a$. (385, 5)

Here the lights are infinitely close to each other, and the one point of equal illumination is midway, and of course, infinitely close to P and Q, while the other position is indeterminate, since we do not know the laws by which the quantities approach the limits. If the lights are practically equal, and at practically the same point, they will illuminate any and every point on either side alike.

360. Validity of Processes of Solution. There are certain processes in the solution of equations which deserve special consideration lest the student be led into serious error in using them.

1. *In the extraction of the same root of the two members of an equation.*

(1. Thus, while $(5-2)^2 = (\pm 3)^2$, $5 - 2 = +3$ only and not -3; that is, the roots of both members must be taken with the same sign.

(2. From $x^2 - 2ax + a^2 = 0$

$$x - a = \pm 0$$

and there are *two* values of x each equal to a.

(3. The equation $12x - 5 + 2\sqrt{36x^2 - 27x + 5} + 3x - 1 = 27x - 2$, (See Ex. 3, 327)

gives $\sqrt{12x-5} + \sqrt{3x-1} = \sqrt{27x-2}$. (Ax. 6)

As shown in **327**, while $x = \frac{4}{51}$ satisfies the first equation, it does not satisfy the second, except when the signs of the radicals are taken in a certain way. This anomaly is caused by the presence of imaginaries.

2. *An equation obtained by squaring or cubing, etc., the two members of an equation will, in general, have roots which do not satisfy the original equation.*

(1. To prove this for squaring take X and X′ for the two members of the given equation, either or both of them functions of the unknown, but neither zero.

$$X = X' \qquad \text{(Hyp.)}$$
$$X^2 = X'^2 \qquad \text{(Ax. 5)}$$
$$X^2 - X'^2 = 0 \qquad \text{(Ax. 2)}$$
$$X^2 - X'^2 = (X + X')(X - X') = 0.$$

By the theory of equations, to find the roots of this equation the two factors are in turn set equal to zero. **(353)**

$$\therefore X - X' = 0, \text{ and } X + X' = 0.$$

Now the first of these *is the original equation*, while the second is a new equation with new roots.

Thus, Ex. 9, 329.

$$\sqrt{4x+5} - \sqrt{x} = \sqrt{x+3}$$

Squaring and transposing, and afterwards factoring

$4x + 5 - 2\sqrt{4x^2 + 5x} + x - (x+3) = (\sqrt{4x+5} - \sqrt{x} + \sqrt{x+3})(\sqrt{4x+5} - \sqrt{x} - \sqrt{x+3}) = 0.$

the second factor being the original equation, while the first furnishes new roots.

(2. To show the same for cubing an equation.

$$X = X' \qquad \text{(Hyp.)}$$
$$X^3 = X'^3 \qquad \text{(Ax. 5)}$$
$$X^3 - X'^3 = (X - X')(X^2 + XX' + X'^2) = 0. \qquad \textbf{(134}, 2)$$

Here, besides the old equation, there is the factor $X^2 + XX' + X'^2 = 0$ which has new roots of its own. And so for other powers.

To illustrate the principle further, let us solve the equation,

$$\sqrt[3]{x^3 + 5x + 19} = \tfrac{1}{2}(x+3)(\sqrt{-3} - 1).$$
$$x^3 + 5x + 19 = x^3 + 9x^2 + 27x + 27 \qquad \text{(Ax. 5)}$$
$$9x^2 + 22x = -8$$
$$9x^2 + 22x + (\tfrac{11}{3})^2 = \tfrac{49}{9}$$
$$3x + \tfrac{11}{3} = \pm \tfrac{7}{3}$$
$$x = -2, \text{ or } -\tfrac{4}{9}.$$

Upon substituting these values in the original equation, we readily find that neither of them satisfies it. They do satisfy, however, the equation framed as $X^2 + XX' + X'^2 = 0$, viz., $\sqrt[3]{(x^3 + 5x + 19)^2} + \frac{1}{2}\sqrt[3]{x^3 + 5x + 19}\,(x + 3)(\sqrt{-3} - 1) + \frac{1}{4}(x + 3)^2(\sqrt{-3} - 1)^2 = 0$.

Thus, $\sqrt[3]{-8 - 10 + 19} = 1$; $x + 3 = 1$; or, $1 + \frac{1}{2} \times 1 \times 1 \times (\sqrt{-3} - 1) + \frac{1}{4}(-2 - 2\sqrt{-3}) = 0$.

So also for the value $-\frac{4}{9}$.

3. *An equation may not be multiplied or divided through by a function of the unknown which equals zero, or infinity, when the value of x is substituted in it.*

(1. In the equation $ax^3 + 2bx^2 - cx = 0$, x is a factor which shows that one of the roots of the equation is zero (**353**). To proceed with the solution, x is divided out, but zero must appear along with the others as a root of the equation.

(2. In like manner it is not allowable to divide out by one or more factors of the form $(x - a)$, and take no further account of the roots they give.

Thus, in $3(x - a)^2 = 5x(a - x)$

$x = a$ is one of the roots of the equation.

(3. It is not allowable to multiply an equation through by x, thus introducing the root, zero, when this root does not by right belong to the equation.

Thus,
$$\frac{x-3}{5} - \frac{8}{x^2} = \frac{2}{x} - \frac{8 - 6x + x^2}{x^2}$$
$$x^3 - 3x^2 - 40 = 10x - 40 + 30x - 5x^2 \qquad \text{(Ax. 3)}$$
$$x^3 + 2x^2 - 40x = 0.$$

Since the last equation contains x, zero is one of its roots. But, upon simplifying the given equation by transposing $-\frac{8}{x^2}$, cancelling, and afterwards multiplying through by x, we have
$$x^2 - 3x = 10 + 30 - 5x$$
$$\text{or, } x^2 + 2x - 40 = 0$$
and this equation does not contain x as a factor at all. Putting $x = 0$ in the original equation, after cancelling $-\frac{8}{x^2}$, $-\frac{3}{5}$ is the left member, and $\frac{8}{x} + 1 = \infty$, the right member; i.e., an infinite quantity equal to a finite quantity, which is absurd. Hence $x = 0$ does not belong to the original equation.

(4. It is allowable in clearing of fractions to multiply through by functions of x, provided that in so doing no alien values of the unknown are introduced. To multiply or divide by any known finite quantity is therefore always admissible.

$$\frac{5x-a}{x+a}+\frac{x}{x+b}=\frac{4x-2a}{x+a}+\frac{4x-a}{x+b}+1.$$

Clearing and simplifying, we have

$$(x+a)x=(x+a)(4x-a). \qquad \text{(Ax. 3)}$$
$$(x+a)x-(x+a)(4x-a)=0. \qquad \text{(Ax. 2)}$$
$$(x+a)(x-(4x-a))=(x+a)(-3x+a)=0. \qquad (136, 3)$$

Hence $x+a=0$, or $x=-a$ is a root of the equation. But the first, third, and fifth terms of the given equation taken together vanish. Consequently, it was not necessary to multiply through by $x+a$, and when it was done, the extraneous root $x=-a$ was wrongfully inserted.

(5. Clearing the equation $\dfrac{y+\frac{1}{y}}{y-\frac{1}{y}}+\dfrac{1+\frac{1}{y}}{1-\frac{1}{y}}=2$ of fractions we get

$$y+\frac{1}{y}-1-\frac{1}{y^2}+y+1-\frac{1}{y}-\frac{1}{y^2}=2\left(y-1-\frac{1}{y}+\frac{1}{y^2}\right);$$

clearing again, collecting and arranging

$$4y^2+4y=8$$
$$\therefore y=1 \text{ or } -2.$$

But while $y=-2$ satisfies the original equation, $y=1$ *does not.* To find the reason for this, we simplify the left member (See Arts. 183, 162), obtaining

$$1+\frac{2}{y^2-1}+1+\frac{2}{y-1}=2;$$

cancelling, we have left,

$$\frac{2}{y^2-1}+\frac{2}{y-1}=\frac{1}{y-1}\left(\frac{2}{y+1}+2\right)=0.$$

The second factor $\dfrac{2}{y+1}+2=0$ gives $y=-2$ as above. The first factor $\dfrac{1}{y-1}$ equals ∞ when $y=1$. (358, 3). Thus in clearing of fractions the equation was multiplied through by a factor $(y-1)^2$, which for a particular value of y equals ∞. Hence we have no right to expect that this value of y will satisfy the given equation.

The principle stated at the beginning of 3 of the present article is of general application and explains the apparent difficulty in 2, as well. It makes clear also the fallacy in the pretended proof of the arithmetical absurdity that $1=2$. The "proof" is as follows: —

If $a = x$, then $ax - x^2 = a^2 - x^2$

$x(a-x) = (a+x)(a-x)$

$x = a + x$ (Dividing out by $(a-x)$)

$x = x + x = 2x$ (Since $a = x$)

$1 = 2$ (Dividing out by x).

Here the second equation is divided through, according to the hypothesis by *zero*, and it does not follow that the *next* equation is true for the same value of x. Thus, $0 \times$ (any number) $= 0 \times$ (any other number).

It may be said in closing this article that, if a student is at any time unable to tell *a priori* whether a certain value of the unknown belongs to the initial form of the equation, he has this recourse, that he can substitute in that equation and find out.

SECTION III.

Equations of Higher Degrees Solved like Quadratics.

361. Equations containing but one unknown solved like incomplete quadratic equations. — **Binomial Equations.** — Incomplete quadratic equations and similar equations of other degrees are often called Binomial Equations, because when reduced they consist of but two terms.

1. Solve $2x^4 - 9 = 7 + x^4$.

$x^4 = 16$ (Ax. *a*)

$x^2 = \pm 4$ (Ax. 6)

$x = \sqrt{+4} = \pm 2$; or $\sqrt{-4} = \pm 2\sqrt{-1}$.

(Ax. 6, **353**, *a*)

2. Given $x^3 = a^3$.

In this example one sees *prima facie* that $x = a$ is one root. Transposing the a^3 to the left member and dividing through by $x - a$ (**353**, c), we have

$$x^3 - a^3 = (x - a)(x^2 + ax + a^2) = 0.$$

Setting the second factor, $x^2 + ax + a^2$ equal to zero (**353**),

$$4x^2 + 4ax = -4a^2$$
$$4x^2 + 4ax + a^2 = -3a^2$$
$$2x + a = \pm a\sqrt{-3}$$
$$x = \frac{-a \pm a\sqrt{-3}}{2}.$$

These two roots along with a form the three roots of the equation (**353**, a). These values,

$$a, \quad \frac{-a + a\sqrt{-3}}{2}, \quad \frac{-a - a\sqrt{-3}}{2},$$

are sometimes called the *three cube roots* of the quantity a^3 or, if $a = 1$, of unity.

3. $x^{\frac{m}{n}} = a$; required to find x.

$$x^m = a^n \qquad \text{(Ax. 5)}$$
$$x = a^{\frac{n}{m}}. \qquad \text{(Ax. 6)}$$

NOTE. — This is the general case. If the exponent of the unknown is negative, make it positive by transference to the opposite term of its fraction.

4. Given $px^{-\frac{1}{2}} = q$.

SOLUTION. $\frac{p}{x^{\frac{1}{2}}} = q$; $p = qx^{\frac{1}{2}}$; $x^{\frac{1}{2}} = \frac{p}{q}$; $x = \frac{p^2}{q^2}$ *Ans.*

5. $x^6 = 1$.

SUGGESTION. $(x^6 - 1) = (x^3 + 1)(x^3 - 1) = 0$.

6. $x^4 = -a^4$.

SUGGESTION. — Transpose $-a^4$, and then factor $x^4 + a^4$ (**134**, 5). The factors set equal to zero will give the four roots.

7. Find only the real roots of this and the following ·

Given $\frac{9}{x} = \frac{x^2}{24}$.

8. $4x^{\frac{3}{4}} = 500$. 9. $x^{\frac{2}{3}} = 81$.

10. $\sqrt{x^{\frac{3}{5}}} = 2\sqrt{2}$.

11. $\frac{x^3 + x + 8}{x^3 + 4} + \frac{x^3 + x - 8}{x^3 - 4} = 2$.

12. $\frac{x^2 - na}{x^2 - a} = \frac{nx^2 - b}{x^2 - b}$.

13. $\frac{a}{x^{\frac{1}{3}} - b} = \frac{b}{x^{\frac{1}{3}} - a}$.

14. $x^5 - 32 = 0$.

15. $x^8 = 390625$.

16. $(x^2 + 3x + 2)^2 = (x^2 - 5x + 10)^2$.

17. $(x^2 - 3x - 10)^2 = 16(x - 5)^2$.

362. Equations containing but One Unknown Solved like Complete Quadratic Equations.

There are two type forms of these equations:

$ax^{2n} + bx^n = c$, the simple form,

$[f(x)]^{2n} + b[f(x)]^n = c$, the complex form,

$f(x)$ in the latter meaning any *function of* x.

1. Solve $3x^6 + 42x^3 = 3321$.

$$x^6 + 14x^3 = 1107. \qquad \text{(Ax. 4)}$$
$$x^6 + 14x^3 + 49 = 1156 \qquad \textbf{(335)}$$
$$x^3 + 7 = \pm 34$$
$$x^3 = 27 \text{ or } -41$$
$$x = \sqrt[3]{27} \text{ or } \sqrt[3]{41}.$$

If in the three answers of **Ex. 2** of the last article we put $a = \sqrt[3]{27} = 3$, or, $\sqrt[3]{-41}$, we get the *six roots* of the given equation, it being of the sixth degree.

2. $\frac{1}{2}y^4 - 3y^2 + 1 = \frac{8y^2 + 1}{2}$

$y^4 - 14y^2 = -1$ (Axs. 3 and 2)

$y^4 - 14y^2 + 49 = 48$

$y^2 - 7 = \pm\sqrt{48}$

$y^2 = 7 \pm \sqrt{48}$

$y = \pm(2 \pm \sqrt{3})$. (**323**)

3. Given $(3x^4 + 8x^2 - 20)^2 - 9(3x^4 + 8x^2) - 2880 = 0$.

(Cf. **345**, 4 and Ex. 16 in **346**.)

$$(3x^4 + 8x^2 - 20)^2 - 9(3x^4 + 8x^2 - 20) = 3060.$$

To save writing, put z for $3x^4 + 8x^2 - 20$: then,

$z^2 - 9z = 3060$

$4z^2 - 36z = 12240$ (**334**)

$4z^2 - (\quad) + 81 = 12321$

$2z = 9 \pm 111$

$z = 60$ or -51

$\therefore 3x^4 + 8x^2 - 20 = 60$

$9x^4 + 24x^2 + 16 = 256$ (**334, 335**)

$3x^2 + 4 = \pm 16$

$x^2 = 4$ or $-\frac{20}{3}$

$x = \pm 2$ or $\frac{2}{3}\sqrt{-15}$.

The other four values of x may be found by using the second value of z, i. e., from

$$3x^4 + 8x^2 - 20 = -51.$$

4. $5x^{\frac{1}{2}} + 8x^{\frac{1}{4}} = 13$

$25x^{\frac{1}{2}} + 40x^{\frac{1}{4}} + 16 = 81$ (**334, 335**)

$5x^{\frac{1}{4}} + 4 = \pm 9$

$x^{\frac{1}{4}} = 1$ or $-\frac{13}{5}$

$x = 1$ or $\frac{28561}{625}$

5. $x^6 - 7x^3 = 8$.

6. $z^4 - 5z^2 + 4 = 0$.

7. $x^4 - 74x^2 = -1225$.

8. $17y^4 - 1 = 16y^8$.

In this and some of the following only the real roots are given in the answer.

9. $3y^4 - 7y^2 = 25$.

10. $x^3 - 24\sqrt{x^3} = 81$.

11. $x^2 - 3 = 6 + \sqrt{x^2 - 3}$.

12. $5x = 39 + 2\sqrt{x}$.

13. $ax^{2n} + bx^n = c$.

14. $\sqrt{x} + 4\sqrt[4]{x} = 21$.

15. $\sqrt[3]{x} + 7\sqrt[3]{x^2} = 350$.

16. $12x^{-\frac{3}{4}} - x^{-\frac{3}{8}} - 2^{-4} = 0$.

17. $x + \sqrt{x^3} = 6\sqrt{x}$.

18. $\frac{1}{3}\sqrt{8} = 2x\sqrt{1 - x^2}$.

19. $\sqrt{x^2 - 8x + 31} + (x - 4)^2 = 5$.

SUGGESTION. — Solve like Ex. 3.

20. $2\sqrt{x^2 - 3x + 11} = x^2 - 3x + 8$.

21. $\left(x + \frac{8}{x}\right)^2 + x = 42 - \frac{8}{x}$.

22. $2x^2 + 3x - 5\sqrt{2x^2 + 3x + 9} + 3 = 0$.

23. $\frac{1}{(2x - 4)^2} = \frac{1}{8} + \frac{2}{(2x - 4)^4}$.

SUGGESTION. — Put $z = (2x - 4)^2$.

24. Given $x^2 + \frac{1}{x^2} = \frac{17}{4}$.

SUGGESTION. — Add 2 to both members and apply Ax. 6

NOTE. — This equation is one of a class called *reciprocal equations*, whose roots are reciprocals of each other. Thus in this example the roots are 2 and $\frac{1}{2}$, and -2 and $-\frac{1}{2}$.

25. Prove by substitution that $\frac{1}{r}$, $-r$, and $-\frac{1}{r}$ are roots of the equation $x^2 + \frac{1}{x^2} = a$ if r is a root, i.e., if $r^2 + \frac{1}{r^2} = a$.

26. $16\, y^2 + \frac{1}{y^2} = 28$.

27. $(x - a)(x - b)(x - c) + abc = 0$.

28. $(a + b + x)(a + b - x)(a - b + x)(a - b - x) = 0$.

After solving this problem by multiplying it out, solve it by **352**.

29. $x^4 + 5\, x^2 + 4\sqrt{x^4 + 5\, x^2} = 60$.

30. $x^3 - 6\, x - 9 = 0$.

SUGGESTION. — Put $y + z = x$, then $(y + z)^3 - 6\,(y + z) = 9$. Developing this we have $y^3 + z^3 + (3\, yz - 6)(y + z) = 9$. Now, since the original equation contains but *one* unknown, and we have just introduced *two*, we are at liberty to make another assumption, *that is form another equation*, as *two* equations are necessary when there are two unknowns. Let us then suppose further that $3\, yz - 6 = 0$. Whereupon the first equation reduces to

$$(1)\quad y^3 + z^3 = 9$$

$$(2)\quad 3\, yz - 6 = 0 \qquad \text{(Hypothesis above)}$$

$$(2_1)\quad z = \frac{2}{y}$$

$$(1)\quad y^3 + \left(\frac{2}{y}\right)^3 = 9$$

$$y^6 + 8 = 9\, y^3$$

$$y^6 - 9\, y^3 + \frac{81}{4} = \frac{81}{4} - 8 = \frac{49}{4}$$

$$y^3 - \frac{9}{2} = \pm \frac{7}{2}$$

$$y^3 = 8 \text{ or } 1$$

$$y = 2 \text{ or } 1$$

$$(2_1)\quad z = 1 \text{ or } 2$$

$$x = y + z = 3. \quad \textit{Ans.}$$

This is called *Cardan's* solution of a cubic equation. The other two roots may be found from $y^3 = 8$ as in Ex. 2, **361**.

31. $x^4 - 2x^3 - 2x^2 + 3x - 108 = 0.$

SUGGESTION. — Sometimes equations of this kind can be arranged in the quadratic form by extracting the square root.

$$\begin{array}{l|l}
 & x^4 - 2x^3 - 2x^2 + 3x - 108 \;(\underline{x^2 - x} \\
 & x^4 \\
2x^2 - x & -2x^3 - 2x^2 \\
 & -2x^3 + x^2 \\
 & \qquad\quad -3x^2 + 3x - 108.
\end{array}$$

Hence $x^4 - 2x^3 - 2x^2 + 3x - 108 = (x^2 - x)^2 - (3x^2 - x) - 108 = 0.$

To proceed with the solution put $z = x^2 - x$.

32. $x^4 - 6x^3 + 5x^2 + 12x = 60.$

33. $4x^4 + \frac{x}{2} = 4x^3 + 1.$

34. $\frac{x + \sqrt{x}}{x - \sqrt{x}} = \frac{x^2 - x}{4}.$

SUGGESTION. — Divide through by $x + \sqrt{x}$. The roots are then found to be, 1, 4 and $\frac{1}{4}(-6 \pm 2\sqrt{-7})$, notwithstanding that in this instance, the first root, $\sqrt{x} = -1$, makes $x + \sqrt{x}$ zero. (360, 3.)

35. $x^4 - 4x^3 + 6x^2 - 4x + 1 = 16.$

SUGGESTION. — Extract the fourth root.

36. $x(x - 2a) = \frac{8a^4}{x^2 - 2ax} + 7a^2.$

37. $x + 4 - 2\sqrt{\frac{x + 4}{x - 4}} = \frac{3}{x - 4}.$

38. $x = \frac{4\sqrt{x} - 48}{x - 18}.$

SUGGESTION. — Clear, add $4x + 49$ to both sides, and apply Ax. 6.

39. $(3 - x)^4 + (4 - x)^4 = 17.$

SUGGESTION. — Put $3 - x = u$, and $4 - x = v$. Then (1) $u^4 + v^4 = 17$. (2) $u - v = -1$. Raising both members of (2) to the fourth power, and subtracting from (1), we have $4u^3v - 6u^2v^2 + 4uv^3 = 4uv(u^2 - 2uv + v^2) + 2u^2v^2 = 16.$

$$4uv(1) + 2u^2v^2 = 16$$
$$u^2v^2 + 2uv + 1 = 9 \qquad (334,\ 335)$$
$$uv = 2 \text{ or } -4.$$

Having the values of $u - v$ and uv the solution is continued as in 342; whence $u = 1$, $v = 2$, which give $x = 2$; and $u = -2$, $v = -1$, which give $x = 5$. The other value of uv leads to imaginary values of x.

40. $(y - 4)^4 + (1 + y)^4 = 97.$

General Remark. — Examples of higher equations solved like quadratics occur in the most diverse forms and are solved by the greatest variety of methods. One or other of the suggestions of this article will often be found helpful.

363. Simultaneous Equations of Higher Degrees that may be Solved like Quadratics.

a. When two equations are combined the number of sets of roots ought to equal the product of the numbers denoting their degrees. Thus combining a cubic with a quadratic should give *six* sets of roots. For, solving the cubic for one of the unknowns, there would result three roots, *each* of which substituted in the quadratic would make three equations for the other unknown, giving in all six values for it. However, such problems as can be solved are degenerate forms, and very often give a less number of answers.

b. These problems are so varied in character that no satisfactory classification can be given. Notwithstanding the variety of forms of the problems, certain methods of solution are worthy of distinct exemplification. Such are: Substitution; division; introduction of new variables; comparison; solving first for functions of the unknowns; extraction of roots; special methods for symmetric equations. (230, *a*.)

1. Given (1) $x^3 + y^3 = 133$, (2) $xy = 10$. (228)

2. (1) $x^4 + y^4 = 89$ (2) $x^2 + y^2 = 13$. (342)

3. $\begin{cases} 12\,xy = 1 \\ x^2 + y^2 = 25\,x^2y^2. \end{cases}$

4. $\begin{cases} x^4 + y^4 = 82 \\ xy = 3. \end{cases}$

5. $\dfrac{x^3 + y^3}{x^3 - y^3} = 2$; and $xy^2 = 6$.

6. $\begin{cases} x^3 + y^3 = 407 \\ x + y = 11. \end{cases}$

SUGGESTION. — Divide one equation by the other.

7. $\begin{cases} x^3 + y^3 = 152 \\ x^2 - xy + y^2 = 19. \end{cases}$

8. $\begin{cases} x^4 - y^4 = b \\ x^2 - y^2 = a. \end{cases}$

9. $\begin{cases} x^3 - y^3 = 2197 \\ x - y = 13. \end{cases}$

10. $\begin{cases} x^4 + x^2y^2 + y^4 = 2128 \\ x^2 + xy + y^2 = 76. \end{cases}$

SUGGESTION. — Divide as in Ex. 6.

11. $\begin{cases} x^3y^2 + x^2y^3 = -4 \\ x^2y + xy^2 = 2. \end{cases}$

12. $\begin{cases} x^3 + y^3 + 3\,x + 3\,y = 378 \\ x^3 + y^3 - 3\,x - 3\,y = 324. \end{cases}$

13. $\begin{cases} x + y = 35 \\ x^{\frac{1}{3}} + y^{\frac{1}{3}} = 5. \end{cases}$

SUGGESTION. — Put $u = x^{\frac{1}{3}}$, and $v = y^{\frac{1}{3}}$. Then, $u^3 = x$, $v^3 = y$.

14. $\begin{cases} x^{\frac{1}{4}} + y^{\frac{1}{4}} = 5 \\ x^{\frac{1}{2}} + y^{\frac{1}{2}} = 13. \end{cases}$

15. $\begin{cases} \sqrt[3]{x^4} + \sqrt[5]{y^2} = 20 \\ \sqrt[3]{x^2} + \sqrt[5]{y} = 6. \end{cases}$

16. $\begin{cases} x^2 + y^2 + 4\sqrt{x^2 + y^2} = 45 \\ x^4 + y^4 = 337. \end{cases}$

SUGGESTION. — Put $z = \sqrt{x^2 + y^2}$.

17. $\begin{cases} xy^2 + xy = 24 \\ xy^2 + x = 40. \end{cases}$

SUGGESTION. — Factor the left members, and eliminate x by comparison.

18. $\begin{cases} x^2(x - y) = 4. \\ x^2(2\,x + 3\,y) = 28. \end{cases}$

19. $\begin{cases} (1)\ xy + xy^2 = 12 \\ (2)\ x + xy^3 = 18. \end{cases}$

SUGGESTION. — Divide one equation by the other.

20 $\begin{cases} 4\,xy = 96 - x^2y^2 \\ x + y = 6. \end{cases}$

21. $\begin{cases} xy^2 + y = 21 \\ x^2y^4 + y^2 = 333. \end{cases}$

SUGGESTION. — Solve first for $xy^2 = x'$, and y.

22. $\begin{cases} x + y + \sqrt{x + y} = 12 \\ x^3 + y^3 = 189. \end{cases}$

23. $\begin{cases} x^3 - y^3 + 6\sqrt{x^3 - y^3} = 16 \\ x^2 + xy + y^2 = 2. \end{cases}$

24. $\begin{matrix} (x^2 + 1)\, y = xy + 126 \\ (x^2 + 1)\, y = x^2y^2 - 744. \end{matrix}$

SUGGESTION. — Eliminate the left members, and solve for xy.

25. $\begin{cases} (1)\ x^3 + y^3 = 133 \\ (2)\ xy\,(x + y) = 70. \end{cases}$

SUGGESTION. — Add three times (2) to (1), and extract the cube root of the members of the new equation.

26. $\begin{cases} x^3 - y^3 = 56 \\ x - y = \dfrac{16}{xy}. \end{cases}$

27. $\begin{cases} x^2 + y^2 = \dfrac{13}{x - y} \\ xy = \dfrac{6}{x - y}. \end{cases}$

28. $\begin{cases} x + y = 3 \\ x^4 + y^4 = 17. \end{cases}$

SUGGESTION. — These equations are symmetrical in x and y. Let $x = u + v$, and $y = u - v$. Then $x + y = 2u = 3$, and $u = \frac{3}{2}$. Therefore $x = \frac{3}{2} + v$, and $y = \frac{3}{2} - v$, and $x^4 + y^4 = (\frac{3}{2} + v)^4 + (\frac{3}{2} - v)^4 = 17$. Upon developing the terms of the last equation it is readily solved as a quadratic.

29. $\begin{cases} xy = a\,(x + y) \\ x^2y^2 = b^2\,(x^2 + y^2). \end{cases}$

Divide these equations through by xy and x^2y^2 respectively, and solve for $\dfrac{1}{x}$ and $\dfrac{1}{y}$.

30. (1) $x - y = 2\,(a - z)$ (2) $x^2 - y^2 = 5\,(a^2 - z^2)$ (3) $x^3 - y^3 = 7\,(a^3 - z^3)$.

SUGGESTION. — Divide (2) by (1), giving (4); solve for x and y from (1) and (4), and substitute in (3) $\div$ (1).

31. (1) $x+y=15$ (2) $xy+xz+yz=264$ (3) $xyz=756$.

SUGGESTION. — Find $z=\frac{756}{(15-x)}$, and $y=15-x$, and substitute these values in (2).

364. Problems Depending on the Solution of Equations of Degrees other than the First or Second.

1. A bin to contain 100 bu. of grain is made $3\frac{1}{2}$ times as long as wide, and $2\frac{3}{7}$ times as deep as wide. If one cu. ft. contains $\frac{4}{5}$ bu., what must be the dimensions.

2. Two cubic vessels together hold 407 cu. inches. When one vessel is placed on the other the total height is 11 in.; find the contents of each.

3. The product of the sum and difference of two numbers is 8, and the product of the sum of their squares and the difference of their squares is 80. What are the numbers?

4. One number is 8 times another, and the sum of their cube roots is 12. What are the numbers?

5. Find two numbers whose product is 15, and the difference of whose cubes is $\frac{49}{4}$ times the cube of their difference.

6. The sum of two numbers is 6, and the sum of their 5th powers is 1056. Find them. (See Ex. 28, last article.)

7. The product of two numbers is 10, and the sum of their cubes is 133. Required the numbers.

8. A man draws a certain quantity of vinegar out of a full vessel that holds 256 gals.; and then, filling the vessel with water, draws off the same number of gallons as before, and so on for four draughts, when there were only 81 gals. of pure vinegar left. How much vinegar did he draw at each draught?

9. B has 33 more than the square of the number of dollars A has, and the product of the numbers expressing the fortunes of each is 1330. How many dollars has each?

10. A box measure has a square bottom, and a bin has for its respective dimensions the squares of the dimensions of the measure. If now the bin is full, and 7 measures full be taken from it, 1710 cu. in. will remain. Moreover, the number of square inches in one of the sides of the measure diminished by the number of linear inches in one side of the bottom is 12. Find the dimensions of the measure.

11. What two numbers are those whose difference, multiplied by the difference of their squares, is 32, and whose sum, multiplied by the sum of their squares, is 272?

SUGGESTION. — Call u their sum and v their difference.

FIFTH GENERAL SUBJECT. — TOPICS RELATED TO EQUATIONS.

(INEQUALITIES, PROPORTION, EXPONENTIAL EQUATIONS, LOGARITHMS, PROGRESSIONS, AND INTEREST.)

CHAPTER XXIV.

INEQUALITIES.

365. An Inequation, as the word itself suggests, consists of two members, one of which is greater or less than the other but not equal to it.

We now propose to study briefly the inequation in the same way the equation itself has been investigated.

a. The signs $>$ and $<$ indicate that the quantities between which they are placed are unequal, the opening of the character being toward the greater. By *greater* is meant higher up in the scale in the positive direction.

Thus $6>4$, $3>-2$, $-11>-15$, etc. Hence if $a>b$, then $a-b$ is always positive. In the examples just given, $6-4=2$, $3-(-2)=5$, $-11-(-15)=4$, etc.

366. Properties of Inequations. The inequation has several of the properties of the equation, *but not all.*

1. If the same number be added to or subtracted from both members of an inequation, the new inequality will be like the old; i.e., will retain the same sign.

Thus, $7 > 5$, and by adding 4, $11 > 9$;
$2 > -30$, and by adding -12, $-10 > -42$;
$-12 > -13$, and by adding 17, $5 > 4$;
$-6 < 0$, and by adding -18, $-24 < -18$.
and so on.

It is evident from these examples that both members will be changed *alike*, or merely shoved along in the scale whether by adding or subtracting, and the new inequality will be of the same kind as the old. A general demonstration of the theorem is as follows:

$a > b$ (Hyp.)
$a - b$ is positive. (a of last Article.)
$a \pm c - (b \pm c)$ is positive. (the c's cancel.)
$\therefore a \pm c > b \pm c$. Q. E. D.

SCHOLIUM. — We learn from this that a term can be *transposed* from one side of an inequation to the other by changing its sign (**210**).

2. An inequation multiplied or divided through by the same *positive* multiplier or divisor will stand as it did before; whereas if the multiplier or divisor be *negative* the new inequation will subsist in the *contrary sense;* i.e., if the left member were greater before, it will be less now, and conversely.

A little consideration will show that this theorem is true for the positive multiplier or divisor, and then also for the negative, since the greater a negative number is numerically, the less it is algebraically. As a special case of this principle, if the signs of the terms of an inequation be changed throughout, the sign of the inequation must be changed.

3. If the two members of an inequation are both positive numbers, when they are raised to any integral power the inequation will still hold as before. If the two members of an inequation are both negative numbers (or at least the

less number is negative), when they are raised to an odd power the inequation will still hold as before. But if the two members of an inequation are both negative numbers, when they are raised to an even power the new inequation will subsist in the contrary sense. Let the student test these cases by various examples.

Raising the members of an inequation the lesser of which is negative to an even power gives rise to an *ambiguity*.

Thus, $5 > -2$, upon squaring gives $25 > 4$,
while, $5 > -8$, upon squaring gives $25 \ngtr 64$.

Evidently in this last case the greater member must also be the greater numerically.

4. If any root of the two members of an inequality both of which are positive is extracted, or any odd root where both are negative, the resulting inequality will be like the old, providing in the former case the root of the greater be taken as positive.

Test this with examples.

5. Two or more like inequalities may be added, just as equations are, and the resulting inequality will subsist in the same sense.

6. But if one inequality be taken from another like inequality, the result is ambiguous.

Thus,

$$\begin{array}{r} 12 > 5 \\ 2 > 1 \\ \hline 10 > 4 \end{array} \qquad \begin{array}{r} 12 > 5 \\ 11 > 1 \\ \hline 1 \ngtr 4 \end{array}$$

7. Like inequalities whose members are all positive may be multiplied together, and the new inequality will subsist in the same sense.

To the theorems given, still others might be added. The foregoing are the most important, and sufficiently illustrate the elementary treatment.

367. Special Theorems in Inequations.

1. To prove that the sum of the squares of any two real and unequal numbers is greater than twice their product.

Let a and b be the numbers.

Now a square number is always positive. Then

$$(a-b)^2 > 0$$

(except when $a = b$, when $(a-b)^2 = 0$)

$$\therefore \quad a^2 - 2ab + b^2 > 0 \qquad \textbf{(114, 2)}$$

$$a^2 + b^2 > 2ab \qquad Q. E. D. \qquad \textbf{(366, 1)}$$

REMARK. — This theorem is much used in the demonstration of other special ones.

2. Prove that any positive fraction whose terms are unequal plus its reciprocal is greater than 2.

Thus, to prove $\frac{x}{y} + \frac{y}{x} > 2$.

3. Show that $a^3 + b^3 > a^2b + ab^2$, when $a + b$ is positive.

SUGGESTION. — Divide the inequation through by $a + b$.

4. Prove that $a^3b + ab^3 > 2a^2b^2$, when $a > b > 0$.

5. Prove that $x^3 + 1 > x^2 + x$, when $x + 1$ is positive and $x \neq 1$.

6. Show that $x^2 + y^2 + z^2 > xy + xz + yz$, if x, y, and z are unequal.

7. Prove that $\frac{m}{n} < \frac{m+p+r}{n+q+s} < \frac{r}{s}$ if $\frac{m}{n} < \frac{p}{q} < \frac{r}{s}$, the denominators n, q and s being all positive.

8. Which is the greater, $a^2 + 3b^2$ or $2b(a+b)$, the letters being unequal and positive?

9. Which is the greater, $a^4 - b^4$ or $4a^3(a-b)$, when $a > b > 0$.

SUGGESTION. — Factor the left member and divide through by $a^3(a-b)$.

10. Given that a, b, and c, are positive and unequal, to show that $(a+b)(b+c)(c+a) > 8\,abc$.

SUGGESTION. — Divide the three factors of the left member by a, b, and c respectively, thus cancelling the same factors in the right member. Then multiply out and apply the theorem of Ex. 2.

368. Examples and Problems in the use of Inequations.

1. Given $2x + \frac{1}{2}x - 4 > 6$ to find the limit of x.

SUGGESTION. — Solve as an equation by clearing and transposing, whence it will appear that $x > 4$.

2. Given $\frac{5x}{8} + \frac{5}{4} < \frac{11}{6} + \frac{7x}{12}$.

3. Find the limit for both x and y in $\begin{cases} (1)\ 2x + 4y > 30. \\ (2)\ 3x + 2y = 31. \end{cases}$

4. Given $\begin{cases} \frac{x}{b} + b > \frac{x}{a} + x & \text{to find the limit of } x, \\ \frac{x}{b} - \frac{x}{a} > \frac{a-b}{b} & \text{when } a > b > 0. \end{cases}$

5. Show that $x^2 - 8x + 22$ is never less than 6 whatever may be the value of x. (See **357**.)

6. The double of a certain number increased by 7 is not greater than 19, and its triple diminished by 5 is not less than 13; required the number.

7. A certain positive whole number plus 23 is less than 6 times the number minus 12; and nine times the number minus 54 is less than twice the number plus 9. What is the number?

8. Prove that the arithmetic mean of two numbers is always greater than their geometric mean. Or, as a formula it is to prove that $\frac{a+b}{2} > \sqrt{ab}$.

CHAPTER XXV.

RATIO AND PROPORTION.

369. The **Ratio** of two quantities is the quotient of the first divided by the second. The first quantity is called the antecedent, and the second the consequent.

Thus, the ratio 3 to 6 is $\frac{3}{6}$ or $\frac{1}{2}$; the ratio of $a+m$ to

$$m + n^2 \text{ is } \frac{a+m}{m+n^2} = (a+m) \div (m+n^2)$$
$$= (a+m) : (m+n^2).$$

a. Ratios are usually denoted by a colon. Thus, $a:b$ is read the ratio of a to b. The colon is supposed to represent a division sign with the horizontal line omitted.

b. Some writers regard the second quantity as the dividend and the first as the divisor. But this usage is out of harmony with the idea that the colon represents a division sign, and should be abandoned.

370. Reduction of Ratios. Since a ratio is a fraction it can be reduced to other terms without altering the value of the fraction, i.e., the ratio. (**159**.)

Thus, $a:b = \frac{a}{b} = \frac{ma}{mb} = \frac{\frac{a}{m}}{\frac{b}{m}}$. In particular, $12:4 = 6:2$.

QUERY. — Will it change a ratio if the two terms be increased or diminished by any number? How, if they are increased or diminished by like parts of each?

SUGGESTION. — Multiply both terms by $\left(1+\frac{m}{n}\right)$ or by $\left(1-\frac{m}{n}\right)$.

371. Ratios can be compared in Value by reducing them to a common denominator.

Thus, to compare the values of $10:11$ and $23:25$.

$$\frac{10}{11} >, =, < \frac{23}{25};\ \frac{250}{275} < \frac{253}{275}\ \textbf{(366, 2)}\ \therefore \frac{10}{11} < \frac{23}{25}.$$

372. To Compound Ratios is to multiply their corresponding terms together.

Thus, compounding $a:b$, $c:d$, and $e:f$ we have $ace:bdf$.

a. When the terms of a ratio are *squared* it is said to be the *duplicate* ratio of the first; when cubed, the *triplicate*.

b. Examples of the compounding of ratios may be seen in compound proportion in arithmetic.

373. A Proportion is the equality of two ratios.

Thus $\frac{6}{3} = \frac{8}{4}$; or, $a:b = c:d$; or $a:b::c:d$ using a double colon instead of the equality sign.

a. The colon form is read "*a* is to *b as c* is to *d*," while the fractional form $\frac{a}{b} = \frac{c}{d}$ *may be read in the same way, or as an equation,* "*a* divided by *b* equals *c* divided by *d*." Indeed, both forms may be read in either way.

b. The first and fourth terms of a proportion (*a* and *d*) are called the *extremes*, and the second and third terms (*b* and *c*), the *means*.

374. Method of Treatment of Proportion.—To accomplish any desired reduction the proportion is first written as an equation, and then transformed by *axiomatic processes.* As regards the colon form of writing proportions *it need never be used.* In truth, it simply duplicates the other notation. If any advantage at all attaches to it, it is due merely to the fact that the proportion is written in a single line. Nothing of importance would be lost and simplicity would be gained if the colon form were entirely discarded. In the study of proportion we seek to familiarize ourselves with the different forms in which the same proportion can appear, and the new proportions which can be constructed out of it.

375. Transformations of a Proportion. — THEOREMS.

1. *In a proportion the product of the extremes is equal to the product of the means. This is the usual test for a proportion.*

Given $a : b :: c : d$ to prove $ad = bc$.

$$\frac{a}{b} = \frac{c}{d} \qquad \textbf{(374)}$$

$$ad = bc \quad Q.\ E.\ D. \qquad \text{(Ax. 3)}$$

2. CONVERSE THEOREM. — *If the product of two numbers is equal to the product of two others, one pair may be made the extremes and the other pair the means of a proportion.*

For if the equation be divided through by any pair of opposite terms, a proportion results.

Taking the equation $ad = bc$, or $bc = ad$

(1) $\frac{a}{b} = \frac{c}{d}$ i.e., $a : b :: c : d$ (by dividing through by bd)

(1′) $\frac{c}{d} = \frac{a}{b}$ i.e., $c : d :: a : b$ (" " " ")

(2) $\frac{b}{a} = \frac{d}{c}$ i.e., $b : a :: d : c$ (by dividing through by ac)

(2′) $\frac{d}{c} = \frac{b}{a}$ i.e., $d : c :: b : a$ (" " " ")

(3) $\frac{a}{c} = \frac{b}{d}$ i.e., $a : c :: b : d$ (by dividing through by ?)

(3′) $\frac{b}{d} = \frac{a}{c}$ i.e., $b : d :: a : c$ (" " " ")

(4) $\frac{c}{a} = \frac{d}{b}$ i.e., $c : a :: d : b$ (by dividing through by ?)

(4′) $\frac{d}{b} = \frac{c}{a}$ i.e., $d : b :: c : a$ (" " " ")

3. GENERAL THEOREM. — *Since if $a : b :: c : d$, then $ad = bc$ (by the first theorem); and furthermore, since all the forms just given follow from $ad = bc$ by axiomatic processes, we*

learn that if $a : b :: c : d$, then all the above forms of the proportion are likewise true.

Evidently the four primed equations (1′), (2′), (3′), (4′) are essentially the same as the others, the difference being the mere interchanging of their members. In translating these formulae into the theorem form, it is customary to call a the first, b the second, c the third, d the fourth term of the proportion. The formula (2′) would be read thus: If four quantities are in proportion taken in order, then the fourth is to the third as the second is to the first. Each formula might in this manner give rise to a theorem. Two of these changes in the original proportion have been given specific names, viz., (2) and (3), called respectively *Inversion* and *Alternation*.

a. A proportion is said to be taken by *inversion* when the second is to the first as the fourth is to the third.

b. A proportion is said to be taken by *alternation* when the first is to the third as the second is to the fourth.

QUERIES. (1) Will a proportion still hold if all its terms be multiplied or divided by the same number? (370, 158.)

(2) Will a proportion still hold if the first two terms are multiplied or divided by one number and the second two by another?

(3) How can the signs of the terms of a proportion be changed without altering its value? Can any two be changed? (157.)

(4.) Will a proportion still hold if the antecedents be multiplied or divided by one number, and the consequents by another?

(5.) Will a proportion still hold if the same number be added to or subtracted from each of its terms?

(6.) Will a ratio represented by an improper fraction be increased, diminished, or unchanged by adding the same quantity to both of its terms? How will a proper fraction be affected? How unity?

376. New Proportions from a Given One.

1. THE COMPOSITION THEOREM. — *If four quantities are in proportion, the sum of the first and second is to the first or second as the sum of the third and fourth is to the third or fourth.*

If $a : b :: c : d$, then $\begin{cases} a + b : a :: c + d : c \\ a + b : b :: c + d : d. \end{cases}$

For $\frac{a}{b} + 1 = \frac{c}{d} + 1$ (Ax. 1)

(1) $\frac{a + b}{b} = \frac{c + d}{d}$ *Q. E. D.* **(167)**

Of course this new proportion can now be alternated, inverted, or transformed in any of the ways described in the last article. Thus,

(2) $\frac{a + b}{c + d} = \frac{b}{d}$ (**375.** *b*)

(3) $\frac{a + b}{c + d} = \frac{a}{c}$ (Ax. 7. See Eq. (3′), **375**)

(4) $\frac{a + b}{a} = \frac{c + d}{c}$.

Equation (4) can be derived directly from $\frac{b}{a} = \frac{d}{c}$ (2)

$\frac{b}{a} + 1 = \frac{d}{c} + 1$ (Ax. 1)

$\frac{b + a}{a} = \frac{d + c}{c}$ *Q. E. D.* **(167)**

2. The Division Theorem. — *If four quantities are in proportion, the difference of the first and second is to the first or second as the difference of the third and fourth is to the third or fourth.*

If $a : b :: c : d$, then $\begin{cases} a - b : a :: c - d : c \\ a - b : b :: c - d : d \end{cases}$

For $\frac{a}{b} - 1 = \frac{c}{d} - 1$ (Ax. 2)

(5) $\frac{a - b}{b} = \frac{c - d}{d}$ *Q. E. D.* **(167)**

(6) $\frac{a - b}{c - d} = \frac{b}{d} = \frac{a}{c}$ (**375.** *b*. (3′))

(7) $\frac{a - b}{a} = \frac{c - d}{c}$ *Q. E. D.* (**375.** *b*)

3. The Composition and Division Theorem.—*If four quantities are in proportion, the sum of the first and second is to their difference as the sum of the third and fourth is to their difference.*

If $a : b :: c : d$, then $a + b : a - b :: c + d : c - d$

For, taking equations (2) and (6) of this article,

$$\frac{a+b}{c+d} = \frac{a-b}{c-d} \qquad \text{(Ax. 7)}$$

$$\frac{a+b}{a-b} = \frac{c+d}{c-d} \quad Q.\ E.\ D. \qquad \text{(375, } b\text{)}$$

4. The Like Powers Theorem.—*If four quantities are in proportion, like powers of the terms are also in proportion.*

If $a : b :: c : d$, then $a^m : b^m :: c^m : d^m$

For $\frac{x}{b} = \frac{c}{d}$ (Hyp.)

$$\frac{a^m}{b^m} = \frac{c^m}{d^m} \quad Q.\ E.\ D. \qquad \text{(Ax. 5)}$$

5. The Like Roots Theorem.—*If four quantities are in proportion, like roots of the terms are also in proportion. This is proved in the same way as* 4; *instead of* Ax. 5, *use* Ax. 6.

377. Combinations of Two or More Proportions.

1. *In a continued proportion* (*i.e., one in which three or more ratios are all equal to each other*) *the sum of any antecedents is to the sum of their consequents as any one antecedent is to its consequent.*

If $a : b :: c : d :: e : f :: g : h ::$ etc., then $a + c + e + g +$ etc., $: b + d + f + h +$ etc. $:: a : b ::$ etc.

To prove it let us suppose r to be the common ratio; then

$$\frac{a}{b} = r,\ \frac{c}{d} = r,\ \frac{e}{f} = r,\ \frac{g}{h} = r, \text{ etc.}$$

$$\therefore a = br,\ c = dr,\ e = fr,\ g = hr, \text{ etc.} \qquad \text{(Ax. 3)}$$

Adding these equations (or any set of them) member by member

$$a+c+e+g+\text{etc.}=(b+d+f+h+\text{etc.})\,r \qquad \text{(Ax. 1)}$$

$$\frac{a+c+e+g+\text{etc.}}{b+d+f+h+\text{etc.}}=r,=\frac{a}{b}=\frac{c}{d}=\frac{e}{f}=\frac{g}{h}=\text{etc.}\quad Q.E.D. \qquad \text{(Ax. 4)}$$

a. Continued proportions are often written more compactly, thus,

$a:c:e::b:d:f$ instead of $a:b::c:d::e:f$.

The former is therefore not *one*, but two different proportions.

b. The above theorem finds its most important application in geometry, where by means of it the perimeters of similar polygons are found to be proportional to any two homologous sides.

2. *The corresponding terms of two or more proportions may be multiplied together, giving rise to a new proportion which holds if the given proportions hold.*

If $\quad a:b::c:d$

$\quad a':b'::c':d'$

$\quad a'':b''::c'':d''$

then $\quad aa'a'':bb'b''::cc'c'':dd'd''$

For, writing the given proportions in the equation form, and multiplying them member by member,

$$\frac{a}{b}=\frac{c}{d},\ \frac{a'}{b'}=\frac{c'}{d'},\ \frac{a''}{b''}=\frac{c''}{d''}$$

$$\frac{aa'a''}{bb'b''}=\frac{cc'c''}{dd'd''}\quad Q.\ E.\ D. \qquad \text{(Ax. 3, 178)}$$

and so for any number of proportions.

Moreover, it is evident that the terms of two proportions may be divided. Thus, if

$$\frac{a}{b}=\frac{c}{d}\ \text{and}\ \frac{a'}{b'}=\frac{c'}{d'}$$

$$\frac{a}{a'}:\frac{b}{b'}::\frac{c}{c'}:\frac{d}{d'}.$$

378. Special Forms in Proportion. —

1. A *mean proportional* between two quantities is one whose square is equal to their product.

Thus, if $a:b::b:c$, b is a mean proportional between a and c.

For by the first theorem, $b^2 = ac$ (or $b = \sqrt{ac}$).

The third quantity c is called the *third proportional* to a and b.

a. A mean proportional is not at all the same as an arithmetical mean. Thus, the *arithmetical* mean of 6 and 4 is 5, or half their sum; while the *geometrical* mean, the mean proportional, equals $\sqrt{6 \times 4} = \sqrt{24}$, which is manifestly less than 5.

2. Two mean proportionals. — If $a:b::b:c::c:d$, b and c are two mean proportionals between a and d.

Prove (1) $a:c::a^2:b^2$ (2) $a:d::a^3:b^3$

379. Exercises in Ratio and Proportion.

1. Which is the greater ratio, $3:4$ or $3^2:4^2$?

2. Write the ratio compounded of $8:7$ and $5:6$. Which is less and which is greater than the compounded ratio?

3. The last three terms of a proportion being 4, 6, and 8, what is the first term?

4. Find a third proportional to 25 and 400.

5. Find a mean proportional between $2\frac{2}{9}$ and $\frac{5}{12}$.

6. Arrange in order of magnitude $4:3$; $9:8$; $25:23$, and $15:14$.

7. If the ratio of a to b is $2\frac{2}{3}$ what is the ratio of $3a$ to $4b$?

8. Prove that if $a:b::c:d$ that $a-b:c-d::a+b:c+d$.

9. If the ratio of m to n is $\frac{4}{7}$, what is the ratio of $m-n$ to $m+n$?

10. Why is a^2-b^2 the mean proportional between $a^2+2ab+b^2$ and $a^2-2ab+b^2$?

11. If $a : b :: c : d$ prove that $ra^5 : sb^5 :: rc^5 : sd^5$.

12. Find two numbers in the ratio of $3 : 4$, (suggestion $3x$ and $4x$) of which their sum is to the sum of their squares as 7 to 50.

13. Solve $x^2 - 4 : x^2 - 9 :: x^2 - 5x + 6 : x^2 + 4x + 3$.

14. Solve for x in the proportion $(a - x) : (x - b) :: a : b$.

15. Solve (1) $x : y :: 3 : 5$ (2) $x : 4 :: 15 : y$.

16. What quantity must be added to each of the terms of the ratio $m : n$ so that it may become equal to that of $p : q$?

17. Four given numbers are represented by m, n, p, and q; what quantity added to each will make them proportional?

18. If four quantities are already proportional, show that there is no number which being added to each will leave the resulting four numbers proportional.

19. If $a : b :: b : c$ prove that $a + b : b + c :: a : b$.

20. If $a : b :: c : d$ prove that $3a + b : b :: 3c + d : d$.

21. If $a : b :: c : d$ prove that $2a + 3b : 3a - 4b :: 2c + 3d : 3c - 4d$.

22. If $(a + b + c + d)(a - b - c + d) = (a - b + c - d)(a + b - c - d)$ prove that $a : b :: c : d$.

23. If $a : b :: c : d$ is $(ma \pm nb) : (pa \pm qb) :: (mc \pm nd) : (pc \pm qd)$?

24. If $a : b :: 4 : 5$, $d : f :: 5 : 2$, $e : c :: 6 : 7$, $d : b :: 7 : 3$, and $f : c :: 4 : 3$ to what numbers is the continued ratio $a : b : c : d : e ::$ proportional?

25. The product of two numbers is 112, and the difference of their cubes is to the cube of their difference as $31 : 3$. What are the numbers?

26. Given $\sqrt[3]{x} + a : \sqrt[3]{x} - a :: m : n$ to find x.

27. Find two numbers whose sum, difference, and product are proportional to m, n, and p.

380. Variation. — One quantity is said to vary with another when there is some law connecting changes in their values. If y is a function of x, then as x changes in value y also changes. The simplest cases of variation (the only ones the word as used technically in algebra refers to) are closely connected with ratio and proportion. Indeed, direct variation is nothing other than a new aspect of proportion.

1. Direct Variation. — One quantity varies directly with another when as one changes the other changes in the same proportion.

Thus, the distance a man travels in a day varies as the number of miles he travels. If he travel 5 miles per hour he will go over $\frac{5}{3}$ times as much ground as when he travels only 3 miles per hour; and if he travel 8 hours per day, he will journey only $\frac{4}{5}$ as far as when he travels 10 hours per day.

When one quantity varies directly as another, or simply varies as another, they are to each other in a constant ratio. This is involved in the definition.

For, if a varies as b, and a' and b', a'' and b'', etc., are other values of a and b, then $\frac{a}{b} = \frac{a'}{b'} = \frac{a''}{b''} =$ etc.; i.e., the ratio is the same in each case, or is constant. When one quantity varies *jointly* as each of two (or more) others, as in the example given above, it varies as their *product*.

As an illustration: if one train runs 30 miles per hour and 22 hours per day, while another train runs 25 miles per hour, and 23 hours per day, the distances they will pass over in any given time are proportional to the *products* of these numbers, or the rate and time.

2. Inverse Variation. — One quantity varies inversely as another when as one increases the other decreases, so that their product is constant.

Thus, if 5 men can do a job of work in 16 days, 10 men can do the same work in 8 days, 20 men in 4 days, and so on.

If a vary inversely as b, and a' and b' are corresponding values, then $a : a' :: b' : b$, since this gives $ab = a'b'$ which is true by definition.

If one quantity varies directly as a second and inversely as a third, it varies as the quotient of the second divided by the third.

Thus the number of bushels of wheat elevated in a mill varies directly as the amount of work performed, and inversely as the height to which it is raised.

3. Other Kinds of Variation. — One quantity may vary as any function of another. It is shown in mechanics that if s is the distance in feet which a body falls in t seconds, then $s = 16\,t^2$. In other words, the distance varies directly as the square of the time. Newton's law of gravitation states that the attraction of the heavenly bodies for each other varies *inversely* as the square of the distance. Works on elementary algebra usually confine their statements and exercises to the first two kinds of variation.

4. Exercise.

(1. If $x \propto y$ and $y = 5$ when $x = 14$ find x when $y = 20$.

(2. If $x \propto y$ and $x = 7$ when $y = 31$ find y when $x = 20$.

(3. If $A \propto BC$ and $A = 6$ when $B = 4$ and $C = 15$, find B when $A = 100$ and $C = 10$.

(4. If $m \propto \frac{1}{n}$, and when $m = a$, $n = b$, find n when $m = c$.

(5. If the cube of x varies as the square of y, and if $x = 3$ when y equals 5, find the equation between x and y.

(6. If $a + b \propto a - b$ prove that $a^2 + b^2 \propto ab$; and if $a \propto b$ prove that $a^2 - b^2 \propto ab$.

Suggestion. — Let $m =$ the common ratio, then in the first part $a + b = m\,(a - b)$.

CHAPTER XXVI.

EXPONENTIAL EQUATIONS. — LOGARITHMS.

381. Exponential Equations are those in which the unknown appears as an exponent.

Thus, $5^x = 25$, in which plainly, $x = 2$.
$8^x = 560$, in which plainly, $x > 3$ and < 4.
$a^x = b$ is the general form of the equation.

a. Except in the case of exact powers, such as $2^x = 16$ (in which $x = 4$); $(\frac{9}{4})^x = \frac{3}{2}$ (in which $x = \frac{1}{2}$); etc., there is no process like those for the solution of simple and quadratic equations by which this class of problems can be solved. However, tables of logarithms enable us to find *approximate* solutions very readily. We turn, therefore, to the subject of logarithms.

382. The **Logarithm** of a number is the exponent of the power to which a fixed number must be raised to produce it. The fixed number is called the base.

a. Evidently three different numbers are concerned. First, the fixed number; second, any number; third, its logarithm.

Thus, if the base is 10, and the number 105, its logarithm is 2.0212, i.e., a little more than 2. For $10^2 = 100$, and $10^{2.0212} = 105$. And so for any other number. As another example, the logarithm of 865 is 2.9370 nearly.

b. Logarithms, except in a relatively small number of instances, are found *approximately* and not *exactly*.

Thus, the logarithm of 105 is not exactly 2.0212, in which four decimal places of its value are given, nor is it 2.021189, in which six places of its value are given, but these are approximations correct as far as they go.

383. Systems of Logarithms. Tables. Any positive number except unity may be taken as the base of a system of logarithms.

a. The logarithm of 1 in any system is 0; for any number (the base) whose exponent is zero is equal to unity. (128, 1.)

1. Let us take 2 as the base of a system. The numbers are written in the left column and directly opposite them their logarithms.

NUMBER.	LOGARITHM.	REASON.	NUMBER.	LOGARITHM.	REASON.
1	0.	$2^0 = 1$	4	2.	$2^2 = 4$
2	1.	$2^1 = 2$	$\frac{1}{4} = .25$	-2.	$2^{-2} = \frac{1}{4}$
$\frac{1}{2} = .5$	-1.	$2^{-1} = \frac{1}{2}$	5	2.3223	$2^{2.3223} = 5$
3	1.5850	$2^{1.5850} = 3$	$\frac{1}{5} = .2$	-2.3223	$2^{-2.3223} = \frac{1}{5}$
$\frac{1}{3} = .33+$	-1.5850	$2^{-1.5850} = \frac{1}{3}$	etc.	etc.	etc.

The student is not supposed to know how the decimal logarithms are found. One method of obtaining them is by a long process of repeated extraction of roots. See the "Encyclopaedia Britannica" article on Logarithms.

b. Logarithms are exponents, integral or fractional, and as such may be interpreted as all others. (283.)

Thus, $2^{1.5850} = 2^{\frac{15850}{10000}}$, means the 15850th power of the 10000th root of 2; and if these operations were actually performed the result would be 3.

So, also, $10^{.021189} = 10^{\frac{21189}{1000000}}$, means the 21189th power of the 1000000th root of 10; and if these operations were actually performed the result would be 1.05. In other words, the logarithm of 1.05 in the system whose base is 10 is .021189 correct to six decimal places. So for all logarithms.

2. Unity cannot be the base of a system of logarithms, for every power of 1 is 1 continually. Neither can *negative* numbers be used as the base of a system of real logarithms; for we saw in this article that the root to be extracted is even. Hence, this root would be imaginary. Then when

raised to the various powers giving the corresponding numbers, some would be imaginary, some plus, some minus, and all in confusion. Therefore, a minus number cannot be the base of a simple system of logarithms.

c. The base of a system being taken positive and greater than unity, the logarithms of all fractions between 0 and 1 are negative (see table in this article); the logarithms between 1 and the base of the system are between 0 and 1; and the logarithms of numbers greater than the base are greater than 1. Negative numbers, as such, are not supposed to have logarithms, though numerical calculations in which negative numbers occur are often made irrespective of signs, or just as if all were positive.

384. Terms and Notation. The integral part of a logarithm is called the **characteristic.** The fractional part is called the **mantissa.** "Logarithm" is usually abbreviated into "log.," and the base of the system used may be written as a subscript to it.

Thus, $\log._{10}50 = 1.6990$ means that the logarithm of 50 to base 10 is equal to 1.6990. In this logarithm, 1 is the characteristic, and .6990 the mantissa.

385. The Briggsian or Common System of Logarithms.[1] The system in practical use, like the Arabic notation, has 10 for its base.

1. Logarithms of fractions in this system (**383**, *c*) are negative, and grow *larger* as the number grows *smaller.* Thus the logarithm of .001 is -3; that of .0001 is -4, and so on *ad infinitum.* At the limit when the number is infinitely small ($\doteq 0$), the logarithm is an infinitely *large* negative number. Furthermore, it may be seen that the logarithm of a fraction say between .001 and .0001 must lie between -3 and -4, one between .0001 and .00001 between -4 and -5, dependent on the powers of $\frac{1}{10}$; and so on.

[1] Logarithms were invented by *Lord Napier* about 1614 A.D. *Henry Briggs*, a great admirer of *Napier's* logarithms, saw the practical advantage of using 10 as the base, and constructed his tables accordingly.

2. Logarithms between 1 and 10. — The logarithm of 1 is 0, and of 10 is 1; hence the logarithms of numbers between 1 and 10 lie between 0 and 1, i.e., are proper fractions.

3. Logarithms of numbers greater than 10. We have $10^1 = 10$, $10^2 = 100$, $10^3 = 1000$, $10^4 = 10000$, $10^5 = 100000$, $10^6 = 1000000$, etc.

Here we may see that the logarithm of a number between 10 and 100 is between 1 and 2; the logarithm of a number between 100 and 1000 is between 2 and 3, i.e., is 2 + a decimal, and so on, *ad infinitum*. Hence, $10^\infty = \infty$. Here as elsewhere the two ∞'s are not equal.

To further illustrate the preceding,

$$\log.\ 67 = 1.8261 \qquad \log.\ 345.25 = 2.5381$$
$$\log.\ 1069.459 = 3.0292 \qquad \log.\ 5.678 = 0.7542$$

386. Briggsian Mantissas. — In the Briggsian or common system *the value of the* MANTISSA *will not change* if the number be multiplied or divided by some power of 10. In other words, the value of the mantissa is independent of the position of the decimal point.

$$\text{Thus, } \log.\ 1.27 = 0.1038$$
$$\log.\ 12.7 = 1.1038$$
$$\log.\ 12700 = 4.1038$$
$$\text{etc.} \qquad \text{etc.}$$

The reason for this follows from the law of exponents. Let us suppose that the logarithm of 1.27 is known to be 0.1038, or that $10^{0.1038} = 1.27$. We have, (287. 2)

$$(1)\ 12.7 = 1.27 \times 10 = 10^{0.1038} \times 10^1 = 10^{0.1038+1} = 10^{1.1038}$$
$$(2)\ .0127 = 1.27 \div 100 = 10^{0.1038} \div 10^2 = 10^{0.1038-2} = 10^{2.1038}$$
$$(3)\ 1270 = 1.27 \times 1000 = 10^{0.1038} \times 10^3 = 10^{0.1038+3} = 10^{3.1038}$$
$$(4)\ .000127 = 1.27 \div 10000 = 10^{0.1038} \div 10^4 = 10^{0.1038-4} = 10^{4.1038}$$

N	0	D	1	D	2	D	3	D	4	D	5	D	6	D	7	D	8	D	9	D
10	0000	43	0043	43	0086	42	0128	42	0170	42	0212	41	0253	41	0294	40	0334	40	0374	40
11	0414	39	0453	39	0492	39	0531	38	0569	38	0607	38	0645	37	0682	37	0719	36	0755	37
12	0792	36	0828	36	0864	35	0899	35	0934	35	0969	35	1004	34	1038	34	1072	34	1106	33
13	1139	34	1173	33	1206	33	1239	32	1271	32	1303	32	1335	32	1367	32	1309	31	1430	31
14	1461	31	1492	31	1523	30	1553	31	1584	30	1614	30	1644	29	1673	30	1703	29	1732	29
15	1761	29	1790	28	1818	29	1847	28	1875	28	1903	28	1931	28	1959	28	1987	27	2014	27
16	2041	27	2068	27	2095	27	2122	26	2148	27	2175	26	2201	26	2227	26	2253	26	2279	25
17	2304	26	2330	25	2355	25	2380	25	2405	25	2430	25	2455	25	2480	24	2504	25	2529	24
18	2553	24	2577	24	2601	24	2625	23	2648	24	2672	23	2695	23	2718	24	2742	23	2765	23
19	2788	22	2810	23	2833	23	2856	22	2878	22	2900	23	2923	22	2945	22	2967	22	2989	21
20	3010	22	3032	22	3054	21	3075	21	3096	22	3118	21	3139	21	3160	21	3181	20	3201	21
21	3222	21	3243	20	3263	21	3284	20	3304	20	3324	21	3345	20	3365	20	3385	19	3404	20
22	3424	20	3444	20	3464	19	3483	19	3502	20	3522	19	3541	19	3560	19	3579	19	3598	19
23	3617	19	3636	19	3655	19	3674	18	3692	19	3711	18	3729	18	3747	19	3766	18	3784	18
24	3802	18	3820	18	3838	18	3856	18	3874	18	3892	17	3909	18	3927	18	3945	17	3962	17
25	3979	18	3997	17	4014	17	4031	17	4048	17	4065	17	4082	17	4099	17	4116	17	4133	17
26	4150	16	4166	17	4183	17	4200	16	4216	16	4232	17	4249	16	4265	16	4281	17	4298	16
27	4314	16	4330	16	4346	16	4362	16	4378	15	4393	16	4409	16	4425	15	4440	16	4456	16
28	4472	15	4487	15	4502	16	4518	15	4533	15	4548	16	4564	15	4579	15	4594	15	4609	15
29	4624	15	4639	15	4654	15	4669	14	4683	15	4698	15	4713	15	4728	14	4742	15	4757	14
30	4771	15	4786	14	4800	14	4814	15	4829	14	4843	14	4857	14	4871	15	4886	14	4900	14
31	4914	14	4928	14	4942	13	4955	14	4969	14	4983	14	4997	14	5011	13	5024	14	5038	13
32	5051	14	5065	14	5079	13	5092	13	5105	14	5119	13	5132	13	5145	14	5159	13	5172	13
33	5185	13	5198	13	5211	13	5224	13	5237	13	5250	13	5263	13	5276	13	5289	13	5302	13
34	5315	13	5328	12	5340	13	5353	13	5366	12	5378	13	5391	12	5403	13	5416	12	5428	13
35	5441	12	5453	12	5465	13	5478	12	5490	12	5502	12	5514	13	5527	12	5539	12	5551	12
36	5563	12	5575	12	5587	12	5599	12	5611	12	5623	12	5635	12	5647	11	5658	12	5670	12
37	5682	12	5694	11	5705	12	5717	12	5729	11	5740	12	5752	11	5763	12	5775	11	5786	12
38	5798	11	5809	12	5821	11	5832	11	5843	12	5855	11	5866	11	5877	11	5888	11	5899	12
39	5911	11	5922	11	5933	11	5944	11	5955	11	5966	11	5977	11	5988	11	5999	11	6010	11
40	6021	10	6031	11	6042	11	6053	11	6064	11	6075	10	6085	11	6096	11	6107	10	6117	11
41	6128	10	6138	11	6149	11	6160	10	6170	10	6180	11	6191	10	6201	11	6212	10	6222	10
42	6232	11	6243	10	6253	10	6263	11	6274	10	6284	10	6294	10	6304	10	6314	11	6325	10
43	6335	10	6345	10	6355	10	6365	10	6375	10	6385	10	6395	10	6405	10	6415	10	6425	10
44	6435	9	6444	10	6454	10	6464	10	6474	10	6484	9	6493	10	6503	10	6513	9	6522	10
45	6532	10	6542	9	6551	10	6561	10	6571	9	6580	10	6590	9	6599	10	6609	9	6618	10
46	6628	9	6637	9	6646	10	6656	9	6665	10	6675	9	6684	9	6693	9	6702	10	6712	9
47	6721	9	6730	9	6739	10	6749	9	6758	9	6767	9	6776	9	6785	9	6794	9	6803	9
48	6812	9	6821	9	6830	9	6839	9	6848	9	6857	9	6866	9	6875	9	6884	9	6893	9
49	6902	9	6911	9	6920	8	6928	9	6937	9	6946	9	6955	9	6964	8	6972	9	6981	9
50	6990	8	6998	9	7007	9	7016	8	7024	9	7033	9	7042	8	7050	9	7059	8	7067	9
51	7076	8	7084	9	7093	8	7101	9	7110	8	7118	8	7126	9	7135	8	7143	9	7152	8
52	7160	8	7168	9	7177	8	7185	8	7193	9	7202	8	7210	8	7218	8	7226	9	7235	8
53	7243	8	7251	8	7259	8	7267	8	7275	9	7284	8	7292	8	7300	8	7308	8	7316	8
54	7324	8	7332	8	7340	8	7348	8	7356	8	7364	8	7372	8	7380	8	7388	8	7396	8

N	0	D	1	D	2	D	3	D	4	D	5	D	6	D	7	D	8	D	9	D
55	7404	8	7412	7	7419	8	7427	8	7435	8	7443	8	7451	8	7459	7	7466	8	7474	8
56	7482	8	7490	7	7497	8	7505	8	7513	7	7520	8	7528	8	7536	7	7543	8	7551	8
57	7559	7	7566	8	7574	8	7582	7	7589	8	7597	7	7604	8	7612	7	7619	8	7627	7
58	7634	8	7642	7	7649	8	7657	7	7664	8	7672	7	7679	7	7686	8	7694	7	7701	8
59	7709	7	7716	7	7723	8	7731	7	7738	7	7745	7	7752	8	7760	7	7767	7	7774	8
60	7782	7	7789	7	7796	7	7803	7	7810	8	7818	7	7825	7	7832	7	7839	7	7846	7
61	7853	7	7860	8	7868	7	7875	7	7882	7	7889	7	7896	7	7903	7	7910	7	7917	7
62	7924	7	7931	7	7938	7	7945	7	7952	7	7959	7	7966	7	7973	7	7980	7	7987	6
63	7993	7	8000	7	8007	7	8014	7	8021	7	8028	7	8035	6	8041	7	8048	7	8055	7
64	8062	7	8069	6	8075	7	8082	7	8089	7	8096	6	8102	7	8109	7	8116	6	8122	7
65	8129	7	8136	6	8142	7	8149	7	8156	6	8162	7	8169	7	8176	6	8182	7	8189	6
66	8195	7	8202	7	8209	6	8215	7	8222	6	8228	7	8235	6	8241	7	8248	6	8254	7
67	8261	6	8267	7	8274	6	8280	7	8287	6	8293	6	8299	7	8306	6	8312	7	8319	6
68	8325	6	8331	7	8338	6	8344	7	8351	6	8357	6	8363	7	8370	6	8376	6	8382	6
69	8388	7	8395	6	8401	6	8407	7	8414	6	8420	6	8426	6	8432	7	8439	6	8445	6
70	8451	6	8457	6	8463	7	8470	6	8476	6	8482	6	8488	6	8494	6	8500	6	8506	7
71	8513	6	8519	6	8525	6	8531	6	8537	6	8543	6	8549	6	8555	6	8561	6	8567	6
72	8573	6	8579	6	8585	6	8591	6	8597	6	8603	6	8609	6	8615	6	8621	6	8627	6
73	8633	6	8639	6	8645	6	8651	6	8657	6	8663	6	8669	6	8675	6	8681	5	8686	6
74	8692	6	8698	6	8704	6	8710	6	8716	6	8722	5	8727	6	8733	6	8739	6	8745	6
75	8751	5	8756	6	8762	6	8768	6	8774	5	8779	6	8785	6	8791	6	8797	5	8802	6
76	8808	6	8814	6	8820	5	8825	6	8831	6	8837	5	8842	6	8848	6	8854	5	8859	6
77	8865	6	8871	5	8876	6	8882	5	8887	6	8893	6	8899	5	8904	6	8910	5	8915	6
78	8921	6	8927	5	8932	6	8938	5	8943	6	8949	5	8954	6	8960	5	8965	6	8971	5
79	8976	6	8982	5	8987	6	8993	5	8998	6	9004	5	9009	6	9015	5	9020	5	9025	6
80	9031	5	9036	6	9042	5	9047	6	9053	5	9058	5	9063	6	9069	5	9074	5	9079	6
81	9085	5	9090	6	9096	5	9101	5	9106	6	9112	5	9117	5	9122	6	9128	5	9133	5
82	9138	5	9143	6	9149	5	9154	5	9159	6	9165	5	9170	5	9175	5	9180	6	9186	5
83	9191	5	9196	5	9201	5	9206	6	9212	5	9217	5	9222	5	9227	5	9232	6	9238	5
84	9243	5	9248	5	9253	5	9258	5	9263	6	9269	5	9274	5	9279	5	9284	5	9289	5
85	9294	5	9299	5	9304	5	9309	6	9315	5	9320	5	9325	5	9330	5	9335	5	9340	5
86	9345	5	9350	5	9355	5	9360	5	9365	5	9370	5	9375	5	9380	5	9385	5	9390	5
87	9395	5	9400	5	9405	5	9410	5	9415	5	9420	5	9425	5	9430	5	9435	5	9440	5
88	9445	5	9450	5	9455	5	9460	5	9465	4	9469	5	9474	5	9479	5	9484	5	9489	5
89	9494	5	9499	5	9504	5	9509	4	9513	5	9518	5	9523	5	9528	5	9533	5	9538	4
90	9542	5	9547	5	9552	5	9557	5	9562	4	9566	5	9571	5	9576	5	9581	5	9586	4
91	9590	5	9595	5	9600	5	9605	4	9609	5	9614	5	9619	5	9624	4	9628	5	9633	5
92	9638	5	9643	4	9647	5	9652	5	9657	4	9661	5	9666	5	9671	4	9675	5	9680	5
93	9685	4	9689	5	9694	5	9699	4	9703	5	9708	5	9713	4	9717	5	9722	5	9727	4
94	9731	5	9736	5	9741	4	9745	5	9750	4	9754	5	9759	4	9763	5	9768	5	9773	4
95	9777	5	9782	4	9786	5	9791	4	9795	5	9800	5	9805	4	9809	5	9814	4	9818	5
96	9823	4	9827	5	9832	4	9836	5	9841	4	9845	5	9850	4	9854	5	9859	4	9863	5
97	9868	4	9872	5	9877	4	9881	5	9886	4	9890	4	9894	5	9899	4	9903	5	9908	4
98	9912	5	9917	4	9921	5	9926	4	9930	4	9934	5	9939	4	9943	5	9948	4	9952	4
99	9956	5	9961	4	9965	4	9969	5	9974	4	9978	5	9983	4	9987	4	9991	5	9996	4

a. *Mantissas* are always taken *positively*. Otherwise they would change when the characteristic became negative. Thus the logarithm of .127 is 0.1038 − 1 = − 0.8962. But by considering the characteristic alone as negative, *we avoid the introduction of a new decimal.* In such cases the minus sign is written *over the characteristic*, and not before the logarithm.

Thus, $\bar{4}.1038$ means the binomial − 4 + .1038 = − (3.8962).

387. Explanation of the Accompanying Four-Place Table.[1] — The table gives *the mantissas only* of all numbers from 100 to 999. The first two figures of the numbers are found in the column marked "N," the third is one of the ten figures at the top of the page.

Thus 487 is found by taking the 48 in the "N" column, and the 7 from the upper row of figures. The mantissa of the logarithm of 487 is found in the 7 column, opposite 48, and is .6875, At the intersection of lines and columns are found the 900 mantissas corresponding to the 900 numbers, the two first figures determining the line, and the last the column. A decimal point before each mantissa is understood.

The columns of numbers marked D are merely the differences between the mantissas. Thus the difference between .3032 and .3054 is 22; and between .4133 (corresponding to 259) and 4150 (corresponding to 260) is 17.

It may be added (see **386**) that this table gives also the mantissas of all numbers consisting of three *or less than three* figures, *preceded or followed by any number of ciphers, and irrespective of the position of the decimal point.* Thus the mantissa of 31 is the same as that of 310, and is .4914; the mantissa of 8 is the same as that of 800, and is .9031.

388. To Find the Logarithms of Numbers.

1. To find the *characteristic* of the logarithm of a number.

(1. Numbers greater than 1.

(1) The logarithms of all numbers consisting of one

[1] The six-place tables prepared by Professor G. W. Jones, of Cornell University, Ithaca, N.Y., are among the best of those published in this country.

integral figure (i.e., between 1 and 10) have 0 for a characteristic, since they must lie between 0 and 1. (**385**, 2.)

Thus 2, 5, 4.75, 6.978, 9.8943, etc., all have 0 for the characteristic of their logarithms.

(2) The logarithms of all numbers between 10 and 100, consisting of two figures before the decimal point, have 1 for a characteristic, since their logarithms are all between 1 and 2. (**385**, 3.)

Thus 40, 75, 35, 96, 89.746, etc.

(3) The logarithms of all numbers between 100 and 1000 have 2 for a characteristic.

Those between 1000 and 10000 have 3, and so on.

The characteristic of the logarithm of a number greater than unity is always one less than the number of the figures preceding the decimal point.

(2. Numbers less than 1.

(1) The logarithms of all numbers between 1 and .1 ($=10^{-1}$) have -1 for a characteristic, since the values are between 0 and -1, and the mantissa is always *added.*

Thus .2, .35, .6978, .934258, all have -1 as characteristic.

(2) The logarithms of all numbers between .1 ($=10^{-1}$) and .01 ($=10^{-2}$) have -2 for their characteristic, since they must lie between -1 and -2, and the mantissa is added.

(3) The logarithms of all numbers between .01 ($=10^{-2}$) and .001 ($=10^{-3}$) have -3 for their characteristic; those between .001 ($=10^{-3}$) and .0001 ($=10^{-4}$), -4, and so on.

Thus, for .0069 the characteristic is -3; for .0006972 it is -4; for .00002 it is -5, etc.

The characteristic of the logarithm of a proper fraction is always negative, and one greater than the number of ciphers preceding the first significant figure in its decimal value.

In (1) above there were *no* ciphers, and the characteristic was -1. *Evidently common fractions must be reduced to decimals to get their logarithms.*

2. To find the *mantissa* of the logarithm of any number. The rules for finding the characteristic have just been given. It remains to investigate the method for finding mantissas.

(1. The finding of the mantissas of logarithms of numbers consisting of three, or less than three, significant figures was explained in **387**.

Write for practice the logarithms of the following:

(1) 457	(6) 40.0	(11) .0000679	(16) 9000000
(2) 345	(7) 4	(12) 3650	(17) $2\frac{1}{2}$
(3) 367	(8) 259	(13) $\frac{1}{2}$	(18) $37\frac{1}{5}$
(4) 450	(9) .037	(14) $\frac{1}{20}$	(19) $2\frac{3}{4}$
(5) 45	(10) .005	(15) 520000	(20) $1.1\frac{1}{2}$

(2. The mantissas of numbers consisting of more than three figures are found by interpolation.

(1) To get the mantissa of a number consisting of four figures, say of 2568.

By Art. **386** the *mantissa* of 2568 equals the mantissa of 256.8. The reasoning will be a little clearer if the latter number is used. Since 256.8 is between 256 and 257 its logarithm must lie (see table) between 2.4082 and 2.4099 and close to the latter. Now one would naturally suppose that as the number increased from 256 to 257, the logarithm would increase proportionately, or at least nearly so. In truth, it does not increase at precisely the same rate. It is shown in the calculus that the Briggsian logarithm of a number, n, increases $\frac{.434+}{n}$ times as fast as the number itself.

Thus, take the number 126, and suppose it to receive the increment 1; then the increment of the logarithm is $\frac{.434+}{126} \times 1 = .0034+$. Adding this increment to the logarithm of 126, which is 2.1004, the sum is 2.1038, or the logarithm of 127. But in passing from 126 to 127 the multiplier has changed from $\frac{.434+}{126}$ to $\frac{.434+}{127}$.[1] Consequently the above logarithmic increment 0.0034 + is not exactly correct, but sufficiently so for a Four-place table. In the example before us the rate will change from $\frac{.434+}{256}$ to $\frac{.434+}{257}$, which is a very small difference. Consequently intermediate values of logarithms may be found *by proportion* with only very small errors.

We have,

Log. 256.	= 2.4082	(From table.)
Log. 256.8	= 2.4096	(See explanation.)
Log. 257.	= 2.4099	(From table.)

To interpolate the logarithm of 256.8 proportionally we say, the difference of the numbers : its fractional part = the difference of the logarithms : its fractional part.

$$\text{Here, } 1 : .8 = 17 : x \therefore x = 17 \times .8 = 13.6 = 14 -.$$

(Inasmuch as the table being used contains 4 decimal places only, 14 is taken instead of 13.6, since 13.6 is nearer to 14 than to 13.)

Adding 14 to 2.4082 gives the result 2.4096.

(2) For a greater number of figures. Thus, to find the logarithm of 52.7298.

Log. 52.7	= 1.7218	(From table.)
Log. 52.7298	= 1.7220	(By interpolation.)
Log. 52.8	= 1.7226	(From table.)

Tabular difference = 8; 8 × .298 = 2.384 = 2 +;
18 + 2 = 20.

The reasoning is the same as before.

From these examples it is clear that to find the mantissa

[1] In order that this change may be small, logarithmic tables begin with 100 or 1000. Four-place tables include the numbers from 100 to 999. Five and Six-place tables have four figures in their numbers, and extend from 1000 to 9999.

of a number consisting of more than three figures, proceed as follows: —

Find the mantissa corresponding to the first three significant figures of the given number. Multiply the tabular difference following (column D) by the remaining figures of the number regarded as a decimal; and add the product, taken to the nearest unit, to the mantissa of the first three figures.

REMARK. — It is convenient to regard the differences between the mantissas as *whole numbers* instead of retaining their decimal places. No mistakes need arise in so doing.

63. Write for practice the logarithms of

(1) 321.6 (4) 2.5675 (7) .03742 (10) .0004523
(2) 14165 (5) 2379.6 (8) 2905624 (11) 7.0633
(3) 1416.5 (6) .200375 (9) .00001328 (12) 1202800
(13) $11\frac{1}{9}$, carry the decimal to 3 places
(14) $\frac{21}{16}$ (15) $5\frac{4}{9}$.

389. Conversely: To Find the Significant Figures of a Number from its Mantissa.

1. When the given mantissa is the same as one in the table.

The first two figures of the number are found in the same row at the left in the column marked "N." The third is at the top of the column in which the given mantissa is found.

Find the number corresponding to 1.3118, its logarithm. Turning to the table, the mantissa is found opposite 20 and 5.

Hence 20.5 is the number. (**388**, 1.)

2. When the given mantissa lies *between* two in the table. Here the *number* is interpolated in the same way that the logarithm was before.

Thus, given 3.8478 to find the number corresponding. Referring to the table, we find

3.8476 (next less) = log. 7040 (See table.)
3.8478 (given log.) = log. 7043.33 (By interpolation.)
3.8482 (next greater) = log. 7050 (See table.)

Tabular difference $= 6$; $78 - 76 = 2$; $6 : 2 :: 1 : x \therefore x = 2 \div 6 = .333$.

The characteristic being 3, there are 4 integral figures in the number (**388**), which places the decimal after the first 3. And so in all cases.

Hence to find a number when its logarithm is given.

(1. From the given mantissa take the next less found in the table, and divide the remainder by the tabular difference following the latter.

(2. The resulting decimal figures are to be annexed to the three figures corresponding to the lesser mantissa.

(3. The decimal point is then inserted between the figures in accordance with the rules of the preceding article.

390. Exercise in Finding Numbers from their Logarithms.

1. Find for practice the number corresponding to 3.2016. In the table we find the next less mantissa to be .2014, corresponding to 159, and the tabular difference 27. Subtracting and dividing, $2 \div 27 = .074 +$. Annexing these figures to 159, we have $159074 +$. Now, the given characteristic is 3. Hence 4 figures precede the decimal point, and the number is $1590.74 +$.

2. 4.8016	5. 2.0095	8. $\bar{4}.7320$	11. 9.8423 — 10
3. 2.1144	6. 4.2488	9. $\bar{1}.0410$	12. 7.0453 — 10
4. 0.4488	7. 1.9488	10. $\bar{3}.0216$	13. 6.5209 — 10

391. Uses of Systems of Logarithms. — Logarithms were invented to abridge the labor of multiplication and division. They are used to multiply, divide, raise to powers, and extract roots, as well as in the solution of exponential equations.

392. To Multiply Numbers by the Use of Logarithms.

1. Let us find the product, for example, of 155, 207 and 939.

$155 = 10^{2.1903}$ (See table, log. 155)

$207 = 10^{2.3160}$ (See table, log. 207)

$939 = 10^{2.9727}$ (See table, log. 939)

$\therefore 155 \times 297 \times 939 = 10^{7.4790}$ (Ax. 3)

(Adding the exponents of 10 by the rule for the multiplication of monomials.)

By Definition, 7.4790 is the logarithm of the product sought. Referring to the table to find the number corresponding, it is found to be 30128571 +. Hence $155 \times 207 \times 939 = 30128571$, approximately.

(= 30127815 exactly.)

Thus multiplication is made to depend on *addition*, a process much more easily performed.

2. Hence, to multiply together two or more numbers, first find their logarithms and add them, and then find the number corresponding to this logarithm.

3. Multiply together the following numbers by logarithms, obtaining the corresponding approximate results.

(1. Multiply 29 by 39.

log. 29 = 1.4624 (Table, log. from number.)
log. 39 = 1.5911 (Table, log. from number.)
log. 1131 + 3.0535 (Table, number from logarithm.)

$\therefore 29 \times 39 = 1131 +$ approximately (= 1131 exactly).

(2. 19×37.

(3. 43×68.

(4. $85 \times 16 \times 21$.

(5. $122 \times 133 \times 144$.

(6. $12 \times 13 \times 14 \times 15 \times 16 \times 17$.

393. To Divide Numbers by the Use of Logarithms.

1. o find the quotient, e. g., of 207 divided by 1.93.

$207 = 10^{2.3160}$ (Table, log. from number.)

$1.93 = 10^{0.2856}$ (Table, log. from number.)

$\therefore 207 \div 1.93 = 10^{2.0304}$ (Ax. 4)

(Subtracting the exponents of 10 by the rule for the division of monomials.)

Now, 2.0304 is the logarithm of the quotient. Turning to the table, we find for the number corresponding, 107.25. Thus, division is made to depend upon subtraction.

2. Hence, to divide one number by another, first find their logarithms and subtract the logarithm of the divisor from that of the dividend, and then find the number corresponding to the difference of their logarithms.

3. Perform the following divisions by logarithms:

(1. Divide 35 by 7.

log. 35 = 1.5441 (Table, log. from number.)
log. 7 = 0.8451 (Table, log. from number.)
log. 5 = .6990 (Table, number from log.)

(2. $91 \div 13$.
(3. $85 \div 17$.
(4. $198 \div 2.67$.
(5. $2345 \div 163.5$.
(6. $796.325 \div 196275$.
(7. $.00367 \div 2.61$.
(8. $.01917 \div .00021$.

4. Rule for evaluating Compound Expressions, such as

$$\frac{9 \times 13 \times 103}{73 \times 87}$$ by means of logarithms.

(1. Find in turn the logarithms of the numerator and denominator by adding the logarithms of their respective factors.

(2. Subtract the latter sum from the former, and the number corresponding is the quotient sought.

5. Find the value of

(1. $\dfrac{9 \times 13 \times 103}{73 \times 87}$. Model Solution.

$$\begin{array}{ll} \log.\ 9 = 0.9542 & \log.\ 73 = 1.8633 \\ \log.\ 13 = 1.1139 & \log.\ 87 = 1.9395 \\ \log.\ 103 = 2.0128 & \text{denominator} = 3.8028 \\ \text{numerator}\quad 4.0809 & \\ \text{denominator}\quad 3.8028 & \\ 1.897 \quad 0.2781 & \end{array}$$

$$\therefore \frac{9 \times 13 \times 103}{73 \times 87} = 1.897.$$

(2. $\dfrac{336.8 \times 37}{7984 \times 22}$.

(3. $\dfrac{.07654}{83.947 \times 0.8395}$.

(4. $\dfrac{212 \times 6.13 \times 2009}{365 \times 5.31 \times 2.576}$.

(5. $\dfrac{.0062 \times .0007 \times 2}{3.6 \times .00005 \times 9.764}$.

(6. $\dfrac{(2\frac{1}{4} \times 6\frac{3}{4}) \div \frac{13}{20}}{(27\frac{7}{8} \div 206)\, 8\frac{1}{9}}$.

(7. Find x in $28.035 : 3.278 = 3114.27 : x$.

394. To Raise Numbers to Powers by the Use of Logarithms.

1. Raise 13 to the fifth power.

$13 = 10^{1.1139}$

$13^5 = 10^{5.5695}$. (Ax. 5)

(Multiplying the exponent of 10 by 5 in accordance with the rule for raising to powers.)

Finding the number corresponding to the logarithm 5.5695, we have $13^5 = 371090.9$, approximately.

2. To raise a number to a power multiply its logarithm by the index of the power and find the number corresponding.

3. Perform the operations indicated in the following exercise.

(1. Raise 6 to the third power.

log. 6 = 0.7782 (Table, log. from number.)

log. 216 + = 2.3346 (Table, number from log.)

$\therefore 6^3 = 216 +$, approximately.

(2. $7^4 = ?$

(3. 53^5.

(4. $(\frac{73}{61})^4$.

(5. $(\frac{816}{207})^6$

(6. $(1.3672)^{10}$.

(7. $(7\frac{8}{11})^8$.

(8. $(.9975)^{24}$.

(9. $(33.9 \times 43.4 \div 3814)^{11}$.

(10. $(\frac{327}{68})^9$.

(11. $\left(\frac{.000106}{36 \times .07}\right)^5$.

395. To Extract the Roots of Numbers by Means of Logarithms.

1. Since the laws of exponents hold for fractional as well as integral powers, the rule of the preceding article holds good here: hence, multiply the logarithm of the given number by the fractional exponent of the power to which the number is to be raised.

If a root is to be extracted, say the fifth, the logarithm would be multiplied by $\frac{1}{5}$ (or divided by 5) as in the extraction of roots of monomials.

2. Exercise.

(1. $(35021)^{\frac{3}{5}}$

$$\log. 35021 = 4.5444$$
$$\tfrac{3}{5} \times 4.5444 = 2.726 -$$
$$\log. 533. = 2.7267 \;\therefore\; (35021)^{\frac{3}{5}} = 533.$$

(2. $(.0069)^{\frac{1}{5}}$ $\qquad \log. .0069 = \bar{3}.8388.$

The log. $\bar{3}.8388$ is now to be divided by 5. If the division were made *as the logarithm stands*, the result would be the same as when the characteristic is *plus*, which is manifestly wrong. The expression, as was shown, is really a binomial, and must be divided in the binomial form. This is accomplished by *borrowing*, so as *to make the negative characteristic divisible.*

$$\log. (.0069)^{\frac{1}{5}} = \tfrac{1}{5}\,(\bar{3}.8388)$$
$$= \tfrac{1}{5}\,(\bar{5} + 2.8388)$$
$$\log. .3697 \textit{ Ans.} = 1 + .5678 \doteq \bar{1}.5678.$$

(3. $8^{\frac{3}{5}}$ (4. $(\frac{19}{23})^{\frac{2}{3}}$ (5. $(8\frac{3}{4})^{2.3}$ (6. $906.80^{\frac{1}{4}}$

(7. $11^{\frac{1}{8}}$ (8. $\sqrt[6]{\frac{249}{1991}}$ (9. $\sqrt[7]{73567}$ (10. $2.5673^{\frac{7}{11}}$

(11. $\sqrt[11]{\left(\frac{12}{7}\right)^{28}}$ (12. $\sqrt[3]{\frac{496 \times \sqrt{117}}{29.62\frac{3}{7}}}$ (13. $\frac{16^{\frac{1}{6}} \times 15^{\frac{1}{5}} \times 14^{\frac{1}{4}}}{13^{\frac{1}{3}} \times 12^{\frac{1}{2}} \times 11}$

396. Accuracy of Results obtained by using Logarithms. — A little consideration will show that in getting a number from its logarithm (4-place table) one cannot be sure of more than 3 figures. As a general thing the fourth and often the fifth will be correct, but the latter is very uncertain.

Besides, the logarithm itself may be somewhat in error from the circumstances under which it was obtained.

Thus, dividing 771 by 119

FOUR-PLACE.	SIX-PLACE.
log. 771 = 2.8871	2.887054
log. 119 = 2.0755	2.075547
log. 6.480 = 0.8116	log. 6.479 = 0.811507

Here four-place logs. are in error by about a half unit of the fourth place, as six-place values show. These errors are superadded in the quotient logarithm, and, as a result, the four-place table gives 6.480, while the more accurate six-place table gives 6.479.

In raising to powers an error may be considerably increased. Thus, $(\frac{32}{6})^4$ worked out by a four-place table gives rise to an error of about 4 units in the fourth place, the reason being that the log of 32 is too small, and that of 6 too large in the table, and the error doubled by subtraction is then multiplied by 4.

We subjoin a little table of errors

PROBLEM.	ANSWER BY LOGARITHMS.	TRUE VALUE.
155 × 207 × 939	30128571+	30127815
207 ÷ 1.93	107.25	107.254
$\frac{9 \times 13 \times 103}{71 \times 75}$	2.2621+	2.2631
$290^{\frac{1}{6}}$	2.5729	2.5728 nearly.

The student is strongly urged to verify by actual multiplications and divisions all problems which can be so solved without too great an expenditure of time and labor. A fair idea of the accuracy of logarithmic computations, and the advantage derived in their use may thus be gained. If other (six- or seven-place) tables are at hand let them be tested in the same way. Every endeavor should be made to get a practical as well as theoretical understanding of the subject at the start. Constant use is made of logarithms in trigonometry and other branches of mathematics.

The student might be led to suppose that if logarithms give results correct only to a certain number of places (depending on the number of decimal places in the table used), not much *practical* use could be made of them. But such is not the case. In actual measurements, as with a rule, it is not possible to get lengths any closer than hundredths of an inch, and often not so close. Hence results obtained by logarithms may be made correct within the limits of error by choosing a table having a sufficient number of places.

397. * Logarithms to other Bases can be derived from the Briggsian Logarithms. — To show this the following theorem must be proved: —

THEOREM: *If the logarithm of any number be taken to two different bases, the first logarithm equals the second multiplied by the logarithm of the second base in the first system.*

Let n be any number and b and b' two bases.

Put $\log._{b'} n = x$, and $\log._b b' = y$; then by definition, $b' = b^y$, and $n = b'^x = (b^y)^x = b^{xy}$,

i.e., $\log._b n = xy = \log._{b'} n \times \log._b b'$ *Q. E. D.*

Thus, $\log._{10} 5 = \log._2 5 \times \log._{10} 2$.

Referring to the values as given in **383**. we have

$$.6990 = 2.3223 \times .3010$$

which is easily verified by multiplication.

If $\log._{10} n = \log._{b'} n \times \log._{10} b'$

$\log._{b'} n = \log._{10} n \div \log._{10} b'$ (Ax. 4)

Hence, *to find the logarithm of any number to a new base, divide the number's logarithm to base 10 by the logarithm of the new base to base 10.*

The student can easily verify that this is true in the examples of **383**.

398. Solution of Exponential Equations by Means of Logarithms.

1. To solve the equation $a^x = b$.

Since the two members of the equation are equal, their logarithms are equal (general axiom), and since log. $(a^x) = x \log. a$ (**395**), we have

$$x \log._{10} a = \log._{10} b$$

$$\therefore\ x = \frac{\log._{10} b}{\log._{10} a}. \qquad \text{(Ax. 4)}$$

REMARK. — This result may also be derived by means of the theorem at the end of the last article, x, b, and a equaling respectively, $\log._{b'} n$, n, b'.

2. **Exercise.**

(1. Solve the equation $13^x = 129$.

$$x = \frac{\log._{10} 129}{\log._{10} 13} = \frac{2.1106}{1.1139} = 1.895 \quad \textit{Ans.}$$

(2. $2^x = 64$.

(3. $20^x = 100$.

(4. $11^x = 3$.

(5. $3^x = .8$ (**386**, *a*)

(6. $9.1^x = 2467.2$

(7. $3^{2x+1} = 27$.

(8. $.0527^{7x+1} = 396.2$

(9. $10^{2x} + 2 \times 10^x = 80$.

SUGGESTION. — Solve first as a quadratic for 10^x.

(10. Given $a^{mx} b^{nx} = c$

Taking logarithms of both sides

$$\log. a^{mx} + \log. b^{nx} = \log c \qquad \textbf{(392)}$$

or,

$$mx \log. a + nx \log. b = \log. c \qquad \textbf{(394)}$$

$$\therefore\ x = \frac{\log. c}{m \log. a + n \log. b}. \qquad \text{(Ax. 4)}$$

(11. $2^x 3^{4x} = 2000$.

(12. $2^x + \frac{1}{2^x} = 5$.

(13. $3^{x^2 - 4x + 5} = 1200$.

SUGGESTION. — Put $x^2 - 4x + 5 = z$ and solve for z.

CHAPTER XXVII.

ARITHMETICAL AND GEOMETRICAL PROGRESSIONS.

399. An **Arithmetical Progression** is a series of terms which increase or decrease by a common difference.

E.g., in 4, 7, 10, 13, 16, etc., the common difference is 3; and in 8, 6, 4, 2, 0, -2, -4, etc., the common difference is 2.

In the study of progressions there are two problems which stand out prominently; viz., to find the nth term, and to find the sum of n terms.

400. Formula for finding the nth Term of an Arithmetical Progression.

Let a be the first term, d the common difference, and n the number of terms considered. Then, by definition,

$$a,\ a \pm d,\ a \pm 2d,\ a \pm 3d,\ a \pm 4d, \text{ etc.},$$

is the arithmetical progression, the signs being taken all positive or all negative according as it is an increasing or decreasing series; for each term is greater or less by d than the one which precedes it.

We are to find the nth term.

Let us write several terms with the ordinal of each above it.

1st	2nd	3^{d}	4th		n^{th}
$a,$	$a+d,$	$a+2d,$	$a+3d,$	. .	$a+(n-1)d.$
$a,$	$a-d,$	$a-2d,$	$a-3d,$	. .	$a-(n-1)d.$

It will be seen that the multiplier of a is always 1 less than the number of the term. Hence, the nth term is of the form given above.

Calling l the nth or last term,

$$(1)\quad l = a \pm (n - 1)\, d.$$

Hence, to find the last term, multiply the common difference by one less than the number of terms, and add the product to the first term if an increasing, or subtract from the first term if a decreasing progression.

401. Formula for finding the Sum of n Terms of an Arithmetical Progression.

As before, call a the first term, d the common difference, n the number of terms, and l the last term. Denote by s the sum of n terms. If d be added each time, we may write the last term l, the next to the last $l - d$, the second from the last $l - 2\,d$, etc. But if d is subtracted each time, the next to the last term would be d *greater* than l, or $l + d$. the second term from the last $l + 2\,d$, etc. (See **45**, b, and remark on $\mp$ Art. **344**.) Consequently we may write

$$a,\ a \pm d,\ a \pm 2\,d,\ a \pm 3\,d,\ \ldots\ l \mp 3\,d,\ l \mp 2\,d,\ l \mp d,\ l.$$

Trying to sum this series, a little study suggests the plan of adding the first and last terms, the second and next to the last terms, the third and third from the last terms, and so on, since the sum in each case would be the same, viz., $a + l$. It will be a little plainer, however, to add the series to itself, writing the second in reverse order, as this also will give the sum $a + l$ for each addition, and it is clear that there will be just n such additions.

$$s = a + a \pm d + a \pm 2\,d + a \pm 3\,d + \ldots\ l$$
$$s = l + l \mp d + l \mp 2\,d + l \mp 3\,d + \ldots\ a$$
$$\overline{2\,s = a + l + a + l + a + l + a + l +\ \ldots\ a + l}\ (\text{Ax. 1})$$
$$2\,s = a + l \text{ counted } n \text{ times since there are } n \text{ terms.}$$

$$(2)\quad s = \frac{n}{2}\,(a + l).$$

Hence, to find the sum of an arithmetical progression, after calculating the value of the last term add it to the first term and multiply the sum by half the number of terms.

402. Other Problems in Arithmetical Progression. — The progressions furnish excellent exercise in the solution of such equations as were given in 44–50 of Art. **216**. Thus, in equation (1) above, $l = a \pm (n - 1)d$, there are four quantities involved (s not appearing), three of which being given, the fourth may be calculated. This gives four prob lems. Then, similarly, equation (2) gives four more involving a, n, l, and s, (the fifth quantity, d, not appearing). Next, by eliminating n between (1) and (2), we have an equation involving a, d, l, and s. Eliminating a between (1) and (2), the resulting equation contains d, n, l, and s. Last of all, eliminating l, the resulting equation contains a, d, n, and s. Consequently there are five equations, each one of which gives four problems.

As an example, suppose l, d, and s given to find a. Since n is the omitted quantity, and (1) and (2) both contain it, it must be eliminated. From (2)

$$\frac{2s}{a+l} = n. \qquad \text{(Ax. } b\text{)}$$

Substituting this value in (1)

$$l = a + \left(\frac{2s}{a+l} - 1\right)d$$

$$l = a + \frac{(2s - a - l)d}{a+l}$$

in which a is supposed to be the unknown.

$$al + l^2 = a^2 + al + 2ds - ad - ld \qquad \text{(Ax. 3)}$$

$$a^2 - da = l^2 + ld - 2ds \qquad \text{(Ax. } a\text{)}$$

$$4a^2 - 4da + d^2 = 4l^2 + 4ld + d^2 - 8ds \qquad \textbf{(335)}$$

$$2a - d = \pm\sqrt{(2l + d)^2 - 8ds}.$$

$$a = \frac{d}{2} \pm \frac{1}{2}\sqrt{(2l + d)^2 - 8ds} \quad \textit{Ans.}$$

1. To find l, being given (1. a, d, n (2. a, n, s (3. a, d, s (4. d, n, s.

2. To find s, being given (1. a, l, n (2. a, l, d (3. a, d, n (4. l, d, n.

3. To find d, being given (1. a, n, l (2. a, n, s (3. a, l, s (4. l, n, s.

4. To find n, being given (1. a, d, l (2. a, d, s (3. a, l, s (4. l, d, s.

5. To find a, being given (1. d, n, l (2. d, n, s (3 l, n, s.

NOTE. — Of these problems only one is commonly given special consideration. It is that in which a, l, and n are given to find d. It may be stated in words as follows: required to insert $n - 2$ arithmetical means (a and l being the other two terms) between any two quantities a and l. After having found the common difference, it is an easy matter to write down these means. The formula for d, as given among the answers, is $d = \frac{l - a}{n - 1}$.

If 7 means are to be inserted, $n = 9$, and so in general.

403. Exercise in Arithmetical Progression.

1. Find the last term and the sum of the terms in the series 5, 7, 9, etc., to 20 terms.

Here $a = 5$, $d = 2$, $n = 20$.
To find l, $l = a + (n - 1)\,d = 5 + 19 \times 2 = 43$. *Ans.*
To find s, $s = \frac{n}{2}(a + l) = 10\,(5 + 43) = 480$. *Ans.*

2. Find the 17th term in the series 10, $11\frac{1}{2}$, 13, etc.

3. Find the last term and the sum in 5, 9, 13, etc., to 19 terms.

4. Find the 20th term of the series 5, 1, -3, etc.

5. Find the nth term of 2, $2\frac{1}{3}$, $2\frac{2}{3}$, etc.

6. Sum $7 + \frac{29}{4} + \frac{15}{2} +$ etc., to 16 terms.

7. Find the last term and the sum in $-\frac{3}{4}$, $-\frac{5}{8}$, etc., to 55 terms.

8. Sum $2a - 5b$, $7a - 2b$, etc., to 9 terms.

9. Given $a = 5$, $d = 4$, and $l = 201$, required n, and s.

10. Insert 12 means between 12 and 77.

11. Insert 9 means between 2 and 5 and write the progression to 8.

12. Given $a = -5$, $l = -47$, $s = -1118$, find n and then d.

13. What is the arithmetical mean between 4 and 10? between 16 and -4? between $a^2 + ab - b^2$ and $a^2 - ab + b^2$?

14. The first term of a series is 2 and the common difference $\frac{1}{3}$. What term will be 10?

15. Which term of the series 5, 8, 11, etc., is 320?

16. Find the sum of $-3q$, $-q$, q, etc., to p terms.

17. How many terms of the series .034, .0344, .0348, etc., amount to 2.748?

18. Given $d = -3$, $l = -39$, $s = -264$ to find a and n.

19. How many terms of the series $5 + 7 + 9 +$ etc., must be taken in order that the sum may be 480?

REMARK. — Negative values of n are by the nature of the subject excluded.

20. Show that if it is known that the mth term of a series is p, and the nth term is q, that the series is known. Find d, a, and l.

21. Find the series in which the 27th term is 186, and the 45th term 222.

22. The sum of 15 terms of an arithmetical progression is 600, and the common difference is 5. Find the first term.

23. Of two arithmetical series whose first terms are equal, the first has for its last term 39, and for sum 207; while the second has for its last term 124, and for its sum 917. What is the first term in both, and what the number of terms in each?

SUGGESTION. — See Art. 250. Use formula $n = \frac{2s}{l + a}$.

24. Two hundred stones being placed on the ground in a straight line at the distance of 2 feet from each other, how far will a person travel who shall bring them separately to a basket which is placed 20 yards from the first stone, if he start from the spot where the basket stands?

25. A body falls $16\frac{1}{12}$ ft. the first second, and in each succeeding second $32\frac{1}{6}$ ft. more than in the next preceding one. How far will it fall during the 16th second, and what will be the whole distance fallen through in 16 seconds?

26. If a person saves \$100 and puts it at simple interest at 5% at the end of each year, how much will his property amount to at the end of 20 years?

27. A man was paid for drilling an artificial well 3.24 marks for the first meter, 3.29 marks for the second, 3.34 marks for the third, and so on. The well had to be sunk 500 meters. How much was paid for the last meter, and how much for the whole?

28. A travels uniformly 20 miles a day; B starts 3 days later and travels 8 miles the first day, 12 the second, and so on in arithmetical progression. In how many days will B overtake A?

29. Show that if the same quantity be added to every term of an arithmetical progression the sums will be in *a. p.*[1]

30. Show that if every term of an *a. p.* be multiplied by the same quantity, the products will be in *a. p.*

31. The sum of three numbers in *a. p.* is 12, and the sum of their squares is 66; find them.

32. Find four integers in *a. p*, such that their sum shall be 24, and their product 945.

33. The sum of five numbers in *a. p.* is 45, and the product of the first and fifth is five-eighths of the product of the second and fourth. Find the numbers.

[1] For brevity *a. p.* will be used for arithmetical progression and *g. p.* for geometrical progression.

404. A Geometrical Progression is a series in which the terms increase or decrease by a common ratio.

E.g., 3, 6, 12, 24, 48, etc., the common ratio being 2; and, 5, $1\frac{2}{3}$, $\frac{5}{9}$, $\frac{5}{27}$, $\frac{5}{81}$, etc., the common ratio being $\frac{1}{3}$.

405. Formula for finding the *n*th term of a Geometrical Progression.

Let a be the first term, r the ratio, n the number of terms, and l the last term. Then, by definition,

1st	2d	3d	4th	5th	6th, etc.
a,	ar,	ar^2,	ar^3,	ar^4,	ar^5, etc.

Here it is evident that the exponent of r in any term is always one less than the number of that term. Consequently the exponent of r in the nth term would be $n-1$. Hence, we have,

$$(1)\quad l = ar^{n-1}.$$

Or, to find the nth term of a geometrical progression, raise the ratio to a power whose exponent is one less than the number of the term, and multiply the result by the first term.

406. Formula for finding the sum of *n* terms of a Geometrical Progression.

We use the same letters, and as in *a. p.* call s the sum of n terms. To sum the geometrical series an artifice is employed analogous to that used for summing an *a. p.* The value of s is first written down, and then, underneath, the same equation multiplied through by the ratio. By subtracting the first equation from the second, all the intermediate terms on the right go out, leaving only the first and last terms.

$$\begin{aligned} s &= a + ar + ar^2 + ar^3 + ar^4 + \ldots\ldots + ar^{n-1} \\ rs &= \quad\;\; ar + ar^2 + ar^3 + ar^4 + ar^5 + \;.\; + ar^{n-1} + ar^n \\ \hline rs - s &= ar^n - a \quad (\text{Ax. } 2) \qquad\qquad [(\text{Ax. } 3) \end{aligned}$$

$$(r-1)s = a(r^n - 1)$$

$$(2) \qquad s = \frac{a(r^n - 1)}{r-1}$$

Or, to find the sum of n terms of a *g. p.*, multiply the first term by the quotient of the ratio raised to the nth power less one, divided by the ratio less one.

A formula in which n is replaced by l is readily derived from (2) and is convenient for reference.

$$(2) \quad s = \frac{ar^n - a}{r-1} = \frac{(ar^{n-1})r - a}{r-1}$$

$$(3) \quad s = \frac{lr - a}{r-1}$$

By this formula, to find the sum, multiply the last term by the ratio, from the product subtract the first term, and divide the remainder by the ratio less one.

REMARK. — When the ratio is less than 1, the use of negatives can be avoided by changing the signs of both numerators and denominators in (2) and (3).

Then, $s = \dfrac{a(1-r^n)}{1-r}$, and $s = \dfrac{a-rl}{1-r}$.

407. Other Problems in Geometrical Progression. — The three equations found in the last two articles are really equivalent to only two independent ones, since one was derived from the other two. Practically the same statement can be made here as that at the beginning of **402**, d being replaced by r. The actual work of elimination and solution in *g. p.* is far more difficult to accomplish than in *a. p.* Indeed, in four cases the general solution of the problem cannot be obtained.

1. Problems which can be solved without the use of logarithms.

(1. Let it be required to find s in terms of r, l, and n, a being eliminated. From (1).

$$a = \frac{l}{r^{n-1}} \qquad \text{(Ax. 4)}$$

$$(3) \qquad s = \frac{lr - a}{r - 1} = \frac{lr - \dfrac{l}{r^{n-1}}}{r - 1} \qquad \textbf{(228)}$$

$$s = \frac{\dfrac{lr^n - l}{r^{n-1}}}{\dfrac{r - 1}{1}} = \frac{l\,(r^n - 1)}{r^{n-1}\,(r - 1)} \qquad \textbf{(182)}$$

(2. To find l, being given (1) a, r, s, (2) r, n, s.
(3. To find a, being given (1) r, n, l, (2) r, n, s, (3) r, l, s.
(4. To find r, being given (1)[1] a, n, l, (2) a, l, s.
(5. To find s, being given a, n, l.

2. Problems which can be solved by the use of logarithms.

(1. Given a, l, and s, to find n.

Here r must be eliminated. Solving from (3).

$$rs - s = lr - a \qquad \text{(Ax. 3)}$$

$$rs - lr = s - a$$

$$r = \frac{s - a}{s - l}$$

Substituting this value in (1) $\frac{l}{a} = r^{n-1}$

$$\frac{l}{a} = \left(\frac{s - a}{s - l}\right)^{n-1}$$

Solving this equation for n, as in **398**.

$$n - 1 = \log._{10}\left(\frac{l}{a}\right) \div \log._{10}\left(\frac{s - a}{s - l}\right)$$

$$n = \frac{\log._{10} l - \log._{10} a}{\log._{10}(s - a) - \log._{10}(s - l)} + 1.$$

(2) To find n, being given (1) a, r, l, (2) a, r, s, (3) r, l, s.

[1] To insert $n - 2$ geometrical means between a and l, $r = \sqrt[n-1]{\frac{l}{a}}$. Let m be the number of means to be inserted. Then $r = \sqrt[m+1]{\frac{l}{a}}$.

3. Problems which lead to the solution of equations higher than the second.

(1) For example, given a, n, s, to find l. Here r must be eliminated. As shown above, $r = \frac{s-a}{s-l}$. Substituting this value in

$$(1), l = a\left(\frac{s-a}{s-l}\right)^{n-1}, \text{ or } l\,(s-l)^{n-1} = a\,(s-a)^{n-1},$$

which is an equation of the nth degree in l.

(2) Given n, l, s, to find a.

(3) To find r, being given (1) a, n, s, (2) n, l, s.

408. Summation of a Decreasing Geometrical Progression which extends to Infinity.

In a geometrical progression in which the ratio is less than unity, the terms grow smaller and smaller the farther the series is carried, and approach the limit zero. (See **358**, 1.) At the limit, therefore, when an infinite number of terms is taken, the last term equals zero. Putting $l = 0$ in the value of s given in (3)

$$s = \frac{a - rl}{1-r} = \frac{a - r \times 0}{1-r} = \frac{a}{1-r}$$

or, the sum of a decreasing geometrical series which extends to infinity equals the first term divided by one less the ratio.

As an example, sum the series 2, $\frac{2}{3}$, $\frac{2}{9}$, $\frac{2}{27}$, . . . in which the first term is 2 and the ratio is $\frac{1}{3}$.

$$s = \frac{2}{1-\frac{1}{3}} = 3 \quad \textit{Ans.}$$

409. Repeating Decimals as Examples of a Decreasing Geometrical Progression.

A repeating decimal is one which repeats certain sets of figures, as in .92282828 . . . the two figures, 28, are repeated to infinity. (See **290**, *a*.)

Writing the repeated orders as common fractions, the first term and ratio are readily recognized:

$$\frac{28}{100} + \frac{28}{10000} + \frac{28}{1000000} + \text{etc.}$$

Here the ratio is $\frac{1}{100}$, and the first term is $\frac{28}{100}$.

$$s = \frac{\frac{28}{100}}{1 - \frac{1}{100}} = \frac{28}{99}$$

Hence the fraction is $.92\frac{28}{99} = \frac{92}{100} + \frac{28}{9900} = \frac{9136}{9900} = \frac{2284}{2475}$. In the same way any repeating decimal can be reduced to the equivalent common fraction. If the repeating part consists of one figure, the ratio is .1, if of two figures, .01, etc.

410. Exercise in Geometrical Progression. — To find the ratio, divide any term by that preceding it.

1. Find the ninth term and sum of nine terms of 1, 3, 9, . . .

2. Find the ninth term and the sum of nine terms of 6, 3, $1\frac{1}{2}$, . . .

3. Find the tenth and sixteenth terms of the series 256, 128, 64, . . .

4. Find the last term and sum of $\sqrt{2}$, $\sqrt{6}$, $3\sqrt{2}$, . . . to 12 terms.

5. Find the last term and sum of 8.1, 2.7, .9, . . . to 7 terms.

6. Find the last term and sum of $-\frac{2}{5}, \frac{1}{2}, -\frac{5}{8}$, . . . to 6 terms.

7. Given $r = 10$, $n = 9$, and $s = 1{,}111{,}111{,}110$, required a and n.

8. Find the sum of $0.1 + 0.5 + 2.5 +$ . . . to 7 terms.

9. Insert 6 geometric means between 56 and $-\frac{7}{16}$.

10. Insert one geometric mean between a and b.

11. Show that of two unequal positive numbers the arithmetic mean is always greater than the geometric mean.

12. Sum to infinity the series 9, 6, 4, . . .

13. Sum to infinity the series $1 - \frac{2}{5} + \frac{4}{25} - \frac{8}{125}$, . . .

14. Sum to infinity the series .9, .03, .001, . . .

15. Sum to infinity the series .3333. . . . . .

16. Sum to infinity .37878; also .12135135. . . . .

17. Show that the reciprocals of the terms of a *g. p.* are also in *g. p.*

18. Find the sum of $\sqrt{a}$, $\sqrt{a^3}$, $\sqrt{a^5}$, etc., to a terms.

19. Find the series in which the 10th term is 320, and the 6th, 20.

20. If a man, whether by his example or designedly, leads a single fellow-man from the path of rectitude each year during twenty years, and each of these men in turn leads astray a single man each year from the time of his change, and so for every one affected, what will be the total number led astray as the outcome of the first man's bad influence?

21. Achilles pursued a tortoise, which was one stadium ahead of him, with a speed twelve times greater than that of the tortoise. When Achilles reached the place where the tortoise had been when he started, the tortoise was still $\frac{1}{12}$ stadium in advance of him; having traversed this distance, the tortoise was still $\frac{1}{144}$ stadium ahead of him, and so on, *ad infinitum*. Will Achilles, then, never catch up with the tortoise?

Explanation. — This problem is a statement of Zeno's celebrated sophism. Let x equal the number of stadia which Achilles must run to catch up with the tortoise. Then we have the equation,

$$x - \frac{x}{12} = 1 \therefore x = 1\tfrac{1}{11} \text{ stadia.}$$

Also, summing the progression $1 + \frac{1}{12} + \frac{1}{144} + \frac{1}{1728} + \ldots$

$$x = \frac{1}{1 - \frac{1}{12}} = 1\tfrac{1}{11}$$

Hence the sum of the infinite series tends to a definite limit. While it is perfectly allowable to *conceive* of the distance divided up in this way, the sum of the parts is only $1\frac{1}{11}$ stadia.

CHAPTER XXVIII.

INTEREST, ANNUITIES, AND BONDS.

411. The Problems of Interest and Annuities require the employment of the formulæ of progressions, furnish valuable exercise in the use of logarithms, and are besides of much practical importance.

SECTION I.

INTEREST.

412. Interest is a payment for the use of money. The sum lent is called the *principal*, and the number of hundredths the yearly interest is of the principal is called the *rate*.

Three kinds of interest may be distinguished: simple, annual, and compound.

413. Simple Interest is interest on the principal alone. It is supposed to be paid annually. But when it is not, the total amount of interest (i) due at the end of n years is, as was stated in Ex. 47, Art. **216**, $i = prn$.

The problems of simple interest may be summarized as follows: —

1. Given p, r, n, to find i; $i = prn$.
2. Given p, r, i, to find n; $n = i \div pr$.
3. Given p, n, i, to find r; $r = i \div pn$.
4. Given n, r, i, to find p; $p = i \div rn$.
5. If a = amount, $a = p + prn = p\ (1 + rn)$.

From this formula a similar set of cases may be distinguished by solving in turn for the different letters. It is not necessary to go into this further.

414. Annual Interest is simple interest on the principal and on each year's interest as it becomes due.

Let p be the principal, r the rate expressed decimally, and $n+f$ the time, in which n represents the whole number of years the principal runs, and f the extra fraction of a year.

Now, assuming the same rate for the unpaid interest as for the principal, we have pr^2 as a year's interest on one payment, pr. But the first payment withheld will run for $n+f-1$ years; the second, for $n+f-2$ years; the third, for $n+f-3$ years; and so on. The last payment will run for the extra fractional part of a year, so that altogether there will be

$$(n+f-1)+(n+f-2)+(n+f-3)+\ldots+(1+f)+f$$

years during which the unit payment pr will draw interest. Summing this arithmetical progression **(401)**, we have

$$\frac{n}{2}(n+2f-1).$$

Multiplying a single year's interest, pr^2, by this total number of years, and adding the product to the simple interest, we have for the annual interest sought

$$pr(n+f)+\frac{pnr^2}{2}(n+2f-1)=pr\left(n+f+\frac{rn}{2}(n+2f-1)\right).$$

EXAMPLE. — Calculate the amount of $1,000, for 10 years and 2 months at annual interest, supposing the rate to be 8%. *Ans.* $1112.

415. In Compound Interest all unpaid interest is added to the principal as fast as it becomes due.

Call p the principal, a the amount, r the rate expressed decimally, and n the number of years. We may have the following problems: —

1. *To find the amount when the principal, rate, and time are given.*

The amount of the principal at the end of the first year is $p(1+r)$. Putting this amount at interest during the second year, the amount is $p(1+r)(1+r)=p(1+r)^2$. Again, putting this sum at interest during the next year, the amount becomes $p(1+r)^3$, and so on through the n periods. Hence,

$$(1) \qquad a=p(1+r)^n$$

If the interest be compounded q times a year, there will be qn periods and the period rate will be $\frac{r}{q}$ and the formula becomes

$$(2) \qquad a=p\left(1+\frac{r}{q}\right)^{qn}.$$

If i is the interest, then

$$i=p(1+r)^n-p=p[(1+r)^n-1].$$

EXAMPLE.—Calculate the amount of a note for \$400, which has been standing for 8 years, allowing 7 per cent interest to be paid semi-annually.

$$\begin{aligned} \log. a &= \log. 400+\log. (1+\tfrac{.07}{2})^{16} && \textbf{(392, 2)} \\ &= \log. 400+16 \log. 1.035 && \textbf{(394, 2)} \\ &= 2.6021+.2384=2.8405 \\ \therefore a &= \$692.67. \end{aligned}$$

REMARK.—It must be borne in mind that a four-place logarithmic table gives only approximate results. Here the exact amount is \$693.60. In this chapter, in order to secure the proper degree of accuracy, a six-place table should be used.

2. *To find the* PRESENT WORTH *of a sum payable n years hence, allowing compound interest.*

Solving from (1) above

$$p=\frac{a}{(1+r)^n} \quad \text{Letting } \frac{1}{1+r}=v,\ p=av^n.$$

NOTE. — The values of the expression $(1+r)^n$ are given in compound interest tables for different values of the arguments r and n. Using these lessens the labor of calculation. The logarithms of the v's may be found by subtracting those of $(1+r)^n$ from 0.

It should be observed that the above formulæ compound the interest in the last fractional period. The rules commonly given direct to calculate simple interest for the odd fraction of a year. Compound interest in this case favors the borrower.

3. *To find the time when the amount, principal, and rate are given.*

Taking logarithms of both sides of $a = p(1+r)^n$

$$\log. a = \log. p + \log. (1+r)^n \qquad \text{(Ax. 8)}$$
$$\log. a = \log. p + n \log. (1+r)$$
$$\therefore n = \frac{\log. a - \log. p}{\log. (1+r)}.$$

EXAMPLE. —Find in how many years \$100 will amount to \$1050 at 5 per cent compound interest. *Ans.*, 48.17 years.

4. *To find the rate when the amount, principal, and time are given.*

This is left as an exercise for the student.

SECTION II.

ANNUITIES CERTAIN.

416. An **Annuity Certain** is a sum of money payable at the end of each year for a fixed period of years, or forever. In the one case the annuity is said to be *terminable*, in the other *perpetual*.

a. Contingent annuities (a term in life insurance) are sums payable as long as specified persons remain alive or for other uncertain periods, and their values are governed by the laws of probability. They cannot therefore be taken up at this stage of the student's progress.

The problems of annuities are analogous to those of compound interest. We shall investigate two cases; viz., to find the amount of an unpaid annuity, and to find the cost of an annuity.

1. *To find the amount of an unpaid annuity.*

Let V represent the amount of an annuity of $A which has run n years. Either simple or compound interest may be taken. We shall use the latter, and we will let $R = 1 + r$, where r is the rate expressed decimally.

The first payment of $A should have been made at the end of the first year, and therefore runs for $n - 1$ years, its amount being AR^{n-1}; the next runs for $n - 2$ years, its amount being AR^{n-2}; the third runs for $n - 3$ years, its amount being AR^{n-3}; and so on. The last payment of $A being made at the end of the period bears no interest. Thus we have for the amount of the annuity,

$$AR^{n-1} + AR^{n-2} + AR^{n-3} + \ldots + A.$$

But these amounts form a geometrical progression whose ratio is R. Summing it (**406**, (3)), we have for the value of V

$$V = \frac{AR^n - A}{R - 1} = \frac{A(R^n - 1)}{r}.$$

(1. If the annuity be not paid until m years after the last payment is due, it will amount to

$$\frac{A(R^n - 1)}{r} R^m.$$

Example.—Let it be required to find the amount of an annuity of $10 to run 12 years at the end of 15 years, interest compounded semi-annually at 5 per cent. Substituting in the formula,

$$V = \frac{\$10\,(1.025^{24} - 1)}{(1.025^2) - 1}\,(1.025)^6.$$

Calculating this by logarithms, starting with the log. of $1.025 = .0107$, we have log. $1.025^{24} = 0.2568$ ∴ $1.025^{24} = 1.806$.

$$\begin{aligned} \log.\ 10 &= 1. \\ \log.\ .806 &= \bar{1}.9063 \\ \log.\ (1.025)^6 &= \underline{0.0642} \\ &\quad .9705 \\ \log.\ .0506 &= \underline{\bar{2}.7042} \\ &\quad 2.2663 \text{ giving V} = \$184.62. \end{aligned}$$

(2. Another way of looking at this problem is to consider the amount of the annuity as an obligation to be met at a certain date in the future, and the payments as sums to be set aside annually as a *Sinking Fund* with which we pay off this debt when due.

Solving for A, we have

$$A = \frac{Vr}{R^n - 1}.$$

As an example, suppose a town borrows \$10,000, agreeing to pay it back at the end of 15 years. What sum will have to be raised by taxation every half year to meet this obligation, supposing interest compounded semi-annually at 4 per cent?

$$A = \frac{10000\ (.02)}{1.02^{30} - 1} = \$246.60\,.$$

2. *To find the cost of an annuity.*

Let \$P be the cost of an annuity of \$A to run n years, and $v = \dfrac{1}{1+r}$.

Then the present value of the first payment of \$A due one year hence is Av; the present value of the second payment due two years hence is Av^2, and so on. The present value of the last, due n years hence, is Av^n. Adding, to find the present value of all, we have

$$\begin{aligned} P &= Av + Av^2 + Av^3 + \quad . \quad . \quad . \quad . \quad . \quad + Av^n \\ &= A\,(v + v^2 + v^3 + \quad . \quad . \quad . \quad . \quad + v^n). \end{aligned}$$

Summing the parenthesis as a *g. p.*, whose ratio is v, we have

$$P = A\,.\,\frac{v^{n+1} - v}{v - 1} = Av\,.\,\frac{v^n - 1}{v - 1} = A\,.\,\frac{1 - \dfrac{1}{(1+r)^n}}{r}$$

(1. If the annuity is *perpetual*, $n = \infty$, and $\left(\dfrac{1}{1+r}\right)^n = 0$, (**358**, 2) so that here $P = \dfrac{A}{r}$.

Example. — Required the cost of a perpetual annuity of \$50, supposing the rate of interest to be 5. It is evident that this is calculated in the same way as a problem in simple interest in which the annual interest and rate are given to find the principal.

(2. To find the cost of an annuity to begin at the end of m years and run for n years. Here all that is necessary is to find the present worth of the preceding result due m years hence. This gives

$$P' = A\;\frac{1 - \dfrac{1}{(1+r)^n}}{r\,(1+r)^m}.$$

(3. If A instead of P is regarded as unknown, the above formula gives the value of n annual payments which will reimburse a lender for a loan of \$P.

Example. — A city borrows \$200,000, which it agrees to repay in 20 annual installments, interest and principal together. How much will have to be paid each year, assuming a 5 per cent rate for interest?

$$A = \frac{Pr}{1 - \frac{1}{(1+r)^n}} = \frac{200000 \times .05}{1 - \frac{1}{1.05^{20}}} = \$16044.$$

NOTE. — For more extended information on the subject of interest and annuities certain the student is referred to *Part First of The Institute of Actuaries Text-Book*, published by the Laytons, London.

SECTION III.

BONDS.[1]

417. Bonds, like promissory notes, draw interest at stated intervals, and the principal itself becomes due at maturity.

Let p be the face of a bond, n the number of years it has to run, r the rate (expressed decimally) named in it, and r' the rate the investor desires to realize on his investment (also regarded as the current rate of interest). We proceed to investigate the different forms in which the problem may appear.

1. *To calculate the price x to be paid for a bond at the time of its issue in order to realize a certain per cent on the investment.*

We solve this problem by balancing income and outgo. The price paid for the bond at compound interest at r' per cent will amount at the time of maturity to $(1 + r')^n x$ dollars. The following amounts come back to the investor: first, the first interest payment $\$pr$, which will therefore run $n - 1$ years at r' per cent compound interest, amounting at the maturity of the bond to $\$pr\,(1 + r)^{n-1}$; next, the second interest payment $\$pr$, which will run for $n - 2$ years amounting to $\$pr\,(1 + r)^{n-2}$; and so on for all the payments, the last being paid at the time of settlement.

[1] The author is indebted to Emory McClintock, LL.D., F.I.A., actuary of the Mutual Life Insurance Company of New York, for important practical suggestions.

Then the face of the bond is also paid at maturity. Hence we have

$$(1+r')^n \cdot x = pr\left[(1+r')^{n-1} + (1+r')^{n-2} + (1+r')^{n-3} + \ldots + 1\right] + p.$$

Summing the bracket quantity, which is a geometrical progression whose ratio is $(1+r')$, we get,

$$(1+r')^n \cdot x = pr\left[\frac{(1+r')^n - 1}{r'}\right] + p$$
$$= \frac{p}{r'}\left(r(1+r')^n - r + r'\right).$$

Dividing through by the coefficient of x.

$$x = \frac{p}{r'}\left(r - \frac{r - r'}{(1+r')^n}\right).$$

Example. — What will be the cost of a bond for \$1000 payable in 12 years and bearing 5 per cent interest, if the purchaser desires to realize 4 per cent on his investment?

$$x = \frac{1000}{.04}\left(.05 - \frac{.05 - .04}{1.04^{12}}\right) = 25000\left(.05 - \frac{.01}{1.04^{12}}\right).$$

Calculating the value of $\frac{.01}{1.04^{12}}$ by logarithmic table, it equals .00625. Whence $x = \$1093.75$.

The formula gives the price to be paid for a bond on the assumption that all interest payments are made annually. But they are usually paid oftener than once a year.

2. *If the interest payments are made q times instead of once a year.*

The period rate will now be $\frac{r}{q}$, and the terms in the amount series will be

$$p\frac{r}{q}(1+r')^{n-\frac{1}{q}},\ \frac{pr}{q}(1+r')^{n-\frac{2}{q}},\ \frac{pr}{q}(1+r')^{n-\frac{3}{q}}, \ldots p\frac{r}{q}.$$

Here the ratio is $(1+r)^{\frac{1}{q}}$. Hence we have **(406, 167)**,

$$(1+r')^n \cdot x = \frac{pr}{q}\left[\frac{(1+r')^n - 1}{(1+r')^{\frac{1}{q}} - 1}\right] + p$$

$$= p\left[\frac{r(1+r')^n - r + q(1+r')^{\frac{1}{q}} - q}{q(1+r')^{\frac{1}{q}} - q}\right]$$

$$\therefore x = \frac{p}{q}\left[\frac{r(1+r')^n + q(1+r')^{\frac{1}{q}} - (q+r)}{(1+r')^n\{(1+r')^{\frac{1}{q}} - 1\}}\right]$$

This and the preceding result were calculated on a basis of annual compound interest. But regular investors expect to compound their interest semi-annually. Tables of bond values are usually constructed in accordance with this practice.

3. *If the interest payments are made q times a year and the interest is compounded semi-annually.*

In order to compound the interest semi-annually, instead of $(1+r)$, we substitute $(1+\frac{r}{2})^2 = 1 + r_e$. This latter (r_e), is often called the *effective* annual rate corresponding to the nominal rate r'.

$$x = \frac{p}{q}\left[\frac{r(1+r_e)^n + q(1+r_e)^{\frac{1}{q}} - (q+r)}{(1+r_e)^n\{(1+r_e)^{\frac{1}{q}} - 1\}}\right]$$

Example. — Required the present value of a municipal bond for $2000 to run for 20 years having six per cent coupons, payable quarterly, to net purchaser 4½ per cent.

Here $r' = .045$; whence $r_e = (1.0225)^2 - 1 = .0455 +$; log. $1.0455 = .0193$; $.0193 \times 20 = .3860$, giving $(1+r_e)^n = 2.4322$; ¼ of $.0193 = .0048$, giving $(1+r_e)^{\frac{1}{q}} \ldots = 1.0112$.

$$x = \frac{2000}{4}\left[\frac{.06 \times 2.4322 + 4 \times 1.0112 - 4.06}{2.4322 \times .0112}\right]$$

$$= \frac{500 \times .1307}{2.4322 \times .0112}$$

$$\begin{array}{rl} \log. \ 500 &= 2.6990 \\ \log. \ .1309 &= \bar{1}.1163 \\ \hline & 1.8153 \\ & \bar{2}.4352 \\ \hline \log. \ x &= 3.3801 \\ \therefore x &= 2400. \end{array} \qquad \begin{array}{rl} \log. \ 2.4322 &= 0.3860 \\ \log. \ .0112 &= \bar{2}.0492 \\ \hline & \bar{2}.4352 \end{array}$$

4. *If interest payments have been made on a bond like a promissory note, it virtually becomes a new bond, to run for the remaining time, and a prospective purchaser may calculate from this standpoint.* If bought at some time *between* interest payments, the discount is from the time of sale to maturity. Thus, instead of n as the exponent of $(1 + r')$, or $(1 + r_e)$, *in the denominator* of the value of x, the exact discount interval must be substituted.

EXAMPLE. — Let it be required to find the present value of a bond for $1000, drawing 7 per cent interest, payable semi-annually, dated Jan. 1st, 1895, and due Jan. 1st, 1910, on June 18th, 1898, supposing the purchaser is willing to take 5 per cent for his money (to be compounded semi-annually).

Here we may calculate the bond as if it had been issued on Jan. 1st, 1898. It will therefore have but 12 years to run. Also the exponent of $(1 + r_e)$ in the denominator will be $11\frac{8}{15}$. The effective rate, $r_e = .050625$.

Log. $1.0506 = .0214$; $.0214 \times 12 = .2568$; $(1 + r_e)^n$
$= 1.8062$; $.0214 \times 11\frac{8}{15} = 0.2466$; $\frac{1}{2}$ of $.0214$
$= .0107$; $(1 + r_e)^{\frac{1}{q}} = 1.025$.

$$x = \frac{1000\,(.07 \times 1.8062 + 2 \times 1.025 - 2.07)}{2 \times 1.0506^{11\frac{8}{15}} \times .025} = \$1206.$$

5. *To calculate the rate of interest which can be realized when the cost of a bond is given.*

In order to make this problem soluble by a formula, we distinguish between the current rate of interest and the rate to be realized. By the current rate of interest we mean the rate at which the interest payments can be promptly re-invested. Let us denote this rate by r_e. Then the formula in 2, above, becomes

$$\left(1+\frac{r'}{2}\right)^{2n} x = \frac{p\,\{r(1+r_e)^n + q(1+r_e)^{\frac{1}{q}} - (q+r)\}}{q\,\{(1+r_e)^{\frac{1}{q}} - 1\}}$$

Solving for r', which is now regarded as unknown,

$$r' = 2\left[\frac{p\,\{r(1+r_e)^n + q(1+r_e)^{\frac{1}{q}} - (q+r)\}}{q\,x\,\{(1+r_e)^{\frac{1}{q}} - 1\}}\right]^{\frac{1}{2n}} - 2.$$

EXAMPLE. — A \$500 bond bears 8% interest payable semi-annually, and is due 18 years hence. If the current rate of interest is $3\frac{1}{2}$, what will a purchaser realize who pays \$800 for it?

Here $r_e = .0353$.

Log. $1.0353 = .0150$; $.0150 \times 18 = .2700$; $(1+r_e)^n = 1.862$; $\frac{1}{2}$ of $.0150 = .0075$; $(1+r_e)^{\frac{1}{q}} = 1.0175$

$$r' = 2\left[\frac{500 \times .104}{2 \times 800 \times .0175}\right]^{\frac{1}{36}} - 2 = .035. \qquad \textbf{(395)}$$

REMARK ON THE LAST CASE. — Properly speaking only *two* rates of interest are recognized in bond calculations; that named in the bond itself, and the rate to be realized by the purchaser. However, we may substitute a probable value of the latter on the right side of the above equation, and thus find an approximate value of the rate to be realized. This answer could be used in a second similar calculation to find a second approximation.

However, a simpler and preferable method of solution is to assume two rates for r_e (one too large and the other too small) in the formula in 3, and calculate the corresponding values of x. Then the desired value of r_e may be found from the given value of x by interpolation.

418. Exercise in Interest, Annuities, and Bond Calculations.

1. What will \$1 amount to in 20 years at 5 per cent, interest being compounded semi-annually?

2. What sum of money will amount to $1000 in 8 years and 4 months at 4 per cent compound interest, interest being compounded quarterly?

3. At what rate per cent per annum will $500 amount to $760.17 in 8 years and 6 months, the interest being compounded semi-annually?

4. How many years will it take a note for $100, bearing 4 per cent interest payable quarterly, to amount to $150?

5. In how many years will a note, whose face is $300, amount to $400, allowing 8 per cent, payable quarterly?

6. In how many years will $967.80 amount to $1269.00 at 5 per cent compound interest, the interest being compounded semi-annually?

7. The population of the United States in 1880 was, in round numbers, 50,000,000; in 1890, 62,500,000. What was the annual gain per cent in this decade, supposing it to have been the same throughout?

8. At the same rate of growth how many years would it require for the population to reach 100,000,000?

9. Allowing 6 per cent interest, what will be the cost of a perpetual annuity of $500?

10. Calculate the annual interest and amount of a note for $500 which had remained unpaid 5 years and 3 months. Rate named 6 per cent.

11. A gentleman wishing to establish a free scholarship of $300 to be in force for fifty years desires to know how much it will cost at a 5 per cent rate of interest. What will he have to pay?

12. An annuity of $20 ran for thirty years. How much would it have amounted to in all had the payments been withheld until the last became due? Assume the rate to be 6 per cent.

13. What will an annuity of $100 cost to begin in 10 years and run for 25? Rate 4 per cent.

NOTE. — If any annuity begin, say Jan. 1st, 1901, the first payment on it will not be made till Jan. 1st, 1902.

14. A town borrowed $18,000, agreeing to repay it in 25 annual installments. What sum had to be raised annually, if 5 per cent was the rate named in the contract?

15. An annuity of $50 was to be paid semi-annually, $25 every six months, and to run for eight years. Not being paid until eight years and six months had elapsed, and the current rate of interest being 6 per cent, what was the amount to be paid?

16. A gentleman wishes to purchase a $1000 bond bearing 7 per cent interest, payable quarterly, due in ten years, so as to realize 5 per cent on his investment. What can he afford to pay for the bond if he compound his interest semi-annually?

17. A $1,000 bond bears 6 per cent semi-annual interest and is to run 12 years. It is offered at $1025. What annual rate will be realized from it by a purchaser, allowing semi-annual interest?

SUGGESTION. — Assume .05¾ as the value of r_e.

18. On a bond for $1000 payable by 6 per cent semi-annual coupons, and which was to run 24 years, the sixth payment of interest has just been made. What sum can be paid for it to net the purchaser 4½ per cent, computing all the interest semi-annually?

19. A town issued a series of 12 bonds of $500 each dated May 1st, 1892, and bearing 6 per cent interest payable semi-annually. The first was to become due May 1st, 1900, the second May 1st, 1901, and so on, the last to become due May 1st, 1911. What can a purchaser afford to pay for, say, the first and last of the series on Sept. 15, 1892 desiring to realize 5 %, interest compounded *annually?*

20. A bond for \$2000 bears 5 per cent semi-annual interest and is to run 30 years. What rate (allowing semi-annual compound interest) will be realized if it be purchased 3 months after its issue for \$1911?

SUGGESTION. — Assume .05½ as the value of r_e.

ANSWERS.

ART.

18. 2. $^{+}5$; 3. $^{-}7$; 4. $^{-}20$; 5. $^{+}15$; 6. $^{-}13$; 7. $^{+}4$; 8. $^{-}3$; 9. $^{+}2$; 10. $^{-}5$; 11. $^{+}41$; 12. $^{-}46$.

28. 1. (3. $+ 118$; (4. $- 341$; 2. (3. $- 40$; (5. $- 7$.

30. 1. 327; 2. $- 95$; 3. 623; 4. $- 1927$; 5. $- 301\frac{5}{8}$; 6. 59.228; 7. $+ 34$ cts.

31. 9. $- 7$; 10. 23; 11. $- 56$; 12. $- 582$; 13. $- 2001$; 14. $- 155$.

33. 1. $+ 2222$; 2. $+19$; 3. $- 90$; 4. $- 5346$; 5. $+ 496$; 6. $+ 17°$.

37. 1. $- 16$; 2. $+ 216$; 3. $- 57$; 4. $- 42$; 5. $+ 288$; 6. $- 8\frac{1}{3}$.

41. 1. $- 1650$; 2. $- 38$; 3. 5760; 4. $- 5\frac{1}{3}$; 5. .00216; 6. $- \frac{2}{15}$.

42. 2. (1. 256; (2. 256; (3. 4096; (4. $- 243$; (5. $- 2744$; (6. 123.21; (7. $- 2.197$.

44. 1. $- 5$; 2. $- 5$; 3. $- 10$; 4. $- 1$; 5. $- .775$; 6. $- .11\frac{7}{9}$; 7. .032.

46. 1. ± 6, ± 13, ± 7, ± 15, ± 11; 2. 2, $- 3$, $- 4$, 5, $- 7$; 3. ± 2, ± 3, ± 4; 4. ± 2, ± 3.

80. 1. $a + b + c$; 2. $2b$; 3. $a + \frac{b}{c}$; 4. $a - 5$; 5. $a^3 + 3c^2 + bcd$; 6. $\frac{cd}{b} + \frac{4b}{3a} - \frac{cd}{24}$.

82. 1. 12; 2. 111; 3. $- 178$; 4. 4; 5. ± 12; 6. ± 22, or ± 16, or ± 6, or 0.

83. 1. 144; 2. 840; 3. 198; 4. 200.

85. 2. $x = 7$; 3. 12; 4. 6; 5. 10; 6. 5; 7. 4; 8. 5; 12. 24, 16 ft.; 13. \$7, 70, 28; 14. 50; 15. 9, 14, 27; 16. 45; 17. 4, 6, 9 cts.

89. 1. (1. $9a$; (2. $7am$; (3. 0; (4. $-7a^2$; (5. $-2m^2n$; (6. $2axy$; 2. $20ab - 2a^2b$; 3. $x - 6a + 3b - 2$;
4. $5a + 4b - 8c - e$; 5. 6; 6. 0;
7. $6ab + 7ax^2 - 4a^2x + ax^3 - 5a^2b$;
8. $5x^4 + 4x^3 + 3x^2 - 4x - 9$.

93. 1. (1. $2a$; (2. $5b$; (3. $-17c$; (4. $-x^2$; (5. $31a^2b^2$; (6. $a^2b - ab^2$; (7. $-15ad^3$; (8. $24a^2b^4$; (9. $6m$;
2. $x^2 - 8a^2x^2 + 12$; 3. $2ax - 7by + 8cz$;
4. $19ab - 2c - 7d - 2x^3 + 3x^2 - x - 1$; 5. $2ax + 2x^2$;
6. $x^3 - 2x^2y - 2x^2y^2 + y^3$; 7. $2x^2y^3 - 6cz + 6m$;
8. $2a^3 - 2a^2c + 8ac^2$.

96. 2. $6(a+b)$; 3. $-2(x+z)$; 4. $16(a+b)^2$;
5. $5(x+y) + (x-y)$; 6. $-9(a+2b)^2 - 8(a-m)$.

97. 2. $(a+3)xy$;
3. $(a+c+e+g)x + (b+d+f+h)y$;
4. $(a+2c+4d)x$.

98. 2. $(a-b)(x+y+z)$;
3. $(a+m)(x+y) + (b-n)(x-y)$.

101. 1. $3x^2 - 2y + 2x - 1$; 2. $2x + 3y$; 3. $4xy$;
4. $2b - 4c$; 5. 0; 6. $3m$.

102. 2. $3a + 3c$; 3. $5a - 6x$; 4. $x^2 - x$; 5. $2a - 4x$;
6. $-a + 8b + 7x - 6y - 2$; 7. $12x - 8y$.

103. 1. $(a+b) + (c+d)$;
2. $(a^2 + 2ab + b^2) - (c^2 + 2cd + d^2)$; 3. $(a+b) + c$.

108. 1. -16; 2. $-12a^2$; 3. $45m^2n^2$; 4. $-24abc$;
5. $6a^3b^5x^3$; 6. $-2ax^3y$; 7. $14a^6bc$; 8. $6a^6b$; 9. a^4b^{2c};
10. $4x^3y^3$; 11. $-20a^2b^2x^3y$; 12. $x^6y^8z^3$; 13. $2x^{m+1}$;
14. $-y^{a+b}$.

112. 1. $-288\,a^6b^6$; 2. $8\,ac + 12\,bc$;
3. $-\frac{3}{7}\,abxy + \frac{9}{7}\,acx^2 - 2\frac{1}{7}\,ax$;
4. $78\,a^8m^8n^2y - 66\,a^6b^6m^4y + 30\,a^8m^4x^4y^3$;
5. $m^2 + mn + ma + mc$; 6. $-14\,ac + 28\,bc + 42\,c^2$;
7. $2\,a^2bc - 2\,ab^2c - 2\,abc^2$;
8. $-8\,a^4b^3 + 24\,a^3b^4 + 40\,a^2b^5$; 9. $x^2 + 5\,x + 6$;
10. $x^2 + 6\,x - 16$; 11. $p^2x^2 + q^2x^2 - p^2y^2 - q^2y^2$;
12. $x^4 - y^4$; 13. $x^3 + x^2 - 4\,x + 80$; 14. $x^2 - 14\,x + 45$;
15. $x^2 + 2\,ax + a^2$; 16. $m^2 + (a + b)\,m + ab$;
19. $3\,x^2 + 4\,xy + y^2 - 2\,yz - 3\,z^2$.

115. 3. $49 + 70 + 25 = 144 = 12^2$; 7. $9\,a^2 + 12\,ab + 4\,b^2$;
8. $25\,c^4d^2 - 40\,c^3d^3 + 16\,c^2d^4$; 9. $\frac{m^4}{4} + \frac{m^2n^2}{3} + \frac{n^4}{9}$;
16. $a^{2x} - 2\,a^xb^y + b^{2y}$.

116. 7. (1. $x^2 + 7\,x + 12$; (2. $x^2 + 15\,x + 54$;
(4. $x^2 - 4\,x - 5$; (10. $x^2 + x - 240$; 9. (1. $x^2 + 24\,x + 143$;
(5. $a^2 + b^2 + 1 + 2\,ab + 2\,a + 2\,b$;
(6. $a^2 + b^2 + 9 - 2\,ab + 6\,a - 6\,b$;
(9. $9\,m^2 + n^2 + 4\,p^2 + 1 + 6\,mn + 12\,mp + 6\,m + 4\,np + 2\,n + 4\,p$; (13. $x^3 + 6\,x^2y + 12\,xy^2 + 8\,y^3$.

117. 1. (3. (2). $9\,x^2y^4$, b^2x^8, $9\,c^2x^6$, $4\,a^2b^2c^4$, $16\,a^8b^{10}x^4$, a^{2m}, x^{4p}; (3). $8\,a^3b^6n^9$, $-27\,a^9b^3$, $-8a^3b^6m^3n^9$, $343\,p^3q^6n^9$, $a^{3m}b^{3n}$
(4). $768\,x^{16}$, $-32\,x^{15}y^5$, $a^{12}x^6$, p^6q^6, $-2\,b^7$, a^{24}.

121. 1. 27; 2. -4; 3. $24\,a^2b$; 4. a^2x^2; 5. $-4\,a^3b^2$; 6. $5\,ex^5y$;
7. $-\frac{1}{4}\,cx^3c^2$; 8. $-\frac{3\,a^2}{2\,xy}$; 10. $a^{n-1}b$; 12. $3\,(a - b)^2$.

127. 2. $3\,m^2\,(a + c)^2$; 3. $3\,x^{2n}$; 4. a^{s-t};
5. $-5\,m^2 - 4\,my + 2\,y^2$; 6. $3\,y^3 - 4\,ay^2$;
7. $-3\,prs^3 + 6\,prt^2 + 7\,rx^2$; 8. $2\,a + 4\,c + 6\,b$;
9. $3\,(a - b)^4 - 2\,(a - b)^2 + (a - b)$; 10. $2\,x + 3\,y$;
11. $c - d$; 12. $3 - a$; 13. $2\,a^2 + 3\,ab + b^2$; 14. $x + y$;
15. $x^2 - \frac{x}{2} + \frac{3}{4}$; 16. $3\,x^2 - x + 2$; 17. $a^3 + a^2x + ax^2 + x^3$.

128. 3. (1. 1; (2. 1; (4. 81, 27, 9, 3, 1, $\frac{1}{3}$, $\frac{1}{9}$, $\frac{1}{27}$, $\frac{1}{81}$;
(5. $\frac{3b^2c^4}{a^2}$, $\frac{y}{2x^3}$, 1.

129. 1. (3. (1) $\pm 4ab^2$; (2) $\pm 3ab^3$; (3) $2xy^3$; (4) ± 4;
(5) $-3m$; (6) $-2ac^2$; (7) $\pm 25p^4q^6$; (8) $\pm 3x^4$.

131. 1. $a+b$; 2. (m^2+mn+n^2);
3. $5^2+5\times 2+2^2=39$; 4. $1+y+y^2$;
5. $(2a+3b)(4a^2-6ab+9b^2)$; 6. $x^{12}-x^6a^{12}+a^{24}$;
7. $(x^4+yz^3)(x^4-yz^3)$; 8. $(a^4-b^3)(a^8+a^4b^3+b^6)$;
11. $(4a+10b)(16a^2-40ab+100b^2)$;
12. $(2a-b)(32a^5+\ldots+b^5)$, $(2a+b)$
$(32a^5-\ldots-b^5)$; 14. $(m^2+n^2)(m^4-m^2n^2+n^4)$;
17 – 20. Not divisible.

133. 1. (2. $2\times 7\, m\cdot m\cdot m\cdot n$;
(3. $7\times 13\, a\cdot a\cdot a\cdot a\cdot b\cdot c\cdot c\cdot c$;
(4. $33(a+b)(a+b)(a+b)$; 2. (2. $5x(4x^2-9y^2)$;
(3. $x^2(12a-3b+1)$; (4. $x^2y^2(x^2-xy+1)$;
(5. $7bc^2x(2c^2-3bc+b^2)$; (6. $3bc^2(2x-5c-b)$.

134. 1. (2. $(a^2+b^2)(a+b)(a-b)$;
(3. $(ab+cd)(ab-cd)$; 2. (2. $(x-y)(x^2+xy+y^2)$;
(3. $(x-2)(x^2+2x+4)$;
3. (2. $(x+y)(x^5-x^4y+x^3y^2-x^2y^3+xy^4-y^5)$, $(x-y)$
$(x^5+x^4y+x^3y^2+x^2y^3+xy^4+y^5)$;
(3. $(a+2)(a^5-2a^4+4a^3-8a^2+16a-32)$, $(a-2)$
$(a^5+2a^4+4a^3+8a^2+16a+32)$;
4. (2. $(x+3y)(x^2-3xy+9y^2)$.

135. 1. (2. $(x+5)(x+5)=(x+5)^2$; (3. $(x+4)^2$;
(4. $(x+6z)^2$; 2. (2. $(x+3)(x+2)$; (3. $(z+5)(z+6)$;
(9. $(rs+5z)(rs+18z)$; (12. $(x-5)(x-4)$;
(19. $(a+5)(a-3)$; (26. $(x-6)(x+5)$;
3. (2. $(3x+2)(3x+1)$; (3. $(3x+7)(x+2)$;
(4. $(4x-1)(x+3)$; (5. $(9x+1)(x+7)$;
(6. $(3x-2y)(x+4y)$; (7. $(2x+1)(x-1)$;
(8. $(3x+2)(x-7)$; (9. $(2c-d)(c-6d)$;

(10. $(2m + y)(m - 2y)$; (11. $(12x + 5)(x - 3)$; (12. $(15z + 1)(z - 15)$; (13. $(3x - 4y)(8x + y)$; 5. (2. $(5x^2 - x - 12)(5x + 1)$; (3. $(4x^2 - 2x - 5)(2x - 5)$.

136. 1. (3. (2) $(a + 1)^3$; (3) $(4b^3 + 1)^3$; (4) $(2x - 5y)^3$; (5) $(6x - y)^3$; 2. (2. $(2a + 3b - c)(2a - 3b + c)$; (3. $(2a + 3b - 2c)(2a - 3b + 2c)$; (4. $(2a + c + 3b)(2a + c - 3b)$; (5. $(l + m - n)(l - m + n)$; (6. $(a + b + x)(a + b - x)$; (7. $(a - b + 1)(a - b - 1)$; (8. $(4a^2 - 4b^2)(2a^2 + 2b^2)$; (19. $(4x^2 + 3xy - y^2)(4x^2 - 3xy - y^2)$, or $(4x^2 + 5xy + y^2)(4x^2 - 5xy + y^2)$; (20. $(3x^2 + 2xy + 7y^2)(3x^2 - 2xy + 7y^2)$; (24. $(2x^2 + 6xz + 9z^2)(2x^2 - 6xz + 9z^2)$; 3. (3. (1) $(a + b)(c - d)$; (2) $(a^2 + 1)(a + 1)$; (3) $(x - 3)(y + 2)$; (4) $(3x - 2y)(2a - 7y)$; (5) $(m - 3n)(9a - 4b)$; (6) $(m + n^2)(m^2 - n)$; (7) $(x^4 - y^4)(x - y)$; 4. (3. $(m^2 + 4m + 1)(m + 1)$; (6. $(x^2 - 4x + 6)(4x - 5)$.

137. 2. (2. $(a + b + c + d)(a + b + c - d)$; (3. $(2m - 3n + p - 2q)(2m - 3n - p + 2q)$; 3. (2. $(a - 3b + 5c)^2$.

142. 1. $3a^2$; 2. $2a$; 3. b^2c; 4. $4x^8y^3z^2$; 5. $2(a + b)$; 6. $7xy$; 7. $2a^2xy(3x - y)$; 8. $x - y$; 9. Prime to each other; 10. $3x(x + 4)$; 11. $3(x - 1)$; 12. $x^3 - 6$; 13. $x^m + 6$; 14. $c(a - b)$.

148. 1. 31; 2. 126; 3. 2; 4. $x + 7$; 5. $x - 2y$; 6. $2a + 3x$; 7. $c(a^2 - m)$; 8. $3x + 4a$; 9. $x(x - 1)$; 10. $x(x - 2)$; 11. $3y - 7$; 12. $ax - by$; 13. $2x + 1$; 14. $x^2(3x + 2)$; 15. $x^2 - 5x + 6)$; 16. $(a + x)^2$.

153. 1. 648; 2. 720; 3. 2160; 4. $252\,ax^2y^2$; 5. $x^2(a + x)$; 6. $x(x^2 - 1)$; 7. $36\,a^2b^2c^2d^2$; 8. $x^2 - y^2$; 9. $a^4b^8c^3x^4$; 10. $30\,x^2y^3z^3$; 11. $54\,a^3b^2c^2$; 12. $72\,a^4b^3c^3$; 13. $210\,a^2b^2x^3$; 14. $12\,xy^2(x^2 - y^2)$; 15. $8(1 - x^2)$; 16. $ab(a + b)$;

17. $12\,x^2(x^2+2)$; **18.** $xy\,(4\,x^2-1)$; **19.** $x^3y^3(x^2-y^2)^2$; **20.** $24\,ab\,(a^2-b^2)$; **21.** $abc\,(x-a)\,(x-b)\,(x-c)$.

160. **2.** $\frac{1}{3}$, $\frac{9}{16}$, $\frac{3}{2}$; **3.** $\frac{7}{11}$, $\frac{809}{1069}$; **4.** $\frac{3\,x}{8}$; **5.** $7\,by$; **6.** $\frac{43}{9\,abx^2}$; **7.** $\frac{1}{a}$; **8.** $\frac{mx}{n}$; **9.** $\frac{4\,(a+b)}{5\,(a-b)}$; **10.** $\frac{x-1}{2\,y}$; **11.** $\frac{x}{c^2}$; **12.** $\frac{b}{c}$; **13.** $\frac{a^3}{a^2+b^2}$.

162. **2.** $3\frac{1}{8}$, $7\frac{14}{15}$, $5\frac{10}{13}$, $12\frac{93}{111}$; **3.** $b+\frac{b^2}{a}$; **4.** $x+y$; **5.** $x+\frac{y^2}{x}$; **6.** $2-\frac{2}{2\,x^2-x+1}$; **7.** $a^2x+x^3-\frac{2\,x^5}{a^2-x^2}$; **8.** $x+\frac{2}{x+3}$; **9.** $2\,a+\frac{4\,b^2}{a-b}$; **10.** $2\,ab^2-3\,a^2b^3+\frac{7\,x}{11\,ab}$.

164. **2.** $\frac{a}{5\,bc}-\frac{b}{10\,ac}+\frac{c}{3\,ab}$; **3.** $\frac{1}{d}+\frac{1}{a}+\frac{1}{b}+\frac{1}{c}$.

166. **2.** $5\,ab^{-1}c^{-2}$; **3.** $ab^{-1}x^3y^2$; **4.** $\frac{5}{3}a^{-6}cd^{-8}$; **5.** $\frac{x}{4}(a-b)^{-1}$.

168. **3.** $\frac{119}{25}$, $\frac{31}{4}$, $\frac{98}{23}$; **4.** $\frac{7\,a-6\,b-x}{3}$; **5** $\frac{a^3+b^3}{a}$; **6.** $\frac{17\,x-17}{3\,x}$; **7.** $\frac{2\,a}{a+b}$; **8.** $\frac{11\,x^2+5}{5\,x}$.

170. **3.** $\frac{5\,a^3b^3d}{ab^2d}$; **4.** $\frac{49}{56}$; **5.** $\frac{3\,a^2-2\,ab-b^2}{a^2-b^2}$; **6.** $\frac{5\,(a^2-b^2)^2}{(a-b)^2}$.

172. **2.** $\frac{128}{180}$, $\frac{135}{180}$, $\frac{108}{180}$, $\frac{126}{180}$; **3.** $\frac{ax^2}{x^3}$, $\frac{bx}{x^3}$, $\frac{1}{x^3}$; **4.** $\frac{9\,ac}{21\,bc}$, $\frac{5\,b^2}{21\,bc}$; **5.** $\frac{adf}{bdf}$, $\frac{bcf}{bdf}$, $\frac{bde}{bdf}$; **6.** $\frac{6\,a^2}{30\,abc}$, $\frac{3\,b^2}{30\,abc}$, $\frac{10\,c^2}{30\,abc}$; **7.** $\frac{15\,anx}{10\,a^2n}$, $\frac{4\,nx}{10\,a^2n}$, $\frac{10\,a^2m}{10\,a^2n}$; **8.** $\frac{567\,a}{504\,x}$, $\frac{98\,b}{504\,x}$, $\frac{198\,ax}{504\,x}$, $\frac{882\,(a+b)}{504\,x}$.

175. **2.** $2\frac{1}{3}$; **3.** $\frac{13\,x-1}{10}$; **4.** $\frac{29}{24}\,x$; **5.** $\frac{adn+bcn+bdm}{bdn}$;

6. $\frac{19x-23}{6}$; 7. $\frac{3bx+2ay}{ab}$; 8. 0; 9. $2c$; 10. $x^2+\frac{332}{63}x$;
11. $8x-\frac{9}{20}a$; 12. $9a^2+\frac{23x-3}{21}$.

177. 1. $\frac{393}{346}$; 2. $\frac{2x-11}{15}$; 3. $\frac{2a}{x+a}$; 4. $\frac{2(bc+ad)}{(a-b)(c+d)}$;
5. $\frac{2x}{x^2-a^2}$; 6. $\frac{1-x^2}{1+x^2}$; 7. $\frac{276x+6}{91}$; 8. $a+2x+\frac{a^2-x^2}{ax}$;
9. $\frac{2x^2}{1-x^4}$; 10. $\frac{y-1}{y}$; 11. $\frac{2a}{x^2-a^2}$; 12. $\frac{2ax}{8x^3-a^3}$;
17. $\frac{b^2-ax}{b^2-x^2}$; 21. 0.

179. 2. x; 3. $\frac{3a^2b}{cd}$; 4. $-5ab$; 5. $\frac{3a^2y}{4bc}$; 6. $\frac{a^2-c^2}{1-x}$;
7. $\frac{9b^2cz^2}{4x^3y}$; 8. m^2-n^2; 9. $\frac{17ax^2-4a}{2ax+1}$; 10. $\frac{ay-x^4y}{b+c}$;
11. $\frac{10x}{3}$; 12. $\frac{a+b+c}{3(x-y)}$; 13. $\frac{ab}{x}+b$; 14. $\frac{c^2-d^2}{10a-5a^2}$.

181. 2. $-\frac{2a^2b^2}{3d}$; 3. $-\frac{m^2}{4a^2b}$; 4. $\frac{9b}{a}$; 5. $\frac{7bd^2}{4a}$; 6. $\frac{p^2q^2y}{10x^2}$;
7. $\frac{y^2}{p}$; 8. $\frac{1}{a-1}$; 9. $\frac{a+b}{x+y}$; 10. $\frac{1}{a+b}$.

183. 2. $\frac{b(a+b)}{x(y-a)}$; 3. $\frac{b(ax+b)}{a(bx-a)}$; 4. $\frac{21cy^2}{10ab}$; 5. $\frac{7ab^2}{4}$;
7. $\frac{bx}{a+by}$; 8. $\frac{an}{m+bn}$; 9. $\frac{m(a+1)}{a(m^2+1)}$; 10. $\frac{1}{m+1}$;
11. $\frac{9(7-4x)}{10(3x-1)}$.

209. 2. 4; 3. 16; 4. $2\frac{7}{11}$; 5. $\frac{16}{a}$; 6. $\frac{7c}{3a}$; 7. 7; 8. $\frac{m}{n}$;
9. $2\frac{1}{3}$; 10. $\frac{3}{17}$; 17. $\frac{d+e+f}{a+b+c}$; 19. $a-b$; 21. 4.

211. 3. 1; 4. $3\frac{4}{9}$; 5. $\frac{n}{m-a}$; 6. 2; 7. $\frac{d}{a+b+c}$;

8. $\frac{b-a}{m+b+p}$; 10. 1; 11. 1; 12. 4; 13. 1.

213. 2. 12; 3. 7; 4. 7; 5. $275\frac{35}{36}$; 6. 420; 7. 7; 8. $\frac{22}{7}$; 9. $\frac{4}{7}$; 10. $\frac{1}{2}$; 12. $\frac{a+b}{2}$.

214. 2. 6; 3. 25; 4. 6.

216. 1. $4\frac{5}{6}$; 2. 4; 3. 45; 4. $\frac{1-a}{9}$; 5. 4; 6. 7; 7. 12; 8. 7; 9. $\frac{3}{2}$; 10. 3; 12. $\frac{mn-ab}{a+b-m-n}$; 17. $\frac{4}{13}$; 28. 4; 33. -9; 49. $a=\frac{(b-c)d}{d-b}$, $b=\frac{d(a+c)}{a+d}$.

224. 4. 7; 5. 106, 75; 6. 60; 7. $58\frac{2}{13}$; 8. 6; 9. 10; 11. 25; 12. 22, 10; 13. 6, 8; 14. 14, 12; 15. 15, 5; 16. 3, 9, 15, 21 yrs.; 17. 21, 7, 14 cts.; 18. 25, 75; 19. 20 of \$100 . . . 1280 of \$1; 20. 34, 17 gals.; 21. 35, 90; 22. \$42, 28, and 18; 23. 175 mi.; 24. 30 ft.; 25. $6\frac{2}{3}$ oz.; 26. 5; 27. 54; 29. 54; 30. $143\frac{19}{32}$ mi.; 32. \$12000; 33. \$25200, \$3000; 34. 10%; 35. 30; 36. \$45; 37. 3; 38. $37\frac{1}{2}$ ft.; 40. 10 da.; 41. 4 min.; 42. 40 min.; 43. 21; 44. $12\frac{12}{13}$ min.; 45. 60; 46. 120.

230. 3. $x=2$, $y=1$, or 2, 1; 4. 3, 5; 5. 8, 2; 6. 25, 15; 7. $-\frac{19}{13}$, $-\frac{69}{13}$; 8. 1, 3; 9. 8, 6; 10. 12, 8; 11. 6, 15; 12. 7, 3; 14. $\frac{f-d}{fc-de}$, $\frac{e-c}{de-fc}$; 15. $\frac{bc}{a^2+b^2}$, $\frac{ac}{a^2+b^2}$.

232. 2. 8, 7; 3. 8. 1; 4. 6, 3; 5. 21, 35; 6. 5, 5; 7. 3, 5.

234. 4. 5, 2; 5. 20, 15; 6. 4, 6; 7. -6, 80; 8. 20, 60; 9. 3, 5; 10. $7\frac{1}{2}$, 5; 11. 3, 4; 12. $9\frac{1}{3}$, 7; 13. .4, .1; 14. $\frac{bq-pd}{a(b-d)}$, $\frac{p-q}{b-d}$; 15. $-\frac{bn}{(m+an)}$, $\frac{bm}{m+an}$.

236. 2. 30, 9; 3. $33\frac{3}{4}$, $21\frac{1}{4}$; 4. 6.3, 39.2; 5. $\frac{a-b^2}{a^2-b}$, $\frac{ab-1}{a^2-b}$.

237. 2. 15, 3; 3. 11, -1; 4. 2, -1; 5. 7, 11.

239. 1. 6, 3; 2. 10, 7; 3. .4, .1; 4. 20, 4; 5. 57, 103; 6. 6, 10; 7. 12, 30; 8. 2, $6\frac{1}{3}$; 9. 1, 4; 10. 19, 3; 11. $\frac{n(c-d)}{an+bm}$, $\frac{m(c-d)}{an+bm}$; 12. 5, 4; 13. $\frac{2ab+6}{7-2b}$, $\frac{ab+2b-4}{7-2b}$.

240. 2. 90, 60 cts.; 3. 714285, 142857; 4. \$420, \$260; 5. 180, 145; 6. 92800000, 67100000 mi.; 7. \$12, 16; 8. \$250, 320; 9. 2.322, 11.03.

244. 1. 1, 2, 3; 2. 7, 8, 9; 3. 98.7, 65.4, 32.1; 4. 24, 9, 5; 5. 6, 2, 1; 6. 3, 9, 15; 7. 4.5, 10.8, 11.7; 8. 7, 4, 3; 9. $\frac{1}{11}a$, $\frac{5}{11}a$, $\frac{7}{11}a$; 10. $\frac{b^3-a^2b+ac^2}{(a^2+b^2)c}$, $\frac{a^3-ab^2+bc^2}{(a^2+b^2)c}$, $\frac{2ab-c^2}{a^2+b^2}$.

245. 1. 1, 3, 5; 2. \$200, 360, 840; 3. 1, 2, 3, 4 cts.; 4. \$40, 30, 24, 26; 5. \$$122\frac{1}{2}$, $97\frac{1}{2}$ horses; $32\frac{1}{12}$, $12\frac{1}{12}$ saddles; 6. 65, 9, 49 years; 7. $3a-2c$, $2b+2c-3a$, $3a-2b$.

249. 1. 11, 1; 8, 3; 5, 5; 2, 7; 2. 9, 1; 2, 6; 3. 5, 1, 1; 2, 2, 1; 3, 1, 2; 1, 1, 3; 7. 7, 8, 9; 9. No values.

250. 1. 55, 16; 15, 56; 2. 8, 2; 15, 4; 22, 6, etc.; 7. 123, 224, 325, etc. 729.

255. 2. $\frac{c}{l+m+n}$; 3. $\frac{nb-a}{n-1}$; 6, 5, $-\frac{1}{2}$; 4. 45, 19; $104\frac{1}{2}$, $90\frac{1}{2}$; 306, -6; 5. $\frac{na-nc+b}{n+1}$; 3; 6. $\frac{a}{m+n}$; 6; 7. $\frac{b'c-bc'}{ab'-a'b}$, $\frac{ac'-a'c}{ab'-a'b}$; $112\frac{3}{5}$, $52\frac{3}{5}$; 8. $\frac{abc}{b+c}$; 9. $\frac{pqr}{pq+pr+qr}$; 5; 10. $\frac{a(m-n)}{2m}$, $\frac{a(m+n)}{2m}$; $1\frac{5}{12}$, $15\frac{7}{12}$; 11. $\frac{n-r}{q+1}$, $\frac{nq+r}{q+1}$; 6, 479.

257. 1. a^6c^2, $121b^4c^6$, $16a^8b^{10}x^4$, $\frac{4}{9}a^4x^8$, $\frac{16}{9x^4y^2}$, $\frac{49a^{2m}}{4}$, $a^{2m}b^{2m}$, $\left(\frac{49a^4}{9b^2}\right)^m$, $a^{2x}b^{-8}$, $\frac{1}{216}$, $\frac{16}{225}$, $(-a^2x)^{12}$, $(-a)^4(-b)^6(-c)^8$, 2^{16}, $\frac{1}{25}a^6x^{2m+4}$, y^{2p-2}, $y^{2(m+n)}$;

2. $-216\,a^{18}$, a^{8x}, $b^{12}y^{8}$, a^{48}, $-8\,a^{21}c^{6}$, $\left(\frac{3\,ab}{5\,cd}\right)^{3n}$, $\frac{m^{24a+12b}}{n^{12a-21b}}$, c^{18a-8b}, $\left(\frac{x^2y^8z^4}{2}\right)^{12}$; 3. x^8y^{12}, $\frac{243\,y^5}{32\,a^5b^{10}}$, $729\,x^6y^{18}$, $a^{2n}b^nc^{8n}$, 9, 729, $\frac{4096}{729}$, $(-1)^m\,a^{2m}b^{5m}c^{3m}$.

260. 1. $p^2q^2-2\,pqr+r^2$; 2. $8\,m^3-12\,m^2p+6\,mp^2-p^3$;
3. $x^6+12\,x^4y^2+48\,x^2y^4+64\,y^6$;
4. $125\,a^3-75\,a^2bc+15\,ab^2c^2-b^3c^3$;
5. $16\,a^8+32\,a^7x+24\,a^6x^2+8\,a^5x^3+a^4x^4$;
6. $1-8\,b+24\,b^2-32\,b^3+16\,b^4$;
7. $a^9+9\,a^8+36\,a^7+84\,a^6+126\,a^5+126\,a^4+84\,a^3+36\,a^2+9\,a+1$;
8. $a^4-\frac{8}{3}\,a^3b+\frac{8}{3}\,a^2b^2-\frac{32}{27}\,ab^3+\frac{16}{81}\,b^4$;
9. $625-2000\,x+2400\,x^2-1280\,x^3+256\,x^4$;
10. $8\,r^3-72\,mr^2+216\,m^2r-216\,m^3$.

261. 3. $x^2y^2+y^2z^2+x^2z^2+2\,x^2yz+2\,xy^2z+2\,xyz^2$;
4. $1+3\,x^2+3\,x^4+x^6+2\,x+4\,x^3+2\,x^5$;
5. $a^3+3\,a^2b+3\,ab^2+b^3-3\,a^2c-6\,abc-3\,b^2c+3\,ac^2+3\,bc^2-c^3$; 6. $1-3\,a+5\,a^3-3\,a^5-a^6$;
7. $1-6\,x+15\,x^2-20\,x^3+15x^4-6\,x^5+x^6$;
8. $8\,a^{3m}-12\,a^{2m}b^n+6\,a^mb^{2n}-b^{3n}+12\,a^{2m}c^p-12\,a^mb^nc^p+3\,b^{2n}c^p+6\,a^m\,c^{2p}-3\,b^nc^{2p}+c^{3p}$;
10. $a^4+16\,b^4+c^4+4\,(2\,a^3b+8\,ab^3-a^3c-ac^3-8\,b^3c-2\,bc^3)+6\,(4\,a^2b^2+a^2c^2+4\,b^2c^2)+12\,(2\,abc^2-2\,a^2bc-4\,ab^2c)$.

266. 1. $\pm\,a^3b^2$, $\pm\,x^2$, c^4, no real root, -4, $\pm\,3\,x^3$;
2. $\frac{2}{3}$, $\pm\,\frac{6}{a^{18}}$, $\pm\,\frac{9\,x^5}{5\,c^2}$, $-x^3$, imaginary, $\pm\,\frac{3\,a^9}{2\,b^6}$;
3. $\frac{5\,ab^2}{6\,x^2y^3}$, -6; 4. $-7\,a^4x^3$, $-\frac{3\,x^9}{4\,y^{21}}$, no real root, no real root:
5. $-\frac{m^2}{x^3y}$, $-2\,a^3$, $\frac{2}{a^9b^8}$, $-12\,c^2d^4x$.

269. 2. $f^3+3\,x^4$; 3. $\frac{3}{2}\,a^4+\frac{2}{3}\,n^3$; 4. $a-\frac{1}{a}$;

5. $\frac{ax + b^2x^2}{a^m + x^n}$; 6. $x^2 - x + 1$; 9. $x + \frac{1}{x} - 2$;

14. $a + \frac{x^2}{2a} - \frac{x^4}{8a^3} + \frac{x^6}{16a^5} +$, etc.

272. 1. 51, 217, 42.1, 20.82; 2. 6.42, 31.08, 4.164, .0321;
3. $\pm \frac{1}{2}$, $\pm 1\frac{7}{8}$, $\pm \frac{85}{133}$, $\pm \frac{53}{63}$.

275. 1. $2a - 3b$; 2. $2a - 7x$; 3. $x^2 + x + 1$;
4. $a^2 - ab + b^2$; 6. $\frac{x}{y} - y$; 9. $1 - \frac{x}{3} - \frac{x^2}{9} -$, etc.

278. 1. 23, 234, 11.4, 5.51; 2. .503, .206, $\frac{2}{7}$; 3. 8.026 −, 2.755 +, 1.710 −, .585 −, 3.332 +.

281. 1. $2a - 3x$; 2. $1 - x + x^2$; 3. $5x - 8y$; 4. $2 - x$;
7. $2x - 1$.

288. 1. $11 \pm 7 = 18$, or 4; 2. $-a^{\frac{1}{2}}bc^2$; 3. $15\,m^{\frac{3}{4}}n^{\frac{5}{6}}$;
4. $6ax^{-m}y^{\frac{1}{2}} + 5bc + 6a^{\;2}x^{-\frac{3}{4}}$; 5. 0;
6. $6ax^{\frac{1}{2}} - 5(x + y)^{\frac{1}{3}} + 6(a - b)^{\frac{1}{2}}$;
7. $7ax^2 - 7bc - 2m^{\frac{1}{2}} + 6c^{\;n} + 5xy$;
8. $8c^{\frac{5}{6}}d^{\frac{2}{3}} + 10c^{\frac{3}{4}}d^{\frac{3}{4}} + 3c^{\frac{1}{2}}d^{\frac{1}{2}} + 11c^{\frac{1}{4}}d^{\frac{3}{4}}$; 9. -32764.
10. 216, 823543, $\frac{1}{128}$, $\frac{1}{128}$, $\frac{1}{3}$;
11. $-18y^{\frac{2n+1}{n}}$, $18a^{\frac{3}{m}}b^{m+n}$; 12. 2401; 13. $a^{3\frac{1}{6}}$, $a^{\frac{3}{4}}b^{3\frac{1}{2}}$;
14. $x^{\frac{1}{2}}m^{2\frac{1}{2}}$; $n^{1\frac{7}{12}}$; 15. $a^{\frac{1}{2}} + a^{\frac{1}{3}} - a^{-\frac{1}{6}} + a^{-\frac{1}{8}} - a^{1\frac{7}{12}} - a^{1\frac{3}{4}}$;
16. $a^{-\frac{3}{4}}b^{\frac{1}{20}}c^{\frac{11}{12}}d^{-\frac{11}{12}}$; 19. $m - n$; 21. $x^{\frac{1}{2}}$; 25. $a^{\frac{1}{2}} - 3a^{\frac{10}{9}}$;
26. $a^n + a^{\frac{n}{2}}b^{-\frac{n}{2}} + b^{-n}$; 31. $a^m b^{\frac{2m}{3}}$, $a^2b^{\frac{3}{2}}c^m$, $\frac{(x+y)^{\frac{2}{p}}}{(x-y)^{\frac{2m}{q}}}$;
33. $x^{\frac{3}{2}}y^{-1}$, $a^{-\frac{q}{2}}b^{-\frac{3q}{4}}$, $a^{\frac{3}{5}}b^{\frac{12}{5}}c^{\;\frac{3}{5}}$; 36. $1 + 2x^{\;\frac{1}{2}} - 3x^{\;\frac{3}{4}} + 4x^{\;1}$.

295. 1. $(2abc)^{\frac{1}{2}}$, $(3ad^2)^{\frac{1}{3}}$, $(50)^{\frac{1}{2}}$; 2. $(9a^2xy^3)^{\frac{1}{4}}$, $(10xy^3)^{\frac{1}{4}}$,
$(\frac{2}{3}x^2y^4z)^{\frac{1}{5}}$.

298. 1. $12(2)^{\frac{1}{2}}$, $10(3)^{\frac{1}{2}}$, $4(4)^{\frac{1}{3}}$, $9(3)^{\frac{1}{3}}$; 2. $3a^2(x)^{\frac{1}{2}}$, $6a(a)^{\frac{1}{2}}$,
$12ab^2(3ab)^{\frac{1}{2}}$, $5bc(2a)^{\frac{1}{2}}$, $4ax^2(5ax)^{\frac{1}{2}}$; 3. $4a^2b(2ab^2)^{\frac{1}{3}}$,
$-3xy(4x)^{\frac{1}{3}}$, $5a(1 + ab)^{\frac{1}{3}}$, $-8a^2y^3$, $x(a + bx^2)^{\frac{1}{3}}$,
$10xy(5y^3)^{\frac{1}{4}}$;

301. 1. $\frac{1}{3}(3)^{\frac{1}{2}}$, $\frac{1}{5}(10)^{\frac{1}{2}}$, $\frac{1}{2}(6)^{\frac{1}{2}}$, $\frac{1}{21}(21)^{\frac{1}{2}}$, $\frac{1}{5}(22)^{\frac{1}{2}}$, $\frac{3}{5}(2)^{\frac{1}{2}}$;
2. $\frac{c}{b}(6\,abc)^{\frac{1}{2}}$, $\frac{x}{4}(6\,ax)^{\frac{1}{2}}$, $\frac{a^2b}{2}(b)^{\frac{1}{2}}$, $\frac{1}{b}(3\,ab)^{\frac{1}{2}}$, $\frac{3}{2}(10)^{\frac{1}{2}}$.

304. 1. $(-125\,a^6b^3)^{\frac{1}{3}}$; **2.** $(36)^{\frac{1}{2}}$, $(4\,a^4b^2)^{\frac{1}{2}}$, $\left(\frac{9}{16}\right)^{\frac{1}{2}}$, $(49\,m^4n^2)^{\frac{1}{2}}$;
3. $\left(\frac{27}{125}\right)^{\frac{1}{3}}$, $((a-x)^3)^{\frac{1}{3}}$, $(216\,a^6x^{-9})^{\frac{1}{3}}$, $\left(\frac{27\,a^3m^8}{8\,n^8y^3}\right)^{\frac{1}{3}}$;
4. $(x^4)^{\frac{1}{6}}$, $(x^{6a})^{\frac{1}{6}}$, $(a^3)^{\frac{1}{6}}$, $(a^6)^{\frac{1}{6}}$.

307. 1. $(350)^{\frac{1}{2}}$, $(\frac{1}{3})^{\frac{1}{2}}$, $\left(\frac{14}{11}\right)^{\frac{1}{2}}$, $(864)^{\frac{1}{3}}$, $(40)^{\frac{1}{3}}$, $(\frac{1}{2})^{\frac{1}{3}}$;
2. $(5\,b)^{\frac{1}{2}}$, $(12\,a^3x^2)^{\frac{1}{2}}$, $(a^8b)^{\frac{1}{4}}$, $(5^{m+2}x^{m+2})^{\frac{1}{m}}$;
3. $\left(\frac{c}{b^2}\right)^{\frac{1}{2}}$, $\left(\frac{a+x}{a-x}\right)^{\frac{1}{2}}$, $\left(\frac{a+b}{a-b}\right)^{\frac{1}{2}}$.

310. 1. $(64)^{\frac{1}{6}}$, $(81)^{\frac{1}{6}}$, $(6)^{\frac{1}{6}}$; **2.** $(a^6)^{\frac{1}{10}}$, $(a^5)^{\frac{1}{10}}$; **3.** $(25)^{\frac{1}{6}}$, $(64)^{\frac{1}{6}}$;
4. $(a^6)^{\frac{1}{3}}$, $(b)^{\frac{1}{3}}$; **5.** $(6561)^{\frac{1}{12}}$, $(512)^{\frac{1}{12}}$, $(15625)^{\frac{1}{12}}$;
6. $(a^{18})^{\frac{1}{12}}$, $(a^8)^{\frac{1}{12}}$, $(a^9)^{\frac{1}{12}}$.

313. 1. $48\,(5)^{\frac{1}{2}}$, $107\,(3)^{\frac{1}{2}}$; **2.** $55\,ay\,(3\,a)^{\frac{1}{2}} - 33\,ay\,(2\,a)^{\frac{1}{2}}$;
3. $(20\,a^2 + 15\,b^2 + 4\,c^2)(7\,x)^{\frac{1}{2}}$; **4.** $16\,(11)^{\frac{1}{2}}$, $20\,(3)^{\frac{1}{2}} - 13\,(2)^{\frac{1}{2}}$;
5. $12\frac{1}{2}\,(3)^{\frac{1}{2}}$; **6.** $a\,(3\,a)^{\frac{1}{2}}$; **7.** $3\,(2)^{\frac{1}{3}}$; **8.** $9\,(2\,a)^{\frac{1}{3}}$.

316. 1. $14\,(6)^{\frac{1}{2}}$, $12\,(3)^{\frac{1}{2}}$, $10\,a\,(3)^{\frac{1}{2}}$, $14\,(9)^{\frac{1}{3}}$;
2. $ab^3\,(ab)^{\frac{1}{2}}$, $\frac{1}{2}\,(2)^{\frac{1}{2}}$, $30\,xy^2$; **3.** $\frac{x}{a}\,(ab)^{\frac{1}{2}}$, $2\,a^2$, 140;
4. $2\,(5)^{\frac{1}{2}}$, $6\,(3)^{\frac{1}{2}}$, $\frac{1}{21}\,(4)^{\frac{1}{3}}$; **5.** $4\,(80000)^{\frac{1}{12}}$, $(648000)^{\frac{1}{12}}$;
7. $(2\,b^2c)^{\frac{1}{3}}$, $\frac{3}{4}\,(12)^{\frac{1}{3}}$; **8.** $\left(\frac{256}{2187}\right)^{\frac{1}{24}}$, $10 + 4\,(35)^{\frac{1}{2}} + 30\,(2)^{\frac{1}{2}}$;
9. $5\,(9\,a^3)^{\frac{1}{6}}$, $20\,b\,(a^5b)^{\frac{1}{6}}$, $(432\,a^7c^7)^{\frac{1}{12}}$; **10.** $6\sqrt[12]{x^9y^6}$;
11. $(x^{np-mq})^{\frac{1}{mn}}$, $y^{\frac{11}{12}}$; **12.** $xy\,(a^6b^4c^3x^6y^4z^3)^{\frac{1}{12}}$;
13. $\frac{a}{b}\left(\frac{a^5}{b^5}\right)^{\frac{1}{12}}$, $2^{\frac{1}{6}}$, $\left(\frac{a+x}{a-x}\right)^{\frac{1}{2}}$; **14.** 4, $-8 - \frac{37}{8}\,(2)^{\frac{1}{2}}$;
15. $3\,(20)^{\frac{1}{3}} - 12\,(3)^{\frac{1}{3}} - (180)^{\frac{1}{3}} + 12$, $ab + \frac{ad}{c}$
$+ \left(\frac{a}{b} + \frac{bd}{c^2}\right)(abc)^{\frac{1}{2}}$.

319. 1. $\frac{m(5)^{\frac{1}{2}}}{5}, \frac{2(3)^{\frac{1}{2}}}{3}, \frac{(2a)^{\frac{1}{2}}}{2a^2}, \frac{3(2)^{\frac{1}{2}}+4}{12}, \frac{a(2b)^{\frac{1}{2}}+(10b)^{\frac{1}{2}}}{2ab}$;

2. $\frac{6(125)^{\frac{1}{2}}}{5}$, 3, $\frac{(25000)^{\frac{1}{2}}}{10}, \frac{(a^3b^4)^{\frac{1}{2}}}{b}$, 15;

3. $5+2(6)^{\frac{1}{2}}, \frac{18+5(10)^{\frac{1}{2}}}{2}, 1+2^{\frac{1}{2}}, 7^{\frac{1}{2}}+2^{\frac{1}{2}}$;

4. $3[(9)^{\frac{1}{2}}+(6)^{\frac{1}{2}}+(4)^{\frac{1}{2}}]$; 5. $\frac{1+3^{\frac{1}{2}}}{2}, \frac{1-(1-x^4)^{\frac{1}{2}}}{x^2}, (x^2+a^2)^{\frac{1}{2}}-x$;

9. $-(17+2\cdot 2^{\frac{1}{2}}3^{\frac{1}{2}}+8\cdot 3^{\frac{1}{2}}+6\cdot 2^{\frac{1}{2}}+4\cdot 3^{\frac{1}{2}}+2\cdot 2^{\frac{1}{2}}3^{\frac{1}{2}})$.

320. 1. $ax^3(a)^{\frac{1}{2}}, -10(-10)^{\frac{1}{2}}$; 2. $\frac{4}{9}, \frac{32}{27}(3)^{\frac{1}{2}}$;

3. $-\frac{x^3y}{4}(2xy^2)^{\frac{1}{2}}, -\frac{54}{49}(14)^{\frac{1}{2}}$; 4. $-c(3)^{\frac{1}{2}}, (\frac{4}{3})^7, (\frac{5}{8})^6, 2a$;

5. $125(5)^{\frac{1}{2}}, (4)^{\frac{1}{2}}, (2a)^{\frac{n}{m}}$; 6. $3(1+2x+x^2), 26-15(3)^{\frac{1}{2}}$;

7. $x-2x^{\frac{1}{2}}y^{-\frac{1}{2}}+y^{-1}, 49-20(6)^{\frac{1}{2}}$;

8. $x^{\frac{3}{2}}-3xy+3x^{\frac{1}{2}}y^2-y^3, \frac{(a-x)(a^2-x^2)^{\frac{1}{2}}}{a^{\frac{1}{2}}}$;

9. $(x+27y)x^{\frac{1}{2}}+(9x+27y)y^{\frac{1}{2}}, -a^{\frac{1}{2}}+(bc)^{\frac{1}{2}}$; 10. $(a^3)^{\frac{1}{2}}, a^{\frac{1}{2}}$;

11. $2a^2(2c)^{\frac{1}{12}}, (a^2b)^{\frac{1}{2}}$; 12. $2ax^{\frac{1}{2}}(12)^{\frac{1}{2}}, (7x)^{\frac{1}{2}}$.

324. 1. $(2)^{\frac{1}{2}}-1, 5-2(6)^{\frac{1}{2}}, 3(7)^{\frac{1}{2}}-2(6)^{\frac{1}{2}}, 6^{\frac{1}{2}}-2$,
$2(7)^{\frac{1}{2}}+14^{\frac{1}{2}}$; 2. $3(7)^{\frac{1}{2}}+2(3)^{\frac{1}{2}}, 2-\frac{1}{3}(3)^{\frac{1}{2}}, 1+\frac{1}{2}(2)^{\frac{1}{2}}$;

3. $(x-1)^{\frac{1}{2}}-1, x^{\frac{1}{2}}-(xy)^{\frac{1}{2}}, (1+a)^{\frac{1}{2}}-(1-a)^{\frac{1}{2}}$,
$a^{\frac{1}{2}}[(x-a)^{\frac{1}{2}}-a^{\frac{1}{2}}], (x+y)^{\frac{1}{2}}+z^{\frac{1}{2}}$;

4. $1+3^{\frac{1}{2}}, 1-5^{\frac{1}{2}}, 2^{\frac{1}{2}}-7^{\frac{1}{2}}, 4-7^{\frac{1}{2}}, 2^{\frac{1}{2}}+3, 5-3^{\frac{1}{2}}$;

5. $1+2^{\frac{1}{2}}, 1+5^{\frac{1}{2}}, \frac{1}{2}(1+5^{\frac{1}{2}}), 50^{\frac{1}{2}}+18^{\frac{1}{2}}$.

326.[1] 1. $14(-3)^{\frac{1}{2}}-18(-2)^{\frac{1}{2}}$; 2. $5-7i$;

3. $3(-1)^{\frac{1}{2}}, (2i)^{\frac{1}{2}}+(3i)^{\frac{1}{2}}$; 4. $(-2)^{\frac{1}{2}}, (b^{\frac{1}{2}}-c^{\frac{1}{2}})i$;

5. $-8(6)^{\frac{1}{2}}, -216$; 6. $-60(42)^{\frac{1}{2}}i, -12$; 7. $-a^3i, a^4$;

8. $\frac{3}{2}(3)^{\frac{1}{2}}, 2-\frac{1}{2}(2)^{\frac{1}{2}}i$; 9. $4i+5^{\frac{1}{2}}, \frac{1}{18}(3)^{\frac{1}{2}}$;

10. $i, -2-2(-3)^{\frac{1}{2}}$;

11. $70-3(30)^{\frac{1}{2}}-[10(5)^{\frac{1}{2}}+21(6)^{\frac{1}{2}}]i, -46-43(-3)^{\frac{1}{2}}$.

[1] The letter i is often written for $\sqrt{-1}$.

329. **1.** 14; **2.** 3; **3.** 4; **4.** 10; **5.** 7; **6.** 144; **7.** $\frac{1}{3}$; **8.** 5; **9.** 1; **10.** $\frac{a^2+b^2}{a}$; **11.** 1; **12.** 1; **13.** $\frac{(a-b)^2}{2a-b}$; **14.** $\frac{17}{8}$.

332. **1.** ± 2; **2.** ± 8; **3.** ± 3; **4.** $(62)^{\frac{1}{2}}$; **5.** ± 4; **6.** ± 7; **7.** $\frac{b}{(a^2+c)^{\frac{1}{2}}}$; **8.** $\pm \frac{1}{2}$; **9.** $(ab)^{\frac{1}{2}}$; **10.** $\pm \frac{24}{5}$.

338. **1.** 6, and -10; **2.** 3, and -25; **3.** 9, -8; **4.** 8, 2; **5** 1, -5; **6.** 6, 0; **7.** 31, -11; **8.** 15, 8; **9.** 14, -2; **10.** $-\frac{1}{2}$, $-\frac{1}{2}$; **11.** 6, $-5\frac{1}{3}$; **12.** 70, 50; **13.** 2, .48; **14.** $\frac{1}{2}(3 \pm (5)^{\frac{1}{2}})$; **15.** $\frac{1}{10}(-19 \pm 581^{\frac{1}{2}})$; **16.** $\frac{1}{4}(5 \pm 67^{\frac{1}{2}})$; **17.** $\frac{1}{15}(8.6 \pm 451.96^{\frac{1}{2}})$; **19.** $2\frac{1}{6}$, $-7\frac{5}{6}$; **20.** $\frac{5}{6}$, $\frac{9}{14}$; **21.** $\frac{5}{12}$, $-\frac{3}{8}$.

339. **1.** ± 2, ± 10; **2.** ± 7; **4.** 7, 3; **5.** 225, 121; **6.** 27; **7.** 112 and 76; **8.** 4 mi. per hour; **9.** 3, 4, 5, and -1, 0, 1; **10.** 259, 481 yds.; **11.** 8; **12.** ± 6; **13.** \$7950; **14.** 15, 17, 8 ft.; **15.** 12; **16.** 140, 120; **17.** 160, 90; **18.** 9 pence.

343. **1.** 4, 3, and $5\frac{1}{3}$, $1\frac{2}{3}$; **2.** 3, 5, and -72, $23\frac{3}{4}$; **3.** 4, 3; $-\frac{3}{5}$, -20; **4.** 4, 1; 7, 10; **5.**[1] 12, 3; 3, 12; **6.** 5, $6\frac{1}{2}$; $-\frac{13}{3}$, $-7\frac{1}{2}$; **7.**† 71, 13; **8.**† 43, -51; **9.**† 1, 1; **10.** 2, 3; $\frac{3}{2}$, 4; **11.**† 12, 4; **12.** 7, 4; -4, -7; **13.** 5, 3.

346. **1.**† ± 13, ± 1; **2.**† ± 1, ± 1; **3.**† ± 2, ± 1;
4. $\pm 10\frac{1}{2}$, ± 4; ± 6, ± 7.
5. $\frac{1}{2}a(\pm 5^{\frac{1}{2}} \pm 1)$, $\frac{1}{2}a(\pm 5^{\frac{1}{2}} \mp 1)$;
6. $\frac{1}{a}(\pm(6ab)^{\frac{1}{2}} \pm (ab)^{\frac{1}{2}})$, $\frac{1}{5b}(\pm(6ab)^{\frac{1}{2}} \mp (ab)^{\frac{1}{2}})$;
7. 11, 5; 6, 10; **8.** $9\frac{1}{4}$, $3\frac{1}{2}$; -2, -4; **9.** $\frac{a}{2}(1 \pm 5^{\frac{1}{2}})$, $\frac{1}{2a}(1 \pm 5^{\frac{1}{2}})$:
10. ± 2, ± 1; $\pm \frac{4}{7}(7)^{\frac{1}{2}}$, $\mp \frac{1}{7}(7)^{\frac{1}{2}}$;
11. ± 2, ± 1; $\pm \infty$, $\pm \infty$, (See **358**, 2); **12.**† $\pm \frac{3}{2}$, $\pm \frac{1}{2}$;
13. ± 7, ± 2; $\pm 3^{\frac{1}{2}}$, $\mp 3(3)^{\frac{1}{2}}$.

347. **1.** ± 36, ± 16; **2.** ± 77, ± 91; **3.** $\frac{1}{2}(1 \pm 5^{\frac{1}{2}})$; $\frac{1}{2}(3 \pm 5^{\frac{1}{2}})$; **4.** 36; **5.** 64, 36; **6.** \$2025, \$900;

[1] This problem is symmetrical in x and y; i.e., they can change places without altering the equations (**230**, a). This the answer indicates, since the values for x and y will satisfy the equation when interchanged. Hereafter, instead of writing these double results, the answer will be marked with a *dagger* to show that the values can be taken the other way.

7. $\pm 10, \pm 2$; $\pm 6\,(2)^{\frac{1}{2}}, \pm 4\,(2)^{\frac{1}{2}}$; 8. \$2, \$3; 60, 40;
9. 4 ft., 13 ft.; 10. 16 yds., 2 yds.; 32 yds., 1 yd.;
11. 8 hrs., 6 hrs.; 12. 36, $38\frac{1}{2}$; 28, $22\frac{1}{2}$.

351. 3. 3, 4; 4. 4, 5; 5. $-5, -6$; 6. $-1, -12$;
7. $-10\,b, -11\,b$; 8. $+9, -3$; 9. $-10, +30$; 10. 3, 5;
11. 21, -10; 12. $\frac{3}{7}, \frac{1}{2}$.

354. 1. $-5\,a, -18\,a$; 2. $0, -\frac{3}{2}, -\frac{4}{3}$; 3. $\frac{3}{4}, \pm\frac{1}{3}(6)^{\frac{1}{2}}$;
4. $-1, \pm i$; 6. $\pm 2, -4$; 7. $0, \pm 4, 2$; 8. $1, 2, \frac{1}{2}$.

355. 2. $x^3 - 7\,x + 6 = 0$; 3. $x^4 + 4\,x^3 - 7\,x^2 - 10\,x = 0$;
4. $x^3 - 15\,x^2 + 60\,x - 46 = 0$.

357. $y = 2$, a minimum, $y = -\frac{10}{9}$, a maximum.

361. 5. $1, \dfrac{-1 \pm (-3)^{\frac{1}{2}}}{2}, -1, \dfrac{1 \pm (-3)^{\frac{1}{2}}}{2}$;
6. $a\,2^{-\frac{1}{2}}(\pm 1 \pm i)$; 7. 6; 8. 625; 9. 18.72; 10. 32; 11. ± 2.

362. 5. $2, -1$; 6. $\pm 2, \pm 1$; 7. $\pm 7, \pm 5$; 8. $\pm 1, \pm\frac{1}{2}$;
9. $[\frac{1}{6}(7 \pm 349^{\frac{1}{2}})]^{\frac{1}{2}}$; 10. $9, 9^{\frac{1}{3}}$; 11. $12^{\frac{1}{2}}, 7^{\frac{1}{2}}$; 12. 9, 6.76;
13. $\left(\dfrac{-b \pm (b^2 + 4\,ac)^{\frac{1}{2}}}{2\,a}\right)^{\frac{1}{n}}$; 14. 81, 2401; 15. $343, -\frac{125000}{343}$;
16. $256, (-24)^{\frac{4}{3}}$; 17. 0, 4, 9; 18. $(\frac{2}{3})^{\frac{1}{2}}, (\frac{1}{3})^{\frac{1}{2}}$; 19. $5, 3, 4 \pm 10^{\frac{1}{2}}$;
20. $2, 1, \frac{1}{2}(3 \pm (-31)^{\frac{1}{2}})$; 21. $4, 2, \frac{1}{2}(-7 \pm 17^{\frac{1}{2}})$.

363. 1.† 5, 2; 2.† $\pm 2\,(2)^{\frac{1}{2}}, \pm 5^{\frac{1}{2}}$; 3.† $\pm\frac{1}{3}, \pm\frac{1}{4}$;
4.† $\pm 3, \pm 1$; 5. $1944^{\frac{1}{2}}, 72^{\frac{1}{2}}$; 6.† 7, 4; 7.† 5, 3;
8. $\pm\left(\dfrac{b + a^2}{2\,a}\right)^{\frac{1}{2}}, \pm\left(\dfrac{b - a^2}{2\,a}\right)^{\frac{1}{2}}$; 9. 13, 0; 10.† $\pm 6, \pm 4$;
11.† $+1, -2$; 12.† 7, 2; 13.† 27, 8; 14. 81, 8;
15. 8, 32; $2^{\frac{2}{3}}, 4^5$; 16.† $\pm 4, \pm 3$; 17. 32, $\frac{1}{2}$; 4, -3; 18. 2, 1;
19. 2, 2; 16, $\frac{1}{2}$; 20. 4, 2; $3 \pm 21^{\frac{1}{2}}, 3 \mp 21^{\frac{1}{2}}$; 21. 2, 3; $\frac{1}{108}$, 18;
22.† 5, 4; $8 \pm \frac{1}{12}(-2505)^{\frac{1}{2}}, 8 \mp \frac{1}{12}(-2505)^{\frac{1}{2}}$.

364. 1. $8\frac{3}{4}, 2\frac{1}{2}, 5\frac{4}{7}$ ft.; 2. 343, 64 cu. ft.; 3. $\pm 3, \pm 1$;
4. 64, 512; 5. $\pm 5, \pm 3$; 6. 4, 2; $3 + (-19)^{\frac{1}{2}}, 3 - (-19)^{\frac{1}{2}}$.

367. 8. The first; 9. The second.

368. 2. $x < 14$; 3. $x < 8$, $y > 3\frac{1}{2}$; 4. $x > a$, and $(ab - a + b)\,x < ab^2$; 6. 6; 7. 8.

379. 1. $\frac{3}{4}$; 2. $\frac{20}{21}$; 2nd, 1st; 3. 3; 4. 6400; 5. $\frac{5}{9}\,(3)^{\frac{1}{2}}$; 6. $\frac{4}{3}$, $\frac{9}{8}$, $\frac{25}{23}$, $\frac{15}{14}$; 7. 2 : 1; 9. $-\frac{3}{11}$; 12. 6, 8; 13. $\frac{7}{9}$; 14. $\dfrac{2\,ab}{a+b}$; 15. ± 6, ± 10; 16. $\dfrac{mq - pn}{p - q}$; 17. $\dfrac{pn - mq}{m - n - p + q}$;

380. 4. (1. 56; (2. $88\frac{4}{7}$; (3. 100; (4. $\dfrac{ab}{c}$; (5. $25\,x^3 = 27\,y^2$.

388. 2. (1. (1). 2.6599; (2). 2.5378; (3). 2.5647; (4). 2.6532; (5). 1.6532; (6). 1.6021; (7). 0.6021; (8). 0.4133; (9). $\bar{2}.5682$; (10). $\bar{3}.6990$; (11). $\bar{5}.8319$; (12). 3.5623; (3. (1). 2.5073; (2). 4.1512; (3). 3.1512; (4). 0.4095; (5). 3.3765; (6). $\bar{1}.3018$; (7). $\bar{2}.5731$; (8). 6.4632; (9). $\bar{5}.1232$.

390. 2. .63329; 3. .013015; 4. 2.8107; 5. 102.2; 6. 17733; 7. 88.88; 8. .0005395; 9. .1099; 10. .001051 · 11. .6955; 12. .00111; 13. .0003318.

392. 3. (2. 703; (3. 2924; (4. 28556; (5. 2337200 (2336544, exactly).

393. 3. (2. 7; (3. 5; (4. 74.167; (5. 14.342; (6. .004057: 5. (2. .07094; (3. .001086; (4. 523; (5. .004939.

394. 3. (2. 2401 +; (3. 418333333; (4. 2.051; (5. 534.1; (6. 22.8; (7. 429.6; (8. .941.

395. 2. (3. 5.656; (4. .8806; (5. 146.76; (6. 5.4875: (7. 1.6155; (8. .70717; (9. 4.957; (10. 1.8217.

398. 2. (2. 6; (3. 1.537; (4. .4581; (5. $-$ .2031; (6. 3.537; (7. 1; (8. $-$ .4319; (9. .9031; (11. 3.011; (12. $\pm$ 2.26; (13. 4.336, $-$ .336.

402. 1. (2. $\dfrac{2s}{n} - a$; (3. $-\frac{1}{2}d \pm \{2\,ds + (a - \frac{1}{2}d)^2\}^{\frac{1}{2}}$;

(4. $\frac{s}{n}+\frac{(n-1)d}{2}$; 2. (1. $\frac{(l+a)n}{2}$; (2. $\frac{l+a}{2}+\frac{l^2-a^2}{2d}$;
(3. $\frac{1}{2}n\{2a+(n-1)d\}$; (4. $\frac{1}{2}n\{2l-(n-1)d\}$;
3. (1. $\frac{l-a}{n-1}$; (2. $\frac{2(s-an)}{n(n-1)}$; (3. $\frac{l^2-a^2}{2s-l-a}$; (4. $\frac{2(nl-s)}{n(n-1)}$;
4. (1. $\frac{l-a}{d}+1$;

(2. $\frac{1}{2d}\{\pm((2a-d)^2+8ds)^{\frac{1}{2}}-2a+d\}$; (3. $\frac{2s}{l+a}$;

(4. $\frac{1}{2d}\{2l+d\pm[(2l+d)^2-8ds]^{\frac{1}{2}}\}$; 5. (1. $l-(n-1)d$;

(2. $\frac{s}{n}-\frac{(n-1)d}{2}$; (3. $\frac{2s}{n}-l$.

403. 2. 34; 3. 77, 779; 4. -71; 5. $\frac{1}{3}(n+5)$; 6. 142; 7. $-2\frac{1}{4}$, -165; 8. $198a+63b$; 9. 50, 5150; 10. $d=5$; 11. 2, 2.3 . . . 7.7, 8; 12. 43, -1; 13. 7, 6, a^2; 14. 25th; 15. 106th.

407. 1. (2. (1) $\frac{a+(r-1)s}{r}$; (2) $\frac{(r-1)sr^{n-1}}{r^n-1}$;
(3. (1) $\frac{l}{r^{n-1}}$; (2) $\frac{(r-1)s}{r^n-1}$; (3) $rl-(r-1)s$; (4. (1) $\left(\frac{l}{a}\right)^{\frac{1}{n-1}}$;

(2) $\frac{s-a}{s-l}$; (5. $\frac{l^{\frac{n}{n-1}}-a^{\frac{n}{n-1}}}{l^{\frac{1}{n-1}}-a^{\frac{1}{n-1}}}$; 2. (2. (1) $\frac{\log. l-\log. a}{\log. r}+1$;

(2) $\frac{\log. (a+(r-1)s)-\log. a}{\log. r}$;

(3) $\frac{\log. l-\log. (lr-(r-1)s)}{\log. r}+1$;

3. (2. $a(s-a)^{n-1}=l(s-l)^{n-1}$;

(3. (1) $r^n-\frac{s}{a}r+\frac{s-a}{a}=0$;

(2) $r^n-\frac{s}{s-l}r^{n-1}+\frac{l}{s-l}=0$.

410. 1. 6561, 9841; 2. $\frac{3}{128}$, $11\frac{125}{128}$;
3. $\frac{1}{2}$, $\frac{1}{128}$, 4. $243\,(6)^{\frac{1}{2}}$, $364\,(6^{\frac{1}{2}}+2^{\frac{1}{2}})$; 5. $\frac{1}{90}$, $12\frac{13}{90}$;
6. $\frac{625}{512}$, $\frac{38443}{7680}$; 7. $a=10$, $l=1000000000$; 8. 1953.1;
9. $-28, 14, -7, \frac{7}{2}, -\frac{7}{4}, \frac{7}{8}$; 10. $(ab)^{\frac{1}{2}}$; 12. 27; 13. $\frac{5}{7}$; 14. $\frac{27}{29}$.

418. 1. $2.68; 2. $719; 3. 5%; 4. $10\frac{1}{4}$, nearly;
5. 3 yrs. 7 mos. 17 da.; 6. $5\frac{1}{2}$; 7. $2\frac{1}{4}$, nearly;
8. A little more than 21 years from 1890; 9. $8333;
10. Int. $177.75; 11. $5477.40; 12. $1580.40; 13. $1055;
14. $1276.76; 15. $516.22; 16. $1160; 17. .0576; 18. $1183.

INDEX.[1]

[1] Can be used to advantage in reviews by ignoring all references in advance of the student's progress.

www.ingramcontent.com/pod-product-compliance
Lightning Source LLC
Chambersburg PA
CBHW020906210326
41598CB00018B/1790

* 9 7 8 3 7 4 4 7 4 5 7 3 4 *